# Applications of Mössbauer Spectroscopy

## Volume I

# Contributors

Richard L. Cohen
Peter G. Debrunner
P. K. Gallagher
N. H. Gangas
V. I. Goldanskii
B. Keisch
L. A. Korytko
A. Kostikas
George Lang
Henry Leidheiser, Jr.
W. T. Oosterhuis
L. H. Schwartz
Gary W. Simmons
A. Simopoulos
K. Spartalian

# APPLICATIONS OF MÖSSBAUER SPECTROSCOPY

## Volume I

Edited by

### Richard L. Cohen

Bell Laboratories
Murray Hill, New Jersey

ACADEMIC PRESS   New York  San Francisco  London   1976

A Subsidiary of Harcourt Brace Jovanovich, Publishers

ACADEMIC PRESS, INC.
111 Fifth Avenue, New York, New York 10003

*United Kingdom Edition published by*
ACADEMIC PRESS, INC. (LONDON) LTD.
24/28 Oval Road, London NW1

Library of Congress Cataloging in Publication Data

Main entry under title:

Applications of Mössbauer spectroscopy.

      Bibliography:    v. 1, p.
      Includes index.
      1.   Mössbauer spectroscopy.   I.   Cohen, Richard Lewis,
(date)
QC491.A66        537.5'352       75-26349
ISBN 0−12−178401−0 (v. 1)

# Contents

## BIOLOGICAL STUDIES

## SOLID STATE CHEMISTRY

# List of Contributors

Numbers in parentheses indicate the pages on which the authors' contributions begin.

Richard L. Cohen (1), Bell Laboratories, Murray Hill, New Jersey

Peter G. Debrunner (171), Physics Department, University of Illinois, Urbana, Illinois

P. K. Gallagher (199), Bell Laboratories, Murray Hill, New Jersey

N. H. Gangas (241), University of Ioannina, Ioannina, Greece

V. I. Goldanskii (287), Institute of Chemical Physics, USSR Academy of Science, Moscow, USSR

B. Keisch (263), National Gallery of Art Research Project, Carnegie-Mellon Institute of Research, Pittsburgh, Pennsylvania

L. A. Korytko (287), Institute of Chemical Physics, USSR Academy of Science, Moscow, USSR

A. Kostikas (241), Nuclear Research Center Democritos, Athens, Greece

George Lang (129), Department of Physics, The Pennsylvania State University, University Park, Pennsylvania

Henry Leidheiser, Jr. (85), Center for Surface and Coatings Research, Lehigh University, Bethlehem, Pennsylvania

W. T. Oosterhuis (141),[†] Physics Department, Carnegie-Mellon University, Pittsburgh, Pennsylvania

L. H. Schwartz (37), Materials Science and Engineering Department, Northwestern University, Evanston, Illinois

Gary W. Simmons (85), Center for Surface and Coatings Research, Lehigh University, Bethlehem, Pennsylvania

A. Simopoulos (241), Nuclear Research Center Democritos, Athens, Greece

K. Spartalian (141),[‡] Physics Department, Carnegie-Mellon University, Pittsburgh, Pennsylvania

[†] Present address: Division of Materials Research, National Science Foundation, Washington, D.C.

[‡] Present address: Physics Department, The Pennsylvania State University, University Park, Pennsylvania.

# *Preface*

The technique of Mössbauer spectroscopy, now 15 years old, is beginning to be used in many analytical and engineering research areas. Many of these efforts in applied science have been highly successful, and we can expect them to form an increasing proportion of the research performed using Mössbauer spectroscopy. The utilization and propagation of these results in new areas has been somewhat hampered by the fact that the experimental techniques, the interpretation of the results, and the significance of the findings are relatively foreign to, for example, an archaeologist or a protein chemist.

The main goal of this series will be to make available to scientists and engineers in varied disciplines, in their own terms, a discussion of what results have been achieved by Mössbauer spectroscopy to date, and what additional advances are likely. The emphasis will be on fields often considered in the grouping "materials science." Basic physics and chemistry studies performed by Mössbauer spectroscopy have been extensively reviewed elsewhere and will be described here only to the extent necessary for the understanding of the applied science.

In this first volume, the emphasis is on metallurgy, solid state and interface chemistry, and structure of iron-containing proteins. In the second volume, the topics will be extended to include catalysis, studies of disordered systems, and diffusion.

# 1

# Elements of
# Mössbauer Spectroscopy

**Richard L. Cohen**

Bell Laboratories
Murray Hill, New Jersey

## I. Introduction

"Mössbauer spectroscopy" is the name given to a technique of studying the absorption of $\gamma$ rays by the nuclei of atoms. The nuclear processes producing this effect were first observed and reported by Rudolf L. Mössbauer in 1958 (Mössbauer, 1958). This work immediately attracted wide interest because of the unprecedented sharpness of the resonance observed, which held out great promise for studies of gravitation, relativity, and certain areas of nuclear physics. The technique would have been relatively little used, however, if those were the only possible topics for study—the vast majority of experiments performed are in the fields of solid state physics and chemistry, metallurgy, geochemistry, and biophysics. These topics also form the basis of most of the applications-oriented work reviewed in this book.

It is easy to describe the processes occurring in nuclear $\gamma$-ray-resonance spectroscopy. Nuclei are the heavy cores of atoms and are generally considered to be composed of protons and neutrons. The number of protons in the nucleus is the atomic number of the atom, and it determines the chemical properties of the atom. For each element (atomic number), a number of different isotopes, corresponding to different numbers of neutrons in the nucleus, may be stable. These are the naturally occurring isotopes of that element, identified by their mass number (sum of protons and neutrons). Unstable (radioactive) nuclei undergo decay, or transformation, with the emission of various kinds of radiation.

Even stable nuclei, however, have excited states, configurations in which the nucleus has some discrete, well-defined quantity of added energy over that present in the stable, or ground state, configuration. These excited states often decay to the nuclear ground state, with the extra energy being emitted in the form of a $\gamma$ ray. Gamma rays are electromagnetic radiation, identical in properties to x rays. They have no electric charge and cannot be deflected by electric or magnetic fields. When $\gamma$ rays pass through matter, they are absorbed or scattered primarily by occasional energetic collisions with electrons. Thus, a beam of $\gamma$ rays that initially have the same energy loses its intensity primarily due to the absorption of individual $\gamma$ rays. Those $\gamma$ rays that are not absorbed or scattered out of the beam continue to propagate with their original energy.

These facts are important for the present discussion because the basis of the Mössbauer effect is the emission of $\gamma$ rays by radioactive nuclei, and the subsequent reabsorption of these gamma rays by other nuclei of the same type. The nuclear emission and absorption energies are slightly affected by the solids in which the nuclei are incorporated. Using the Mössbauer effect, these tiny energy changes can be measured, and used to deduce information about the surroundings of the nucleus.

Figure 1 shows a schematic drawing of the nuclear decay and excitation process. It is worthwhile to emphasize here a number of features of the $\gamma$-ray resonance absorption process that are implicit in the above description and provide some of the distinctive advantages of the technique.

1. Since nuclear energy levels in the range involved here are so narrow and sharply defined (see Section II,E,1), $\gamma$ rays from any nucleus (e.g., $^{57}$Fe) can only be reabsorbed by nuclei of the same type, since any other isotope will have absorption energies (corresponding to excited states) in a different

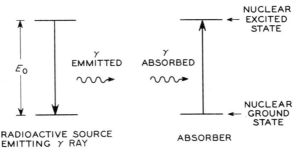

**Fig. 1.** Schematic indication of the events occurring in Mössbauer spectroscopy. The horizontal lines represent the nuclear states. The diagram shows (left) the source nucleus going from the excited state to the ground state, emitting a $\gamma$ ray. The $\gamma$ ray is subsequently absorbed (right), raising the absorber nucleus to its excited state. The resonance absorption can be detected either by the decreased transmission of the absorber, or by the subsequent decay of the absorber nucleus out of the excited state.

energy region. Thus, experiments are absolutely specific to the particular isotope involved, and no cross-interference from other isotopes or elements ever arises.

2. It is possible to incorporate the radioactive source atoms in the material to be studied and thus combine the advantages of radioactive tracer experiments with those of the Mössbauer technique. Samples containing as few as $\sim 10^{12}$ probe atoms can thus be studied.

3. Since the resonance absorption is a purely nuclear process, its existence is inherently independent of the properties of the host (e.g., symmetry, metallic character), which sometimes interfere with the use of other resonance techniques.

4. The nuclear energy level perturbations observable using the Mössbauer effect arise only from the first few nearest-neighbor shells of an ion. Thus, short-range order, over as little as 10–15 Å, is adequate to provide sharp Mössbauer spectra. Glassy materials, disordered alloys, and very finely divided samples can all produce well-defined spectra.

5. Sample preparation is usually very simple—single crystals are not normally necessary, and no special polishing or surface treatment is required. Power samples can be readily utilized.

6. The dependence of the recoil-free fraction (Debye–Waller factor, discussed in Section II,E,2) on the properties of the host lattice allows investigation of the Debye temperature and anharmonic binding forces via the temperature dependence of the resonance intensity.

7. Although the technique is in principle limited to studying nuclei in solids, it is often possible to investigate dissolved molecules and complexes by freezing the solutions and making measurements on the resulting solid.

8. The existence of chemically, crystallographically, or magnetically inequivalent sites is generally revealed by the appearance of distinct components, arising from the different sites, in the Mössbauer spectrum.

## II. Interpretation of Mössbauer Spectra

### A. What Does Mössbauer Spectroscopy Measure?

A Mössbauer spectrum is normally produced by varying the source $\gamma$-ray energy and measuring the (nuclear) resonance absorption as a function of $\gamma$-ray energy, as shown in Fig. 2. At $\gamma$-ray energies that match the possible excitation energies in nuclei in the absorber, the nuclear resonance will result

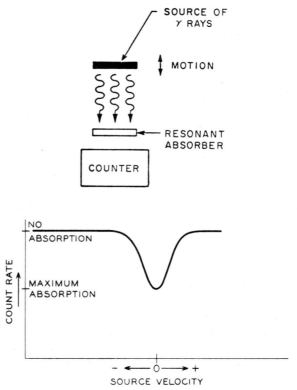

**Fig. 2.**   Basic arrangement for measuring a Mössbauer spectrum in transmission geometry. The source is moved to Doppler modulate the $\gamma$-ray energy. When the $\gamma$ rays have the proper energy to be resonantly absorbed, the increased absorption produces a decrease in the number of $\gamma$ rays transmitted through the absorber, and the counting rate decreases.

in increased absorption, and an absorption "line" will occur. This dip (or series of dips) is the Mössbauer spectrum.

Chapters 3 and 9 describe resonance scattering experiments in which backscatter geometry (as shown in Fig. 3) is used, rather than transmission

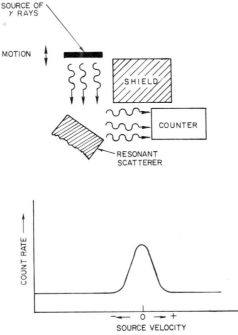

**Fig. 3.**   Mössbauer spectrometer for backscatter experiments. When the $\gamma$ rays incident on the absorber are the proper energy to be resonantly absorbed, absorber nuclei are raised into their excited states. They subsequently decay, emitting radiation that is detected by the counter.

geometry. There are two advantages to this approach. (1) It is possible to study thick samples *in situ* without the special preparation of a thin layer of the material; the advantages for the study of a valuable painting or locomotive boiler should be obvious. (2) If the scattering is observed via the emission of secondary radiation of low penetrating power, only the atoms near the surface of the sample will contribute to the observed resonance effect. Depending on the isotope used and the sample material, surface layers from 0.1 to 1 $\mu$m thick can be studied. The backscatter technique is also useful in cases where the absorber consists of only a very thin layer of material, as in corrosion research.

There are basically two types of information that can be derived from the Mössbauer spectrum. The relative and absolute line energies are determined by electronic effects on the nuclear energy levels. These effects are generally lumped together as "hyperfine parameters," and are the isomer shift (Section II,B), (electric) quadrupole hyperfine interaction (Section II,C), and (magnetic) dipole hf interaction (Section II,C). Figure 4 shows how these

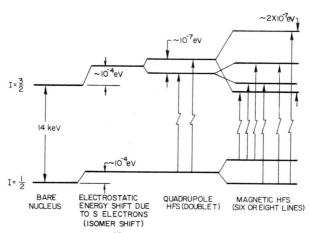

**Fig. 4.** Energy level diagram for $^{57}$Fe showing the origins of the isomer shift and hyperfine splittings. Each of the vertical broken arrows represents a possible absorption line. The observed isomer shift arises from a very small difference between the size of the electrostatic energy shift in the ground state and in the excited state.

effects split the nuclear levels of $^{57}$Fe, the isotope most commonly used for Mössbauer effect research. The overall intensity (amplitude) of the resonance spectrum depends on the recoil-free fraction (Debye–Waller factor), normally termed f. The recoil-free fraction is dependent on the binding of the Mössbauer-active atom in the lattice and can thus be used for studies of lattice vibrations (see Section II,E).

In addition to the hyperfine parameters and recoil-free fraction, which form the principal experimental results determinable from a Mössbauer spectrum, there are two other, less important, features. The first is the relative intensity of various lines of the spectrum, which can be analyzed to yield information on the direction between the crystal and hyperfine field axes and the direction of the γ-ray beam. The second effect, called the second-order Doppler shift, or thermal red shift, is a displacement of the entire spectrum arising from the thermally excited vibrations of the Mössbauer ion. It is discussed in Section II,E,3.

For the applications work discussed in this book, a detailed understanding of these interactions is often unnecessary—the information sought may be found simply by comparing the Mössbauer spectrum of an unknown (e.g., the product of a chemical reaction) with the spectra of various known materials. Even in such cases, however, it is useful to know which systems might have very similar hyperfine parameters (and therefore, similar spectra), and which differences between known and unknown spectra might arise from unimportant extrinsic effects, such as particle size or orientation differences. One of the major aims of this chapter is to allow the reader who is not a specialist in the interpretation of Mössbauer spectra to understand the basic approaches used in analyzing the spectra in these experiments, and the important interactions and their origins. Toward this end, a discussion of all of the hyperfine and lattice-related parameters is presented below. Table I shows the major fields in which useful information has been gained using Mössbauer spectroscopy and the particular significance of the results obtained.

Throughout this section, the characteristics of $^{57}$Fe will be emphasized. This isotope is the one most widely used in both basic research and applications-oriented work, and demonstrates essentially all of the phenomena encountered in Mössbauer studies. Table II shows the commonly used isotopes and their salient features.

### B. Isomer Shift

The isomer shift (also called the chemical shift) arises from the interaction energy of the part of the electronic cloud inside the volume of the nucleus with the nuclear charge. The isomer shift provides direct information about the electron density at the nucleus, and this can often be interpreted to give unequivocal information about the valence state of the ion under study.

For a uniformly charged spherical nucleus of radius $R$, in an electron density $\rho$, the interaction energy can readily be shown (Wertheim, 1964) to be

$$(2\pi/5)Ze^2\rho R^2, \tag{1}$$

with $Z$ the atomic number and $e$ the electronic charge. This interaction energy shifts the nuclear energy levels slightly, as shown in Fig. 4. Since $R_e$, the nuclear excited state radius, is generally slightly different from the nuclear ground state radius $R_g$, the energy change described by Eq. (1) will be different for the two states.

Thus,

$$(2\pi/5)Ze^2\rho[R_e^2 - R_g^2] \tag{2}$$

is the net change in $\gamma$-ray emission or absorption energy arising from the electronic density at the nucleus. All of the terms except $\rho$ are constant for a given nuclear transition. In actual experiments, it is not the total energy perturbation as written (Eq. 2) that is measured, but rather the difference in transition energies between a source (electronic density $\rho_s$) and an absorber

**TABLE I**

Relationship between Variables Measured Using Mössbauer Spectroscopy and Various Research Fields

| Measured variable | Nuclear physics | Solid state physics | Chemistry | Metallurgy | Biology | Analytical |
|---|---|---|---|---|---|---|
| Isomer shift | Nuclear radius change | Electronic structure | Valence state and covalency | Electronic structure | Valence changes | Chemical or phase identification by comparison with spectra of known materials |
| Magnetic hyperfine structure | ⎫ Nuclear moments | Magnetic structure, electronic configuration of magnetic ions | Electronic configuration of magnetic ions | ⎫ Order–disorder, precipitate identification | ⎫ Ligand conformation | Particle size information via superparamagnetism |
| Electric quadrupole hyperfine structure | ⎭ | Electronic configuration | Ligand symmetry, bonding orbitals | ⎭ | ⎭ | |
| Recoil-free fraction | | Phonon spectrum, anisotropic binding | Anisotropic binding | Force constants and anharmonic binding | Free or bound complex | |

**TABLE II**

Elements that Can Be Studied with Mössbauer Spectroscopy

| K | Ca | Sc | Ti | V | Cr | Mn | Fe | Co | Ni | Cu | Zn | Ga | Ge | As | Se | Br | Kr |
|---|----|----|----|---|----|----|----|----|----|----|----|----|----|----|----|----|----|
| Rb | Sr | Y | Zr | Nb | Mo | Tc | Ru | Rh | Pd | Ag | Cd | In | Sn | Sb | Te | I | Xe |
| Cs | Ba | La | Hf | Ta | W | Re | Os | Ir | Pt | Au | Hg | Tl | Pb | Bi | Po | At | Rn |
| Fr | Ra | Ac | | | | | | | | | | | | | | | |

| Ce | Pr | Nd | Pm | Sm | Eu | Gd | Tb | Dy | Ho | Er | Tm | Yb | Lu |
|----|----|----|----|----|----|----|----|----|----|----|----|----|----|
| Th | Pa | U | Np | Pu | Am | Cm | Bk | Cf | Es | Fm | Md | No | Lw |

▨▨▨ EASY TO STUDY, EXTENSIVE RESEARCH

▨▨▨ MORE DIFFICULT TO STUDY, SOME RESEARCH

▨▨▨ VERY DIFFICULT OR LIMITED RESULTS

($\rho_a$). The isomer shift as observed in an experiment thus is equal to

$$\delta_{IS} = (2\pi/5)Ze^2[R_e^2 - R_g^2]\{\rho_a - \rho_s\}. \tag{3}$$

Since the change between $R_e$ and $R_g$ is generally very small ($\sim 0.01\%$), an alternate convenient formulation, using $\delta R = R_e - R_g$, is

$$\delta_{IS} = (4\pi/5)Ze^2[R\,\delta R]\{\rho_a - \rho_s\}. \tag{4}$$

In situations where it is not realistic to consider the nucleus as a uniformly charged sphere, the isomer shift is written

$$\delta_{IS} = (4\pi/3)Ze^2[\langle r^2 \rangle_e - \langle r^2 \rangle_g]\{\rho_a - \rho_s\}, \tag{5}$$

where $\langle r^2 \rangle$ is the mean-square charge radius. For nuclear physics, where the charge radius change between the excited and ground states can be related to nuclear structure models, the term in the square brackets, containing only parameters characteristic of the nucleus, is the significant one. For all other research and analytical applications, the term in the curly brackets is the interesting one, and the first term is considered a constant determined from other experiments.

There are two basic mechanisms by which the valence of the ion under study influences $\rho$ and effects change in the isomer shift energy. The first is a direct change in $\rho$ due to the presence or absence of valence s electrons. This is the primary contributor to electron density changes in the widely studied isotope $^{119}$Sn. Figure 5 shows the way in which measured isomer shift values for a number of inorganic tin compounds are related to the formal valence state of the ion.

In ions of the transition elements, where valence changes occur via a change in d or f electron number, no direct change in $\rho$ at the nucleus is

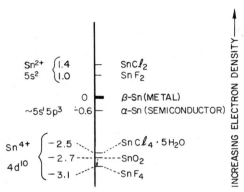

**Fig. 5.**  Isomer shift (mm/sec) and electron configuration for assorted tin compounds. All values are referred to $\beta$-tin, which is ordinary tin metal.

produced, because the charge densities of those electrons are negligible at the nucleus. There is, however, a substantial indirect change in the total $\rho$ arising from the altered shielding of the outermost s electron shell. For example, in iron, if a 3d electron is removed (increasing the valence by 1), the 4s electrons will be pulled closer to the nucleus, and the charge density due to them will increase. Note that in this case, an increase in valence, corresponding to removal of a non-s electron, results in an increase in total electron density at the nucleus.

Actual behavior in compounds that are not simple ionic insulators is more complex than that described above and results from a combination of these two effects plus additional phenomena, such as back-donation and formation of conduction bands. A classic analysis of isomer shift systematics in iron compounds is shown in Fig. 6 (Walker *et al.*, 1961). This approach laid the groundwork for many of the subsequent analyses in this field.

Despite the fact that the basic interactions underlying the electronic density changes are well established, it is extremely difficult to make "first principles" calculations that accurately represent real solids. Most analysis of Mössbauer isomer shift results is still based on systematics and semiempirical calculation using comparison with compounds of relatively well-established electronic structure.

## C. *Quadrupole Coupling and Magnetic Hyperfine Interaction*

In addition to the changes in nuclear energy levels produced by the isomer shift, the levels may be split by the electric quadrupole and magnetic dipole hyperfine interactions. This splitting leads to a number of possible absorption

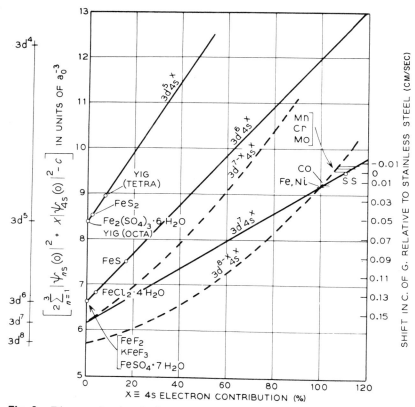

**Fig. 6.** Diagram showing the isomer shift as a function of electronic configuration for various iron compounds. (After Walker et al., 1961.)

energies, and thus a number of lines in the absorption (or emission) spectrum. Figure 4 shows how, for $^{57}Fe$, the nuclear excited state ($I = \frac{3}{2}$) and ground state ($I = \frac{1}{2}$) are split to produce two absorption lines (a doublet) under the quadrupole interaction, and then (normally) six lines under the magnetic hyperfine interaction.

The quadrupole coupling arises from the fact that the nucleus is not perfectly spherical, but may be ellipsoidal, either elongated (prolate, cigar shaped) or flattened (oblate, pancake shaped). As seen in Fig. 7, the electrostatic forces between the surrounding ligands—assumed to be negatively charged—and the nonspherical part of the nuclear charge tend to make the nuclear axis point toward the ligands. This state is then the lower energy state of the quadrupole split doublet, and the state with the nuclear axis perpendicular to the ligand axis is the higher energy state.

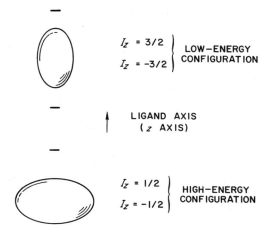

**Fig. 7.** Coupling of the nuclear quadrupole moment with nearby charges. For the prolate nucleus shown ($Q > 0$) the low energy configuration (top) is obtained with the nuclear axis pointing toward the negative charges. The difference in energy between the two configurations is the quadrupole splitting, $\Delta_{EQ}$. The drawing is oversimplified to show the $I_z = \frac{3}{2}$ and $I_z = \frac{1}{2}$ spin states exactly along and perpendicular to the $z$ axis; the correct quantum mechanical solution is more complex.

In the quantum mechanical analysis of the situation, the nucleus is normally considered to have $2I + 1$ ($I$ is the nuclear spin) orientations relative to an external axis, and these are described by values $I_z$, running in steps of one unit from $-I$ to $+I$. Large positive and negative values of $I_z$ correspond to the nuclear spin pointing along the axis; small values of $I_z$ correspond to the nuclear spin perpendicular to the axis. Since the two "ends" of the nucleus have the same charge distributions, the coupling energies of the nucleus with the external charges are identical in the "spin up" $I_z = +\frac{3}{2}$ and "spin down" $I_z = -\frac{3}{2}$ states, and these are termed "degenerate" (two or more distinct states with the same energy). For a nuclear spin of $I = \frac{1}{2}$, the states available are $I_z = +\frac{1}{2}$ and $I_z = -\frac{1}{2}$; following the same analysis, these states will both have the same energy, and there will be no quadrupole splitting. Thus, the doublet observed in $^{57}$Fe, $^{119}$Sn, $^{197}$Au, $^{125}$Te, and a number of other Mössbauer nuclides with $I = \frac{3}{2}$ in one state and $I = \frac{1}{2}$ in the other arises from the splitting of the $\frac{3}{2}$ state into two levels ($\pm\frac{3}{2}$ and $\pm\frac{1}{2}$), while the $\frac{1}{2}$ state remains unsplit.

For nuclear states with $I = \frac{3}{2}$ a mathematical analysis leads to the expression

$$E_Q = \mp\tfrac{1}{4}eV_{zz}(1 + \tfrac{1}{3}\eta^2)^{1/2}Q = \pm\tfrac{1}{4}e^2qQ(1 + \tfrac{1}{3}\eta^2)^{1/2} \qquad (6)$$

for the quadrupole coupling energy $E_Q$, where the $\pm$ term corresponds to the fact that the $I_z = \pm\frac{3}{2}$ and $I_z = \pm\frac{1}{2}$ state energies are perturbed equally, but in opposite directions. Note that the measured quadrupole splitting, sometimes called $\Delta$, or $\Delta_{EQ}$, the observed energy difference between the lines of a quadrupole-split doublet, is twice the value of $E_Q$ as defined above. In Eq. (6), $Q$ is the nuclear quadrupole moment, $V_{zz} = -eq$ is the *electric field gradient* (along the $z$ direction), and $\eta = (V_{xx} - V_{yy})/V_{zz}$ is the *asymmetry parameter* describing the difference in electric field gradient in the $x$ and $y$ directions. Since axes are normally defined so that $0 \leqslant \eta \leqslant 1$, the value of the asymmetry term is always between 1 and 1.16, so that it is not an important determinant of the spectrum. A detailed discussion of the origins and significance of this term is beyond the scope of this section; the reader is referred to Wertheim (1964) and Goldanskii and Makarov (1968). The discussion of quadrupole coupling here is valid only for $I = \frac{3}{2}$; for higher spin states the analysis is more complicated, particularly if the asymmetry parameter is nonzero.

We note that, as with the isomer shift, quadrupole coupling information can be factored into a nuclear term $Q$ and a term due to the ion and its surroundings, the field gradient, $V_{zz}$. Once again, the latter term contains all of the interesting solid state physics or chemistry information. The field gradient arises from nonsymmetric disposition of the electronic charge in the ion under study and its surroundings, and the principal contributions are:

1. In transition ions, especially iron or the rare earths, if the d or f electronic wavefunction has orbital angular momentum ($L \neq 0$), a large contribution to $V_{zz}$ is obtained. Since d and f wavefunctions can be calculated relatively accurately and $V_{zz}$ can be calculated directly from the wavefunctions, it is sometimes possible to decide if a particular electronic configuration could be the source of the observed field gradient.

2. Non-s electrons participating in bonding also provide large contribution to $V_{zz}$. For example, tin organic compounds have extremely large quadrupole couplings arising from 5p orbitals. In metallic systems, this term would include contributions from conduction electrons.

3. The direct field gradient contribution from ligand ions is negligible in comparison to the first two terms. However, distortion of the electronic charge distribution in the outer shells of the ion under the influence of the ligand field greatly enhances the gradient actually observed at the nucleus. This phenomenon is called *Sternheimer shielding* (Sternheimer, 1963; Freeman and Watson, 1964), and increases the field gradient from the ligands by a factor which is $\sim 7$ for iron and $\sim 80$ for rare earth ions.

The *magnetic hyperfine interaction* (*nuclear Zeeman effect*) arises from the coupling of the nuclear magnetic moment with effective magnetic fields at the nucleus and results in splitting of the nuclear ground and excited states if they have nuclear spin $I > 0$. As with the quadrupole coupling, the magnetic interaction can be written as the product of a nuclear term and a term arising from the surroundings. The latter can be considered to be the sum of components arising from a number of sources:

1. Hyperfine fields as large as 9 MG[†] are produced in rare earth ions (and actinides) by the orbital angular momentum of the 4f (or 5f) electrons. For most rare earth ions [for a detailed discussion, see Ofer et al. (1968)], this component of the hyperfine field is proportional to the magnetic moment of the ion with a known constant. Thus a measurement of the hyperfine field determines the ion moment.

2. If the ion is in a magnetic metal or is itself magnetic, densities of the spin-up and spin-down electrons at the nucleus will not be the same, and there will be a net spin polarization. The interaction of this electron spin polarization with the nucleus is called the *Fermi contact interaction* and can be considered as an effective magnetic field at the nucleus. In ions with a partly filled 3d or 4f shell, the net electronic spin density is produced by *core polarization* the distortion of core level electronic wavefunctions by the interaction with the spin of the d or f electrons. This mechanism leads to hf fields of $\sim -220$ kG/spin for iron (a total of $\sim -550$ kG for high-spin $Fe^{3+}$, $S = \frac{5}{2}$) and $\sim -100$ kG/spin for 4f electrons ($\sim -330$ kG for $Gd^{3+}$, $S = \frac{7}{2}$). The polarization of conduction electrons by magnetic ions produces hf fields as large as $\sim 1500$ kG.

3. In insulating materials, hf fields as large as a few hundred kilogauss can be produced by polarization of the ion's electrons by magnetic ligand ions. A notable example is that of tin as a dilute impurity in a magnetic garnet (Goldanskii et al., 1966), where the iron ions produce a hf field of 200 kG via polarization of the outer s electrons of the tin ion.

4. Externally applied fields are also effective in producing magnetic splitting of the nuclear energy levels.

In most cases, measured hf fields arise from more than one of these sources, and a major part of the interpretation of the results is the analysis of the various contributions. The simplest cases tend to be the rare earths and iron in magnetically ordered systems, where the first and second terms, respectively, are dominant. For many analytical applications, of course, the exact origins of the hf fields may not be important.

---

† MG = megagauss = $10^6$ G. The proper dimension for reporting hf fields is gauss. Many researchers report fields in oersteds (Oe). Gauss and oersteds are numerically the same if the permeability of the medium is 1. A few workers report tesla (T); 1 T = 10,000 G.

The hf field is often called the *internal field*, since it is measured at the nucleus, and is usually termed $B_{int}$, $H_{int}$, $B_{hf}$, or $H_{eff}$. The energy by which the nuclear levels are perturbed by the magnetic interaction alone can be written

$$\Delta E_M = B_{int}\mu(I_z/I), \tag{7}$$

where $\mu$ is the magnetic moment of the nuclear state, and $I$ and $I_z$ are the spin of the nuclear state and its projection along a $z$ axis (as discussed in the above section on the quadrupole interaction), which is chosen to be in the direction of the hf field. Since the possible $I_z$ values are in steps of one unit from $I$ to $-I$, the magnetic splitting leads to $2I + 1$ equally spaced nuclear energy levels.

For simplicity, we have considered the magnetic and quadrupole splittings separately. Often, both interactions are present, and their combination determines the hf splitting. For the special but frequently occurring case in which $V_{zz}$ lies along the magnetic hf field and $\eta = 0$, the energies of the nuclear levels can be determined by simply adding the perturbation arising from the quadrupole interaction (Eq. 6) and magnetic (Eq. 7) perturbations. Solutions for more complex cases have been described by Goldanskii and Makarov (1968).

Up to this point, we have discussed only the hf splitting of the nuclear energy levels, and have not demonstrated how those splittings lead to the hf spectrum. As shown in Fig. 4 for the $\frac{1}{2} \rightarrow \frac{3}{2}$ transition in $^{57}$Fe, in general, both the nuclear ground state and excited state are split by the hf interaction. In principle, one could expect a hf absorption line from each ground state $I_z$ level to each excited state $I_z$ level.

However, the character of the $\gamma$-ray emission, termed the *multipolarity*, allows transitions only between nuclear states with $I_z$ values differing by no more than the order of the multipolarity. Thus, for $^{57}$Fe and $^{119}$Sn, whose $\gamma$ transitions are both M1, if the absolute differences between the $I_z$ of the ground state and the excited state is greater than 1, the transition is forbidden. Thus (Fig. 4), there is no absorption line corresponding to the energy difference between the ground nuclear level with $I_z = -\frac{1}{2}$ and excited level with $I_z = \frac{3}{2}$ (the change in $I_z$ would be 2). Thus, the spectrum is restricted to the six lines shown in Fig. 4. Occasionally, spectra do have lines at the energies appropriate to forbidden transitions. This is the result of interactions in which $V_{zz}$ is not along the magnetic hf field. In such cases, the nuclear levels are no longer "pure" states of a particular $I_z$, but contain admixtures of other $I_z$ values. The $\gamma$-ray transition can then couple to the admixed $I_z$ value, producing lines that are usually relatively weak.

The relative intensities of the transitions connecting the nuclear states are determined by the transition multipolarity, the $I$ and $I_z$ of the initial and final states, and the angle between the $\gamma$-ray direction and the $z$ axis. The transition matrix elements are identical in form to those used in atomic spectroscopy and represent the interaction of the nuclear moments with the electromagnetic radiation field. For the important case of M1 radiation and a $\frac{3}{2}$–$\frac{1}{2}$ transition ($^{57}$Fe and $^{119}$Sn), the relative transition probabilities are

$$P(\pm\tfrac{1}{2}, \pm\tfrac{1}{2}) = \sin^2 \theta, \tag{8a}$$

$$P(\pm\tfrac{1}{2}, \mp\tfrac{1}{2}) = \tfrac{1}{4}(1 + \cos^2 \theta), \tag{8b}$$

$$P(\pm\tfrac{3}{2}, \pm\tfrac{1}{2}) = \tfrac{3}{4}(1 + \cos^2 \theta), \tag{8c}$$

where the first index represents the $I_z$ of the $\frac{3}{2}$ state, the second the $I_z$ of the $\frac{1}{2}$ state, and $\theta$ is the angle between the $z$ axis and the incoming (or outgoing) $\gamma$. The relative transition probabilities are the same for both emission and absorption.

We have indicated above that the $z$ axis for the hf interaction will be defined by the largest component of the electric field gradient $V_{zz}$ or the magnetic hf field direction. For absorbers in which all atoms see the hf axis in the same direction—e.g., a magnetic metal saturated by an applied field—the angle between the $\gamma$-ray beam and the $z$ axis will define a unique value of $\theta$ for all atoms, and observed intensities will be according to Eq. (8). Two special cases are of particular interest:

1. For magnetic hf structure, if the field is along the $\gamma$-ray direction, $P(\pm\tfrac{1}{2}, \pm\tfrac{1}{2}) = 0$, and those lines do not appear in the spectrum. That effect can be used either to simplify the spectrum or provide a sensitive measure of the direction of the internal field.

2. In systems in which the hf splitting is quadrupolar in origin, as shown previously, the four transitions included in Eqs. (8a) and (8b) all have the same energy and form one line of a doublet, with the two transitions in Eq. (8c) forming the other line. The ratio of the relative transition intensities is then $(2 + 3 \sin^2 \theta)/3(1 + \cos^2 \theta)$; measured transition intensity ratios can thus be used to determine the angle between $V_{zz}$ and the $\gamma$-ray axis.

Most samples for Mössbauer experiments are not oriented, however, but consist of powder or polycrystalline materials in which to some approximation all values of $\theta$ are present with equal likelihood. For such samples, the relative transition intensities are obtained by integrating Eqs. (8) over a sphere. For magnetic splitting, $a:b:c$ corresponds to $1:2:3$; this is the origin of the familiar intensity pattern seen in iron metal. In the quadrupole case, $(a + b):c$ is $1:1$, yielding a symmetric doublet. In reality, all samples can be expected to possess a certain degree of orientation, and extreme care is needed to determine quantitative information from line intensity ratios (see Section II,E).

It is worthwhile to sum up here some of the basic features determining Mössbauer spectra:

1. Three important interactions determine the energy of the lines—the isomer shift, which moves all lines in the spectrum by the same energy, and the electric quadrupole and magnetic dipole interactions, which split the nuclear energy levels and determine the splittings of the hf lines.

2. Each of those interactions can be factored into a nuclear term, which is always the same for a given transition, and a "solid state" or chemical term characteristic of the material under study.

3. The nuclear energy levels are determined by the hf interaction parameters; the observed spectrum is then determined by $\gamma$-ray absorption transitions connecting these energy levels, with relative intensities dependent on the transition multipolarities and absorber orientation.

4. The analysis has used implicitly the idea that all atoms in the sample had identical hf parameters. In cases in which inequivalent sites (i.e., classes of nuclei with different hf parameters) are present, each class contributes its own complete spectrum, and the observed spectrum is the sum of those arising from the individual sites.

One of the major advantages of Mössbauer spectroscopy is that it is useful to study systems with a substantial amount of damage or disorder. This capability arises from two factors. First, it is a relatively low-resolution measurement technique in that the size of the hf interaction energies is only moderately large (typically, 10–200 times) the natural linewidth of the resonance. Thus, even if a range of hf parameters exists, lines will not normally be broadened to the point of invisibility. Second, an examination of the origins of the hf parameters themselves shows that many of the most important hf parameters are determined by the ion under study, and most of the remainder by its immediate ligands. For example, any charge defect in the lattice leads to (neglecting screening) a field gradient of order charge/$r^3$ at a radius $r$ from it. Thus, field gradients from charge defects in the second and third coordination sphere will be at least 10 times down from those arising from the ligand ion charges, and the latter gradients are themselves relatively weak.

In contrast, polarization of conduction electrons by magnetic ions tends to be a long-range phenomenon, with the induced polarization dropping off only as $(\sin kr)/kr$ at a radius $r$ from the magnetic impurity ($k^{-1}$ is a wavelength characteristic of the conduction electrons). With even a few percent concentration of magnetic impurities, the net spin polarization of the conduction electrons provides a significant "average" hf field seen by the nonmagnetic probe atoms. Mössbauer studies of a model system of this kind—manganese as a magnetic impurity in Au–Mn random alloys—have shown (Cohen and West, 1971) that the resolution of the hf spectrum is adequate to begin to separate long- and short-range interaction terms. Most of the articles in this volume demonstrate, and some depend vitally on, the fact that hf parameters significant in determining the Mössbauer spectra are dependent almost entirely on the atom under study and its first two nearest-neighbor shells.

### D. Dynamic Effects—Paramagnetic Hyperfine Structure and Superparamagnetism

#### 1. Hyperfine Structure in Paramagnetic Materials

Up to this point, we have assumed that the hf fields determining the splitting of the nuclear energy levels were static, i.e., both magnitude and direction constant in time. In many materials this is a valid assumption. In others, the hf parameters may fluctuate so rapidly that the hf structure is determined by a time average of the fluctuating value; this occurs, for example, in magnetically ordered materials at temperatures at which the sublattice magnetization is significantly less than the saturation value. In fact, bulk samples of magnetically ordered systems containing high concentrations of 3d elements normally satisfy this criterion. The critical parameter determining whether the fluctuation is rapid enough to be considered as a static average field is the ratio $\tau_r/\tau_{hf}$, which we will call $\beta$. Here the mean time between changes of magnitude or direction of the hf field, or fluctuation time, is called $\tau_r$, and $\tau_{hf}$ is the reciprocal of the hf frequency $\nu_{hf}$, which is often called the *Larmour frequency*. The hyperfine frequency is related to the hyperfine splitting of the nuclear energy levels $E_{hf}$ via $\nu_{hf} = E_{hf}/h$, where $h$ is Planck's constant, $4.135 \times 10^{-15}$ eV sec. In general, values of $\beta < 0.01$ correspond to fluctuation rates that are rapid enough (fast relaxation) that the fluctuations will not be observed in the experimental spectra, and the observed hf structure will correspond to the mean value of the hf field. Values of $\beta > 100$ correspond to fluctuation rates so slow that each of the hf field values among which the fluctuations occur will provide a distinct hf spectrum. The spectra observed in systems with intermediate values of $\beta$ tend to be extremely complex.

For $^{57}$Fe, with typical hf coupling energies of about $10^{-7}$ eV, $\beta = 1$ for $\tau_r \approx 4 \times 10^{-8}$ sec; the hf coupling energies for rare earth ions (Ofer *et al.*, 1968) may be 100 times larger, corresponding to $\tau_r \approx 4 \times 10^{-10}$ sec. There are a number of different ways in which these relaxation spectra can provide useful information about the systems under study. Most obviously, spectra taken in the regime of intermediate relaxation time may be analyzed to determine $\tau_r$. Additionally, if $\beta$ is very large, it is possible to determine the hf parameters (including the magnetic coupling energy) even in systems that are not magnetically ordered.

Figure 8 shows the fluctuation process in a simple situation where the (magnetic) hf field inverts suddenly in direction but stays the same size. The case shown is appropriate to represent a material containing high-spin, paramagnetic $Fe^{3+}$, with only the $S_z = \pm\frac{5}{2}$ doublet populated. The hf field "up" would correspond to the $S_z = -\frac{5}{2}$ state, and vice versa.

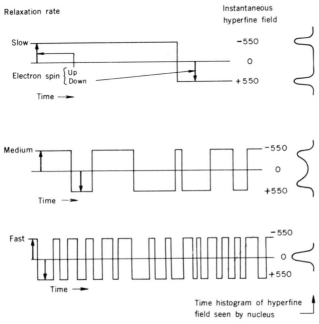

**Fig. 8.** Electron spin and effective instantaneous hf field as a function of time for various electron spin relaxation rates. The slow electron relaxation rate (top) corresponds to the situation obtained at low temperature in Fig. 9.

The actual experimental spectrum of a substance having approximately these properties is shown in Fig. 9 (adapted from Wickman and Wagner, 1969). In this work, $\tau_r$ was varied over a wide range by changing the sample temperature as indicated; changes in the spectra arise primarily from the change in $\tau_r$. These data show many features characteristic of spectra in the range of intermediate relaxation time; particularly evident are the broad shoulders on the 15°K spectrum and the persistence of weak absorption peaks at a splitting corresponding to the full hf field value. At the lowest temperature shown, the material is still paramagnetic (not magnetically ordered).

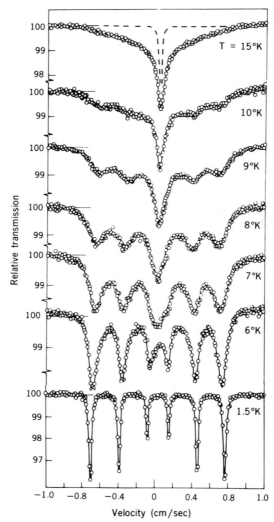

**Fig. 9.** Experimental spectrum of $^{57}Fe$ in a high-spin dithiocarbamate. At 1.5°K, slow relaxation of the $Fe^{3+}$ spin produces a well-resolved hf pattern like that in a static field case. At high temperature, rapid fluctuation of the spin averages out the hf pattern, but the shoulders remain broad. The dashed line shows the pattern that would be obtained if there were no hf broadening. (After Wickman and Wagner, 1969.)

There are two important classes of systems in which relaxation effects are common. The first is exemplified above by the iron dithiocarbamate, where the hf structure arises from a paramagnetic ion, or a small cluster of paramagnetic ions, isolated by virtue of being dilute in a diamagnetic surrounding. Almost all of the biological materials studied by Mössbauer spectroscopy fall into this category. The second type of system in which relaxation effects are important is that of very fine particles of magnetically ordered (either ferro- or antiferromagnetic) material and is discussed in the next section.

### 2. Superparamagnetism

Superparamagnetism is the process of collective reorientation of the magnetic moment direction in fine particles. It is commonly encountered in iron-containing clay minerals, pigments, chemical precipitation reaction products, and catalysts. A few biological systems containing large numbers of iron atoms in a single molecule (such as gastroferrin, see Chapter 5) are also in this category. Superparamagnetism arises in the following way: if a magnetic material is cooled below the magnetic ordering temperature, the spins of the magnetic ions will tend to lock together, producing magnetic ordering. In a large crystallite, once a particular spin orientation is established below the magnetic ordering temperature, it will not change except under the influence of external magnetic fields. If there are only some tens or hundreds of magnetic ions in the particle, however, thermal excitation energy may be enough to invert all of the spins simultaneously, thereby reversing the magnetization of all the sublattices. For the normal situation of no applied field, the overall energy of the two spin states is equal, and the mean time between spin flips is proportional to

$$\exp(KV/kT), \tag{9}$$

where $K$ is the anisotropy energy of the material, $V$ is the particle volume, $k$ is the Boltzmann constant, and $T$ is the temperature. This formula has been extensively used to analyze superparamagnetic hf structure by Kündig *et al.* (1966) and Afanas'ev *et al.* (1970); see also Chapters 5 and 6. Krop and Williams (1971) have studied the size of cobalt precipitates in Cu–Co alloys via the superparamagnetic hf structure.

The effect of the spin flips is the same as that shown above in Fig. 8. The hf field seen by the nucleus is aligned with the ion's spin moment, and when the spin is inverted by the superparamagnetic relaxation, the hf field inverts also. All spins in the particle flip simultaneously, in contrast to the case of paramagnetic relaxation discussed above, where spin flips of different ions occur independently. Spectra obtained are similar to those in Fig. 9, however.

There are three ways in which hf structure in the Mössbauer spectra of fine particles can be useful. First, the observation of relaxation spectra below the magnetic ordering temperature may be evidence that very fine particles are involved. Second, the temperature dependence of the relaxation time may be determined from the spectra and through Eq. (9) used to determine the particle size. Third, at very low temperatures, $\tau_r$ will be very long, and the spectrum will be that obtained in the static case. If the hf parameters are the same as those obtained for the bulk material, that is good evidence that the microstructure of the particles is similar to that of the bulk.

### 3. Paramagnetic Hyperfine Structure in Cases of Slow Relaxation

As mentioned above, most of the molecules of importance for biological studies contain individual iron ions, or a cluster of a few iron ions, relatively isolated from other paramagnetic groups. Especially for $Fe^{3+}$ ions, the spin relaxation arises primarily from the presence of nearby paramagnetic ions, and the spin relaxation time $\tau_r$ thus tends to be long in these systems. Since it is difficult to extract the hf parameters from the complex spectra observed in the regime of intermediate relaxation time ($\beta \approx 1$), biological materials are often studied at very low temperatures, where $\beta \gg 1$. Under these conditions, the spins of the iron ions are essentially static, and, as in the dithio-carbamate case discussed above, well-resolved spectra are obtained and the hf parameters can readily be extracted.

The interpretation of paramagnetic hf structure in many systems is more complicated than indicated above. We have so far used implicitly the *effective field model* for hf structure; i.e., the idea that the nucleus acts as a probe for a hf field that is produced by the ion and its surroundings, and that the effective hf field is independent of the nuclear state and properties. This approximation is a valid one when the hf interaction energy between the nucleus and the ionic electrons is small compared to the coupling energy between the ion and the surroundings. The hf interaction is typically $10^{-7}$ eV for iron and $10^{-5}$–$10^{-6}$ eV for rare earth ions. Thus, concentrated para-magnetic materials, magnetically ordered materials, and materials in applied fields stronger than $\sim 20$ kG, where the interaction energy of the electrons with the surroundings is at least $10^{-3}$ eV, can usually be satisfactorily treated using the effective field approach.

Most of the biological molecules studied, and some systems of dilute rare earths, cannot be adequately treated in this model, and much more complex theories have been developed to analyze the data. Both the general ap-proaches used and the application to specific systems are described in Chapters 5 and 6.

## E. Lineshape and Resonance Intensity

### 1. Resonance Cross Section and Linewidth

The absorption cross section for $\gamma$ rays to produce a transition between nuclear ground and excited states at resonance is given by

$$\sigma_0 = \frac{1}{2\pi} \frac{h^2 c^2}{E_0^2} \frac{2I_e + 1}{2I_g + 1} \frac{1}{1 + \alpha}, \tag{10}$$

where $h$ is Planck's constant, $c$ is the velocity of light, $E_0$ is the transition energy, $I_e$ and $I_g$ are the excited- and ground-state spins, respectively, and $\alpha$ is the internal conversion coefficient. [The value of $\alpha$ describes the relative strength of radiative ($\gamma$ ray) and nonradiative (electron conversion) processes connecting the ground and excited states; $\alpha = 0$ if all the decays from the excited state involve the emission of a $\gamma$ ray.] The fact that this cross section is dependent entirely on nuclear parameters is an important and useful feature of the Mössbauer effect. For a single $^{57}$Fe nucleus, $\sigma_0$ has the value $2.5 \times 10^{-18}$ cm$^2$ for the 14 keV transition. This cross section is about $10^6$ times larger than the actual geometrical cross section of the nucleus and is also very large compared to the photoelectric (electronic) absorption cross section for iron, which is $5.5 \times 10^{-21}$ cm$^2$ per atom at this energy. Thus, the nuclear resonance absorption process is a strong effect.

The size of the observed resonance will depend not only on the cross section per nucleus, but on the *effective thickness $t$* of the absorber,

$$t = nf\sigma_0 \tag{11}$$

where $n$ is the number of atoms of the Mössbauer species per square centimeter, and $f$ is the recoil-free fraction (Section II,E,2). If the nuclear levels are split via hf interactions, the resonance cross section of Eq. (10) is divided among the different transitions, and the effective value of $t$ is decreased accordingly. For small values of $t$ ($\leqslant 0.5$), the intensity of the observed resonance is approximately proportional to $t$. For values more typically used in experiments ($\sim 1 \leqslant t \leqslant \sim 4$), the proportionality fails badly, and corrections must be made for this effect if the intensity of the observed resonance is being used to establish the relative amount of different phases. This problem is discussed in detail by Schwartz in Chapter 2 of this book, and by Shenoy et al. (1975).

Equation (10) gives the maximum value of the resonance absorption cross section. The energy dependent absorption cross section is

$$\sigma = \sigma_0 [1 + 4(E - E_0)^2 / \Gamma^2]^{-1} \tag{12}$$

where $E$ is the incident $\gamma$-ray energy and $\Gamma$ is the uncertainty-principle energy width of the nuclear excited state. This width is defined by $\Gamma = h/2\pi\tau$, where

$\tau$ is the *mean life* ($= 1.44 \times$ the half life) of the excited state. For $^{57}$Fe, with a half life of $10^{-7}$ sec for the 14 keV level, $\Gamma = 4.65 \times 10^{-9}$ eV, and this is called the *natural linewidth* of the level. In an actual experiment, both the source radiation and the absorber cross section have linewidth values of $\Gamma$ associated with them, and the minimum width actually observable in a transmission experiment is $2\Gamma$, a value corresponding to 0.2 mm/sec of velocity Doppler shift for $^{57}$Fe. The corresponding value for $^{119}$Sn is 0.65 mm/sec.

Two effects make the linewidths actually observed in experiments broader than these theoretical minimum values. One, called *inhomogeneous broadening*, occurs when the Mössbauer atoms in the absorber have a small range of, e.g., isomer shift values associated with them. Inhomogeneous broadening thus arises from imperfections in the source or absorber. The other effect is called *thickness broadening*, or *saturation broadening*, and arises from the fact that when the absorber is thick ($t > 1$), the absorption effect at the center of the resonance saturates before that at the wings of the line. This makes the observed spectral line wider, increasing the effective $\Gamma$ by a factor of $\sim(1 + 0.135t)$ for $t < 5$. The cross section shown in Eq. (12) has *Lorentzian shape* and is of the form $1/(1 + x^2)$ about $E_0$. For thin absorbers, the observed resonance curve has this shape, and Lorentzian lineshapes are frequently used to fit Mössbauer spectra. For precise determinations of line areas, or of hyperfine parameters of poorly resolved spectra, the deviations of the observed lines from Lorentzian shape must be taken into account. Shenoy *et al.* (1975) describe the possible origins of these deviations and survey the approaches that can be used to analyze data correctly.

### 2. The Recoil-Free Fraction

Gamma rays have a momentum equal to $E_\gamma/c$, where $E_\gamma$ is the $\gamma$ energy and $c$ is the velocity of light. When a $\gamma$ ray is emitted from an excited nucleus, conservation of momentum requires that the nucleus recoil with an equal momentum in the opposite direction. The kinetic energy of the nucleus corresponding to this amount of recoil momentum is (approximately) $E_r = E_\gamma^2/2Mc^2$, where $M$ is the nuclear mass and $E_r$ is called the recoil energy loss. The value of $E_r$ is typically $10^{-2}$–$10^{-3}$ eV for transitions used in Mössbauer spectroscopy, but may be as large as 1–10 eV for high-energy $\gamma$ rays emitted from light nuclei. The recoil energy is supplied from the initial energy $E_0$ of the nuclear excitation. The nucleus gets a kinetic energy $E_r$, and the $\gamma$ energy is $E_0 - E_r$, slightly less than the intrinsic energy of the nuclear excitation. Although $E_r$ is only a tiny fraction of the $\gamma$-ray energy (one part in $10^7$ for $^{57}$Fe), it is much larger than the width of the nuclear level, typically $10^{-7}$ eV. Thus, the $\gamma$ rays emitted from a freely recoiling nucleus are far too low in energy to be resonantly absorbed by a nucleus of the same type.

In 1958, Mössbauer demonstrated both experimentally and theoretically how to overcome this problem. He showed that for a nucleus bound in a solid, the recoil energy loss could be essentially eliminated for a significant fraction of the γ-ray transitions. This *recoil-free fraction* is called $f$ and has value ranging from almost unity in a few cases to 0.1 or 0.01 in more typical cases. Since only the recoil-free γ rays have the right energy to be resonantly absorbed (see Fig. 10), the maximum size of the resonance effect observed in a transmission experiment can never be larger than $f$, and many experiments are limited in feasibility or precision by small values of $f$.

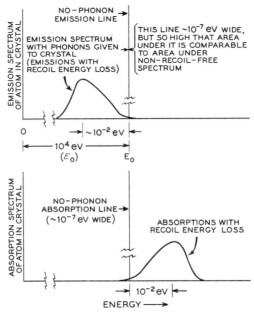

**Fig. 10.**   Energies of no-phonon emission and absorption lines, and emission and absorption spectra in the case of recoil energy loss. Note that only the no-phonon lines overlap in energy. The fraction of the total area under the no-phonon line is $f$, the recoil-free fraction.

In most Mössbauer experiments, where the values of the hf parameters or the identification of substances from their characteristic spectra are the objective, the size and parametric dependence of $f$ are not significant. Increasingly, recently, the Mössbauer effect has been used for studies of diffusion, measurements of the binding forces of atoms in solids, and quantitative analysis; in all of these phenomena, the size of the recoil-free fraction is a primary object of study. We will thus outline the fundamental variables which determine $f$.

Mössbauer's original analysis of the processes underlying the recoil-free $\gamma$-ray emission was based on previous theoretical work (Lamb, 1939) developed to describe neutron scattering from atoms bound in crystals. That approach considers the elementary vibrational excitations (phonons) created in the lattice by the recoil from the $\gamma$-ray emission. In the Mössbauer–Lamb approach, if a Debye model is assumed for the solid,

$$ f = \exp\left(-\frac{6E_r}{k\theta_D}\left[\frac{1}{4} + \left(\frac{T}{\theta_D}\right)^2 \int_0^{\theta_D/T} \frac{x}{e^x - 1}\, dx\right]\right), \tag{13}$$

where $T$ is the temperature, $\theta_D$ is the characteristic Debye temperature of the crystal, $k$ is the Boltzmann constant ($8.616 \times 10^{-5}$ eV/°K), and $E_r$ is the recoil energy loss as defined above. This reduces to the forms

$$ f = \exp\left(-\frac{3}{2}\frac{E_r}{k\theta_D}\right) \quad \text{at} \quad T = 0°\text{K} \tag{14}$$

and

$$ f \approx \exp\left(-\frac{6E_r T}{k\theta_D^2}\right) \quad \text{for} \quad T > \theta_D/2. \tag{15}$$

A useful plot showing the temperature dependence of $f$ was generated by Muir (1962) and reproduced by Cohen and Wertheim (1974).

It should be emphasized that the temperature dependence of $f$ in Eqs. (13–15) is obtained only for a solid for which the Debye model is a valid description of the phonon spectrum. In fact, deviations between the observed temperature dependence of $f$ and Debye model values have recently been used to determine the anharmonic force constants (Howard and Nussbaum, 1974).

An alternate analysis of the recoil-free emission process yields

$$ f = \exp(-4\pi^2\langle x^2\rangle/\lambda^2), \tag{16}$$

where $\lambda$ is the wavelength of the $\gamma$ ray (0.86 Å for the 14 keV $\gamma$ of $^{57}$Fe). The term $\langle x^2\rangle$ is the mean of the square of the displacement of the emitting atom from its equilibrium position under thermal vibration. In cases where the mean-square vibration displacement amplitude is different along different directions in the crystal, $f$ will be a function of the direction of $\gamma$-ray emission. This variation will produce an effective orientation of the source or absorber, often called the Goldanskii–Karyagin effect (Goldanskii and Makarov, 1968).

The recoil-free fraction can be strongly temperature dependent, and the rapid decrease in $f$ at high temperatures limits the range over which the

Mössbauer effect can be a useful tool. For isotopes with low energy $\gamma$ transitions ([57]Fe, [119]Sn, [151]Eu, [161]Dy, [169]Tm, [181]Ta) in materials of moderately high Debye temperature, measurements can be made as high as 700–1500°C before $f$ becomes too small. Most of the other isotopes with $\gamma$ transitions of 60 keV or more can only be used for measurements at temperatures well below 100°K.

### 3. Second-Order Doppler Shift

The *second-order Doppler shift* (SODS), also called the *thermal red shift*, is a small decrease in the energy of the $\gamma$-ray emission or absorption resulting from the relativistic effects of the thermal vibrational velocity of the nuclei. The shift has the value

$$\delta E_{SODS} = -\tfrac{1}{2}(\overline{v^2}/c^2)E_0 \tag{17}$$

where $\overline{v^2}$ is the mean of the square of the velocity of the atom as it vibrates in the lattice, $c$ is the velocity of light, and $E_0$ is the $\gamma$ transition energy. Since the second-order Doppler shift depends only on the vibrational motion of the atoms, it does not usually convey any useful information. It is, however, superimposed on the isomer shift, and must be taken into account, especially if the isomer shift temperature dependence is being studied.

At temperatures high compared to the Debye temperature $\theta_D$, $\overline{v^2}$ is proportional to temperature and produces a shift of $-0.073$ mm/sec (for [57]Fe) and $-0.035$ mm/sec (for [119]Sn) per 100°K. The measured linewidth for iron is often $\sim 0.3$ mm/sec, so that a shift of about one-quarter of a linewidth is produced per 100°K temperature change. For [119]Sn, both the larger linewidth ($\sim 1$ mm/sec) and smaller shift make the thermal shift unimportant except in cases where very precise measurements are made.

It should be emphasized that the thermal shift values cited here are valid only at high temperatures. For temperatures much lower than the Debye temperature of the lattice, the shift becomes vanishingly small and depends on the relationship between the temperature and $\theta_D$. A family of curves showing the low-temperature behavior of the second-order Doppler shift is shown by Tseng et al. (1968).

Precise determination of the true isomer shift between a source and absorber with different $\theta_D$ values requires either measurements at temperatures $\gg \theta_D$ (where source and absorber SODS are the same), or a series of measurements at different temperatures, from which the effective $\theta_D$ values can be determined. In practice, the temperature dependence of the isomer shift (due to lattice expansion or wave function changes) introduces complications that make it difficult to determine these corrections accurately.

## III. Experimental Techniques

### A. *Mössbauer Spectrometers, Detectors, and Data Collection*

There are two extensive reviews of techniques used in Mössbauer experiments (Cohen and Wertheim, 1974; Kalvius and Kankeleit, 1972) (see also Chapter 4). The purpose of this section is to provide an overview of the approaches used in experimental work and familiarize the reader with terminology commonly used in discussing Mössbauer spectroscopy results.

As indicated in Sections I and II, most Mössbauer spectroscopy experiments are transmission experiments, with the energy scan being obtained by a Doppler shift. Generally, the source is held at room temperature and moved by the velocity drive to produce the Doppler modulation scanning of the $\gamma$-ray energy. Electromagnetic velocity scanning drives, which operate like moving-coil loudspeakers, are by far the most commonly used. The state of the art of drive design is such that, except in experiments of unusually high precision, error arising from drive calibration and nonlinearity should be negligible. Spectra are normally displayed in velocity units of millimeters per second or centimeters per second.

Absorbers typically contain a layer of 2–100 mg/cm² of the material under study and vary from $\sim 1$ mm² to $\sim 2$ cm² in area. Metallic samples may be rolled, filed, or ground; crystalline materials are generally powdered and mixed with an inert binder such as wax or plastic resin. When very thin ($\sim 10$ mg/cm² or less) absorbers are made of powdered material, it is common to mix an inert filler, such as graphite or boron nitride powder, with the sample to aid dispersion.

If the experiments are to be carried out above or below room temperature, the absorber is mounted in an oven or dewar. Such equipment is basically similar to that used in other types of research, but has two special features—windows, which are transparent to the $\gamma$ rays, and a rigid mechanical structure to prevent vibration of the absorber. Even very small vibration amplitudes can produce enough velocity to broaden the observed spectral lines.

For transmission experiments, $\gamma$ rays are detected with conventional commercial equipment; most commonly, gas-filled proportional counters are used for $^{57}$Fe, since they are relatively insensitive to an interfering 122 keV $\gamma$ ray, which is ten times as intense as the 14 keV $\gamma$ ray used for Mössbauer experiments. For $^{119}$Sn, sodium iodide scintillation detectors are used with a critical absorber of palladium. The palladium acts as a filter to selectively attenuate the tin K x rays, which are about 1 keV more energetic than the $\gamma$ ray of interest. Germanium and silicon semiconductor detectors are widely used for other isotopes. Detectors used for backscatter experiments tend to be specialized to the particular application involved and sometimes have the absorber under study incorporated as part of the counter volume.

The spectroscopic data are taken in the form of $\gamma$-ray counts versus Doppler velocity. Most commonly, multichannel analyzers are used to collect the data. The analyzer is synchronized with the drive so that each analyzer channel (memory location in which data is stored) corresponds to a particular drive velocity. Thus, the channel number corresponds to the Doppler-shifted $\gamma$-ray energy, and the channel contents corresponds to the number of $\gamma$-rays transmitted through the absorber at that energy.

Normally, data from both transmission and backscatter experiments are plotted as count rate versus source velocity, so that for transmission experiments, absorption lines appear as dips below the baseline, while for scattering experiments, the lines appear as peaks. Unless otherwise stated, *positive velocity* signifies motion of source toward the absorber, so that the effective source $\gamma$-ray energy is increased. Thus, resonance lines at positive velocity result when the transition energy in the source is less than the transition energy in the absorber.

### B. Data Analysis

Precise analysis of Mössbauer data is normally carried out using computer least-squares fitting for two reasons: (1) the spectra contain hundreds of data points, which are already in digital form and are difficult to deal with graphically or by hand; (2) the observed line positions are often not independent, but arise from the possible combinations of hf splittings of the excited and ground nuclear levels, as outlined above. Thus, the relative line positions (and intensities) are constrained to those arising from the hf Hamiltonian describing the interaction of the nucleus with its surroundings. These constraints are simple to insert in a computer program, but are difficult to use in manual data reduction.

Data reduction by constrained least-squares fits allows use of all of the information in the spectrum to determine the hf parameters and therefore produces high precision. More important, however, the positions (and approximate intensities) of all the lines can be specified by just a few hf parameters. Thus, spectra in which some lines are incompletely resolved can be analyzed with confidence. This approach is particularly useful when the sample under study contains inequivalent lattice sites or two or more phases.

When Mössbauer spectroscopy is being used as an analytical tool, the line positions and relative intensities are constrained to those arising from the phases anticipated to be present in the unknown. The least-squares fitting program then contains only the degrees of freedom corresponding to the relative intensities of spectra arising from the various phases present. Chapter 2 contains detailed examples of this approach and brings out the importance of saturation effects. These effects, also called finite absorber thickness effects, make the intensity of the observed spectrum a nonlinear function of absorber thickness (Section II,E,1). Thus, if quantitative absorber thickness measurements are to be obtained from Mössbauer experiments, corrections must be made for this phenomenon.

## IV. Limitations and Future Possibilities

### A. New Isotopes

Mössbauer spectroscopy is necessarily restricted to nuclei having low-lying excited states. Nuclear level schemes are well established for stable isotopes, and the likelihood that new, unexpected Mössbauer states will be found is very small. Over the past few years, there have been many successful attempts to develop techniques for using isotopes with very narrow lines, such as $^{67}$Zn, $^{73}$Ge, and $^{181}$Ta. All of these isotopes have linewidths about one hundred times sharper than that of $^{57}$Fe. The major challenge in the experimental work with such isotopes is to produce sources and absorbers that are pure enough and free enough from defects that line broadening from inhomogeneities does not destroy the resonance. Although this has been accomplished for these isotopes with $\sim 1$ $\mu$sec half life for the Mössbauer state, the care required suggests that basically new approaches will be required to use longer lived nuclear states for Mössbauer spectroscopy. Proposals for eliminating the effects of at least some of the possible sources of inhomogeneous line broadening have been made in conjunction with speculative proposals to build $\gamma$-ray lasers using stimulated emission (via the Mössbauer effect) from nuclear excited states (Baldwin and Khokhlov, 1975).

The lower limit of nuclear lifetimes used in Mössbauer effect experiments is not limited by feasibility, but by the fact that for nuclear state lifetimes of $\sim 10^{-10}$ sec or less, the natural linewidth (Eq. 12) is very broad. Thus, it is difficult to make any measurements on hf interactions, since they are smaller than the natural linewidths.

The maximum $\gamma$-ray energy that can be utilized for Mössbauer spectroscopy is limited to about 160 keV, since the $f$ value (see Eqs. 13–15) decreases very rapidly at higher energies. For low $\gamma$-ray energies, the recoil-free fraction is unity, but the electronic absorption cross section becomes very large compared to the Mössbauer cross section (Eq. 10), and the resonance becomes more difficult to observe. The 6 keV $\gamma$ ray of $^{181}$Ta has been used for a number of experiments despite this difficulty.

Thus, it appears that very few new isotopes will be added to the list of those already used for Mössbauer experiments. The main growth in research using the Mössbauer effect will come from the study of new problems with the currently used isotopes. This projection is borne out by the fact that over the past ten years, research using $^{57}$Fe has consistently produced about half the Mössbauer effect literature, and research with $^{119}$Sn has produced most of the remainder. This emphasis has been carried into the articles in this volume.

## B. Absorber Size and Geometry

As stated above, an average absorber for a transmission experiment consists of a sheet or powdered layer of the material to be studied, a fraction of a millimeter thick and of area 0.1–1 cm$^2$. In general, from a few milligrams to 1 gm of material is used, with the larger quantities needed only if the $f$, isotopic abundance, or concentration of the absorbing element is very low. Using extremely small area sources and absorbers, experiments have been carried out with as little as $\sim 1$ $\mu$g of material.

The extent to which the Mössbauer species can be used as a probe atom, dispersed in a matrix of other elements, is limited primarily by the fact that the host will absorb, via photoelectric processes, the $\gamma$ rays to be studied. Thus, the restriction is that the electronic absorption cross section for the $\gamma$ rays should not be more than ten to a hundred times the Mössbauer cross section, or the resonance will be too weak to observe. For $^{57}$Fe and $^{119}$Sn in hosts of low or moderate atomic number, this condition can be met for a concentration of 0.1–1 at. % of the Mössbauer element. This can be a severe limitation in experiments to study biological materials, in which a few iron atoms may be present in a protein with a weight of 100,000. In such situations, it is frequently possible to make the material with $^{57}$Fe or to enrich it with $^{57}$Fe by exchange. Since normal iron contains only 2.2% $^{57}$Fe, a gain of almost fifty times is attainable by this enrichment process (see Chapter 4).

## C. Source Experiments

In metallic systems, it is often possible to study cases in which the Mössbauer probe atom is very dilute by performing a source experiment. The radioactive parent isotope of the Mössbauer state is doped into the material to be studied, and then the hyperfine perturbations are observed by their effect on the energy of the $\gamma$ ray emitted in the radioactive decay. Since only about $10^{12}$ active atoms are necessary to carry out such an experiment, extremely high dilutions can be obtained. The radioactive decay often produces ionization of the atom, and additional complications can occur in source experiments. If the source is metallic, in general, all of the inner and outer shell holes will be filled long before the Mössbauer $\gamma$ ray is emitted. Thus, the spectrum observed will correspond to that expected in an absorber experiment, where the ion is in its ground state. In insulating sources, however, effects arising from the ionization make the spectrum complex, and not, in general, representative of the corresponding absorber.

### General Bibliography

An excellent introduction to the basic principles of Mössbauer Spectroscopy is given by G. K. Wertheim ("Mössbauer Effect; Principles and Applications," Academic Press, 1964). An encyclopedic book (N. N. Greenwood and T. C. Gibb, "Mössbauer Spectroscopy," Chapman and Hall, London, 1971) contains an extensive survey of the research literature, with emphasis on the chemical aspects. A new book (G. K. Shenoy and F. E. Wagner, "Mössbauer Isomer Shifts," North Holland Publ., Amsterdam, 1975) contains detailed reviews of the isomer shift fundamentals and results, with a more physical emphasis. The *Proc. 1974 Int. Conf. Appl. Mössbauer Effect* have been published as *J. Phys. (Suppl., Colloq.* C–6) (1974). Proceedings of a periodic conference are published as "Mössbauer Methodology" (I. J. Gruverman, (ed.), Vol. 1–9. Plenum Press, New York, 1965–75). The "Mössbauer Effect Data Index" (J. G. Stevens and V. E. Stevens (eds.), IFI/Plenum Press, New York, 1966–73) tabulates and indexes essentially all Mössbauer effect research results published.

### References

Afanas'ev, A. M., Suzdalev, I. P., Gen, M. Ya., Goldanskii, V. I., Korneev, V. P., and Manykin, E. A. (1970). *Zh. Eksp. Teor. Fiz.* **58**, 115.

Baldwin, G. C., and Khokhlov, R. V. (1975). *Phys. Today* **28** (2), 32.

Cohen, R. L., and Wertheim, G. K. (1974). *In* "Methods of Experimental Physics," (L. Marton, ed.) vol. 11, pp. 307–369. Academic Press, New York.

Cohen, R. L., and West, K. W. (1971). *J. Phys. (Suppl. Colloq.* C–1) **32**, 781.

Freeman, A. J., and Watson, R. E. (1964). *Phys. Rev.* **135**, A1209.

Goldanskii, V. I., and Makarov, E. F. (1968). *In* "Chemical Applications of Mössbauer Spectroscopy" (V. I. Goldanskii and R. H. Herber, eds.), pp. 1–113. Academic Press, New York.

Goldanskii, V. I., Devisheva, M. N., Makarov, E. F., Novikov, G. V., and Trukhtanov, V. A. (1966). *Zh. Eksp. Teor. Fiz. Pis'ma Red.* **2**, 63.

Howard, D. G., and Nussbaum, R. H. (1974). *Phys. Rev. B* **9**, 794.

Kalvius, G. M., and Kankeleit, E. (1972). *In* "Mössbauer Spectroscopy and Its Applications," pp. 9–88. Int. Atomic Energy Agency, Vienna.

Krop, K., and Williams, J. M. (1971). *J. Phys. F* **1**, 938.

Kündig, W., Bömmel, H., Constabaris, G., and Lindquist, R. H. (1966). *Phys. Rev.* **142**, 327.

Lamb, W. E. (1939). *Phys. Rev.* **55**, 190.

Mössbauer, R. L. (1958a). *Z. Phys. 151*, 124. Mössbauer, R. L. (1958b). *Naturwissenschaften* **45**, 538.

Muir, A. H., Jr. (1962). Unpublished report.

Ofer, S., Nowik, I., and Cohen, S. G. (1968) *In* "Chemical Applications of Mössbauer Spectroscopy" (V. I. Goldanskii and R. H. Herber, eds.), pp. 427–503. Academic Press, New York.

Shenoy, G. K., Friedt, J. M., Maletta, H., and Ruby, S. L. (1975), *In* "Mössbauer Effect Methodology" (I. J. Gruverman, ed.), Vol. 9, p. 277. Plenum Press, New York.

Sternheimer, R. M. (1963). *Phys. Rev.* **132**, 1637.

Tseng, P. K., Ruby, S. L., and Vincent, D. H. (1968). *Phys. Rev.* **172**, 249.

Walker, L. R., Wertheim, G. K., and Jaccarino, V. (1961). *Phys. Rev. Lett.* **6**, 98.

Wertheim, G. K. (1964). "Mössbauer Effect: Principles and Applications." Academic Press, New York.

Wickman, H. H., and Wagner, C. F. (1969). *J. Chem. Phys.* **51**, 435.

# Metallurgy

# 2

# *Ferrous Alloy Phase Transformations*

**L. H. Schwartz**

Materials Science and Engineering Department
Northwestern University
Evanston, Illinois

## I. Introduction

The growth in activity in Mössbauer spectroscopy has been accompanied by the use of the Mössbauer effect as an analytical tool for increasingly more quantitative studies of metallurgical phenomena. In these studies, the metallurgist shares with his colleagues in biology, chemistry, and magnetism the fact that the most conveniently studied isotope, $^{57}$Fe, is also one of the most interesting. Metallurgical applications of the Mössbauer effect have been extensively reviewed, although critical evaluation of the literature has been rare. Perhaps the first suggestions of the potential of this new technique for metallurgy were noted by Epstein (1963). Subsequently reviews by Flinn (1967a, b, 1971, 1973), Bokshtein, *et al.* (1968), Gonser (1971),

Jones (1973), and Ron (1973) have appeared, all listing some of the many applications in a variety of metallurgical experiments. Descriptions of quantitative data treatment may be found, for example, in Chow *et al.* (1969) and Schwartz (1970).

With such a formidable number of predecessors, yet another review article in this field may seem redundant. However, recent concerns with quantitative analytical results justify an in-depth, critical examination of the field. To keep the discussion within reasonable bounds, a somewhat arbitrary division of the subject has been made. The present article deals with transformations in ferrous alloys, and includes most of the extensive literature in this field.

Metallurgical studies of transformations using the Mössbauer effect usually depend for their interpretation on the differences between spectra and phases exhibiting long-range magnetic order and spectra of phases that are paramagnetic or diamagnetic. In many cases, determination of the metallurgically variable parameters depends on an understanding of the hyperfine fields at the iron nuclei in alloys and how they depend on local atomic environment. This paper will commence with a brief review of the experimental analysis of the Mössbauer spectra of single phase, substitutional iron-based alloys. Emphasis will be given to the importance of considering the possibilities of anisotropic hyperfine interactions and to accounting for the finite thickness of the sample in fitting programs. In subsequent sections, consideration will be given to the applications of the Mössbauer effect to decomposition into two phases via nucleation and growth and via spinodal decomposition and to the related determination of retained austenite in martensitic ferrous alloys. The final sections of the paper will deal with the determination of local and long-range atomic order in ferrous alloys using the Mössbauer effect.

## II. Mössbauer Spectra of Iron Alloys

### A. *Hyperfine Interactions in Iron Alloys*

An example of the complex Mössbauer spectrum for Fe–3.2 at. % Mo (Asano, 1974) is shown in Fig. 1. Note that while six clearly defined resonance envelopes can be seen, the spectrum is significantly modified from the six well-defined resonance lines characteristic of ferromagnetic iron. The presence of the solute atoms evidently perturb the hyperfine fields characteristic of the pure iron.

The basis for most of the analyses of hyperfine fields in iron alloys is the work of Wertheim *et al.* (1964). The Mössbauer spectrum for dilute polycrystalline iron alloys was shown by them to be due to the superposition of many six-line subspectra, each with hyperfine magnetic fields and isomer

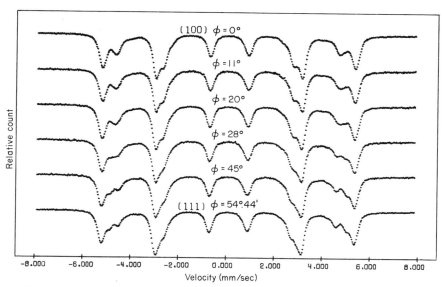

**Fig. 1.**  Mössbauer spectra of single crystal Fe–3.2 at. % Mo. An external field of 2.7 KOe was used to polarize the sample at various angles $\phi$ relative to the crystal axes and reveal the anisotropic nature of the hyperfine fields. (Asano, 1974.)

shift determined by the near neighbor (nn) and next near neighbor (nnn) solute concentration. The intensities of each of these subspectra was taken proportional to the fraction of iron atoms with $i$-nn and $j$-nnn solute atoms, $w(i, j)$. In their analysis the values of $w(i, j)$ were calculated assuming a random distribution of solute atoms on substitutional lattice sites. While this assumption may be valid for very dilute alloys, it is almost certainly violated in more concentrated ordering alloys, for example, $Fe$Al and $Fe$Si studied by Stearns (1966) and Cranshaw (1964) and later by many other workers. Wertheim *et al.* (1964) concluded that the hyperfine magnetic field $H(i, j)$ at a given iron nucleus was changed from that for $^{57}$Fe in iron metal $H_0$ by an amount

$$H(i, j) - H_0(1 + kc) = H_0(ia + jb)(1 + kc) \tag{1}$$

where $a$, $b$, and $k$ are empirically determined for a particular alloy, and $c$ is the atomic concentration of the solute. The essential sufficiency of Eq. (1) in describing various iron alloys has been verified by many workers [see, e.g., Vincze and Campbell (1973) for a partial review of this extensive literature]; however, Stearns (1971) has shown the importance of considering the weak effects of solute atoms more distant than the second neighbor shell in determining the electronic origin of the hyperfine fields. While some controversy continues regarding the origin of these fields, effects due to conduction electron polarization (discussed recently by Stearns, 1971) and local magnetic moment changes due to the introduction of the solute (see Stearns, 1974 and Vincze and Aldred, 1974), both play important roles.

Fortunately, understanding the physical origin of the hyperfine fields in terms of electron redistribution is not crucial to the metallurgical applications. What is important, however, is knowing which subspectrum is associated with which atomic configuration. The most serious shortcoming of Eq. (1) in this context is the assumption that the magnetic hyperfine field is isotropic, i.e., that $H(i, j)$ is determined only by the distance of the solute atom from the iron atom in question. Polycrystalline samples are not useful for delineating anisotropic effects as the sample averages over all crystallographic orientations. Using single crystal samples, Cranshaw (1964, 1972) has shown that strong anisotropic perturbations to the magnetic hyperfine fields exist in many alloys of iron with transition and nontransition elements. This anisotropic interaction partially eliminates the degeneracy for iron atoms in the same neighbor shell, splitting the resonance line associated with these iron atoms. The amount of the splitting depends on the angle between the magnetic spin polarization and the vector from the iron to solute atom, and differs in magnitude for different solutes. Using the assumed magnetic dipole angular dependence, Cranshaw showed that the (1, 0) and (0, 1) satellite positions could be located by taking spectra with magnetic field saturating the sample along the [100] and [111] directions.

In a recent study by Asano and Schwartz (1973), the dipole symmetry of the anisotropic perturbation was confirmed using single crystals of dilute $Fe$Mo alloys. In these alloys, when the magnetic spins are aligned along the easy axis [100] (as is also the case in all experiments on samples in the absence of magnetic fields), the anisotropic effect splits the resonance line associated with iron atoms with one molybdenum in the nnn shell by an amount of the order $\Delta H(0, 1)/2$ as shown in Fig. 2. The circles indicate the data taken in transmission through a one mil thick single crystal absorber of Fe–3.2 at. % Mo. Only the lowest energy resonance envelope is shown. In Fig. 2a, the data have been fit as is most commonly done, neglecting the anisotropic interaction. In Fig. 2b, the anisotropic splitting has been accounted for [for details of the fitting procedure, see Asano (1974) and Asano and Schwartz (1973)]. In this case, neglect of the large anisotropic interaction leads to a totally erroneous identification of the location of the subspectrum due to iron atoms with one molybdenum in the nnn shell. Furthermore, neglect of this peak splitting can lead to errors in the determination of the resonance peak areas [proportional to $w(i, j)$], and hence to errors in the estimate of the local atomic order in the sample. While the magnitude of this anisotropic splitting may be negligibly small for diamagnetic solutes (Stearns, 1971), it is clearly not negligible for transition metal solutes far from iron in the

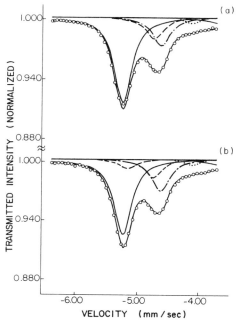

**Fig. 2.**  The lowest energy envelope of the Mössbauer spectrum of single crystal Fe–3.2 at. % Mo taken in transmission with an external magnetic field of 2.7 KOe along the (100) direction. ○: experimental data; ——: (0, 0) peak; –·–: (1, 0) peak; ———: (0, 1) peak; ···: superposition of the (1, 1), (2, 0), (0, 2) peaks.  (a) Analysis neglecting anisotropic hyperfine interaction; (b) analysis including anisotropic hyperfine interaction; no correction for finite sample thickness has been made. (Schwartz and Asano, 1974.)

periodic table (Cranshaw, 1972). Unless evidence to the contrary is present, data neglecting the anisotropic interaction should be considered suspect.

## B. Finite Sample Thickness

As noted in Chapter 1 by R. L. Cohen, the cross-section for emission or absorption of resonant $\gamma$-rays is Lorentzian in the energy. In the most commonly encountered experiment, transmission through an absorber of finite thickness $t_a$ the normalized transmission $P(V)$ is given by

$$P(V) = (1 - f_s) + f_s(\Gamma/2\pi) \int_{-\infty}^{\infty} \mathscr{L}(\Gamma, E, V)\exp[-\sigma\Gamma^2(t_a/4)] \, dE. \quad (2)$$

Equation (2) is adapted from Margulies and Ehrman (1961) and describes the transmitted intensity as a function of Doppler energy $V$ (normalized to unity for $V \to \infty$) in terms of the parameters: $f_s$ = source Mössbauer fraction; $\Gamma$ = full width at half-height of emission or absorption lines centered at $E_i$; $E$ = $\gamma$-ray energy; $\mathscr{L}(\Gamma, E, V) = [(E + V)^2 + (\Gamma^2/4)]^{-1}$, the Lorentzian line shape, assumed centered at $E = 0$; $\sigma$ = the absorption cross-section = $\sum_i h_i \mathscr{L}(\Gamma, E - E_i, V = 0)$, a superposition of Lorentzian lines with the values of $E_i$ determined by the hyperfine interactions; $h_i$ = cross-section of the $i$th Lorentzian line; $\tau_a = \sigma t_a$ = effective thickness of the absorber. It is common to use thin absorbers and to assume that they are sufficiently thin that the exponential in Eq. (2) may be expanded, retaining only the first two terms. This leads to the thin absorber approximation (renormalized to unity):

$$P(V) \cong \Gamma^2 \sum_i h_i \mathscr{L}(2\Gamma, E - E_i, V = 0). \tag{3}$$

Using Eq. (3) the observed spectrum is fit using a digital computer to minimize the residuals between the data and a sum of Lorentzians. Depending on the magnitude of $\tau_a$, this procedure always leads to errors in the determination of the magnitudes of $h_i$ and may often distort a complex spectrum in such a way that even the values of $E_i$ are erroneous (Shenoy, 1974).

Many techniques for correcting for finite sample thickness have been proposed. Dibar-Ure and Flinn (1971) and Asano and Schwartz (1973) have developed techniques for deconvoluting the observed spectrum as described in Eq. (2) to obtain $\sigma$, which may then correctly be analyzed as a sum of Lorentzians. Alternatively, one may use Eq. (2) directly in a computer fit. This procedure involves a lengthy and costly numerical integration, but a recent algorithm developed by Cranshaw (1974) has reduced the time required to the point where this procedure is practical. These techniques for deconvolution and for using the true transmission integral have been discussed in detail by Lin and Preston (1974). A third alternative is to fit the observed spectrum with Lorentzians and then use an analytical expression for the resonance line area that includes the effects of finite sample thickness. This procedure was used by Abe and Schwartz (1974) and will be referred to later in the discussion of retained austenite in Section III,D. Finally, it should be noted that an integral similar to that given in Eq. (2) arises when the Mössbauer experiment is done in scattering geometry, the geometry most convenient for nondestructive testing. Corrections for finite sample thickness for this geometry may also be done by deconvolution, by numerical integration, or by evaluating the analytical expressions for the resonance line areas as was done for example by Chow et al. (1969) and by Swartzendruber et al. (1974). In summary, while the corrections for finite sample thickness may often be small, they can never be completely neglected if a quantitative analysis is to be made.

A convincing demonstration of the importance of making a thickness correction is illustrated using the data taken by Asano (1974) for Fe–3.2 at. % Mo and shown earlier as Fig. 1. The thickness correction was made by deconvoluting the observed spectrum [using a technique developed by Asano (see Asano and Schwartz, 1973)] to obtain the true absorption cross-section. In the hyperfine field analysis, account was taken of the large anisotropic hyperfine magnetic field perturbation discussed in Section II,A. The results for $w(i, j)$ are presented in Table I. The values given on the first

**TABLE I**

Measured and Calculated Values of $w(i, j)$ for Fe–3.2 at. % Mo[a]

|  | $w(0, 0)$ | $w(1, 0)$ | $w(0, 1)$ |
|---|---|---|---|
| Mössbauer–anisotropic interaction | 0.54 | 0.225 | 0.195 |
| Mössbauer–anisotropic interaction + thickness correction | 0.63 | 0.185 | 0.115 |
| Random—calculated from Eq. (13) with $P_1 = P_2 = c = 0.032$ | 0.64 | 0.165 | 0.125 |

[a] Schwartz and Asano (1974).

line represent the best fit when no thickness correction is made, i.e., assuming a linear superposition of Lorentzian lines, while those on the second line are the results of analysis after thickness correction. A comparison of these numbers to those shown in the third line (calculated for a random distribution of the solute atoms as described below in Section IV) is quite revealing. When properly treated, the data agree within experimental error with a random solute model. Failure to correct for the thickness distortion leads to an erroneously low $w(0, 0)$ and high $w(1, 0)$, a result that would be interpreted as a preference for unlike nn pairs or short-range order.

In closing this section, it should be noted that analytical techniques that correctly identify the atomic configurations associated with each subspectrum and account for finite sample thickness determine the quantity denoted $h_i$ following Eq. (2). This cross-section $h_i$ is proportional to the fraction of iron atoms in the sample having the given configuration and to the Mössbauer fraction for the nuclei of those atoms, $f_i$. For different atomic configurations in a single-phase alloy, one may reasonably neglect the small changes in $f_i$ from one configuration to another. However, in general, the values of $f_i$ will differ for iron in two chemically or structurally different phases, and should be taken account of in multiphase quantitative analysis. This involves experimental determination of the Mössbauer fraction in each phase using samples of known thickness and concentration of the two phases or, better yet, using single-phase samples (see Abe and Schwartz, 1974).

## III. Decomposition into Two Phases

### A. Theory of Two-Phase Analysis

In quantitative analysis of samples containing two phases, one must take account of finite sample thickness, polarization effects in the case where one or both of the phases are magnetically ordered, and the effect of granularity in powder specimens. These topics have been described in the literature and are discussed in some detail by Abe and Schwartz (1974). The conclusions regarding sample thickness and polarization effects will be briefly reviewed here.

In these experiments, details of the hyperfine fields in the magnetic phase are often ignored. The spectrum is commonly fit with enough Lorentzian lines to give a good measure of the total resonance associated with each of the two phases (i.e., the area of the Lorentzian absorption curves associated with each phase). The resultant areas are compared to give the fraction of iron atoms in each of the two phases. Again, care should be taken, as the Mössbauer fractions of $^{57}Fe$ in the two phases may differ, and the measured area is related to the product of the number of $^{57}Fe$ atoms and the relevant Mössbauer fraction.

For a single resonance line, the area is obtained from Eq. (2) as

$$A = \int_{-\infty}^{\infty} [1 - P(V)] \, dV. \tag{4}$$

This integral may be evaluated analytically as shown by ByKov and Hien (1963) and reduces to the form

$$A = f_s \tfrac{1}{2} \Gamma \pi K(\tau_a) \tag{5}$$

where $K(\tau_a) = \tau_a e^{-\tau_a/2} [I_0(\tau_a/2) + I_1(\tau_a/2)]$, and $I_0(x)$ and $I_1(x)$ are, respectively, the zero- and first-order Bessel functions of the imaginary argument.

In the derivation of Eq. (4), it was assumed that absorber and source linewidths are the same and equal to the natural line width $\Gamma$ ($= 0.097$ mm/sec for $^{57}Fe$) giving rise to a measured $\Gamma_{app} = 2\Gamma$. (In what follows $\Gamma$, $\Gamma'$, $\Gamma_{app}$ refer to values that would be found upon extrapolation to samples of zero thickness.) Experimentally one often finds $\Gamma_{app} > 2\Gamma$ due to varying local environments at the $^{57}Fe$ nuclei affecting the resonance spectrum. If $\Gamma_s \gtrsim \Gamma$, in the thin-source approximation the area is unaffected, as the normalization process is equivalent to setting

$$\Gamma_s/2\pi \int_{-\infty}^{\infty} [(E - E_0 + V)^2 + \tfrac{1}{4}\Gamma_s^2]^{-1} \, dV = 1. \tag{6}$$

If, however, the absorber linewidth is $\Gamma' > \Gamma$, the expression for the area is modified (Johnson and Dash, 1968) and becomes

$$A = f_s \tfrac{1}{2} \Gamma' \pi K(g\tau_a), \tag{7}$$

where $g = \Gamma/\Gamma'$ must be empirically determined. Thus, through the thickness effect the measured area for a sample of finite effective thickness $\tau_a$ depends on $\Gamma'$.

For a six-line spectrum from a ferromagnetic phase, the expression for the area may be written in a simple form when the six peaks only slightly overlap and the expression reduces to (Abe and Schwartz, 1974)

$$A_{\text{ferro}} = f_s \tfrac{1}{2} \pi \sum_{i=1}^{6} \Gamma' K(g_i q_i \tau_a), \tag{8}$$

where in general the values of $\Gamma'$ and $g_i$ are different for each of the six resonance lines. The transition probabilities $q_i$ depend on the sample polarization. For a powder sample in the absence of a magnetic field, the magnetic domains are random in orientation and

$$q_1 : q_2 : q_3 : q_4 : q_5 : q_6 = \tfrac{3}{12} : \tfrac{2}{12} : \tfrac{1}{12} : \tfrac{1}{12} : \tfrac{2}{12} : \tfrac{3}{12},$$

but in general, for foil specimens of finite thickness, the values of $q_i$ are unknown. Rosencwaig et al. (1969) have discussed this problem in some detail for nonrandom domain orientations including the polarization effects introduced when the absorber is subjected to an external magnetic field.

For a two-phase mixture in a metal foil, one may use an external field to polarize the sample (Rosencwaig et al., 1969) or note that for a small paramagnetic fraction ($\tau_{\text{para}}$ small), comparison of $A_{\text{para}}$ to the area of the two weak innermost lines of the ferromagnetic spectrum minimizes those errors (Abe, 1972). It is never strictly correct to compare the total area of the six ferromagnetic lines ($A_{\text{ferro}}$) to that of the paramagnetic line ($A_{\text{para}}$). In an extreme example of equal mixtures of powders of paramagnetic and ferromagnetic phases, each with $\tau = 9$ (corresponding to a total foil thickness of $\sim 1$ mil), $A_{\text{ferro}}/A_{\text{para}} \approx 1.9$—i.e., a 90% error in relative fraction is made. In most of the studies described in Sections III,B–III,D, no corrections have been made for the effects of sample thickness and polarization, so these experiments must be viewed as qualitative only.

## B. Nucleation and Growth

### 1. Metallurgical Theory

The decomposition of a homogeneous single phase into two homogeneous phases may be classified into one of three types, depending on the mechanism of the decomposition. Diffusionless transformation involving atomic rearrangements over distances less than one unit cell length leave the composition of the new phase the same as that of the parent phase. The most often studied example of such diffusionless transformation is the austenite–martensite transformation, discussed in Section III,D. The remaining two types of transformation are associated with the presence of a miscibility gap in the temperature–composition equilibrium phase diagram. The simplest example of such a miscibility gap is shown in Fig. 3. The solid line defines

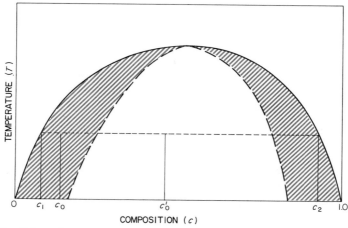

**Fig. 3.** Schematic temperature–composition phase diagram showing miscibility gap (solid line) and chemical spinodal (dashed line). The cross-hatched region is characterized by nucleation and growth, while the decomposition below the dashed line is via the spinodal mechanism.

the equilibrium solid solubility, while the dashed line, the locus of zeroes of the second derivative of chemical free energy with composition, is the spinodal. More accurately, one should locate the coherent spinodal, a curve at somewhat lower temperatures that includes the contributions of the elastic strain energy to the free energy (Cahn, 1968). Alloys quenched below the spinodal temperature decompose by the spinodal mechanism discussed in Section III,C.

Alloys quenched into the cross-hatched region between the miscibility gap and the (coherent) spinodal in Fig. 3 decompose by a process of nucleation and growth. The decomposition theory (see, for example, Fine, 1964) indicates that only near-equilibrium concentration fluctuations can reduce the chemical free energy. Thus, for an alloy of initial composition $c_0$, one expects the precipitates with composition near to the equilibrium value $c_2$ to be formed. Thermally activated composition fluctuations of a certain minimum size (radius $= r^*$) are necessary to overcome the increase in surface free energy, and fluctuations with radius $>r*$ are stable nuclei which may subsequently grow via diffusion. The value of $r*$ is commonly of the order of 50–200 Å, so that one may expect that even at the nucleation stage the iron atoms in the precipitate will have hyperfine fields characteristic of the bulk equilibrium phase of composition $c_2$ (however, one must always be aware of possible superparamagnetic behavior of these fine particles). Thus, the Mössbauer effect may be used to follow the evolution of the precipitate from the earliest stages of decomposition. This is particularly valuable for precipitation of a paramagnetic phase from an iron-rich alloy, as the resonance line for the paramagnetic phase is usually located in the resonance-free portion of the matrix spectrum. The kinetics of the decomposition may thus be studied by measuring the relative fraction of the precipitate with time. Furthermore, the spectrum of the matrix will change as its composition approaches $c_1$, and one may often deduce still further information by a careful study of this spectrum.

### 2. Experimental Results

*a. Maraging Steels.* One of the most extensively studied systems is the group of high alloy steels known as maraging. These steels contain varying amounts of nickel (or manganese), cobalt and molybdenum (or tungsten). They are first transformed from the high-temperature austenite to the low-temperature martensite phase by the diffusionless martensitic reaction, then heat treated to produce a two phase structure including a precipitate containing molybdenum (or tungsten), which further hardens the already hard martensite. The composition of this precipitate, the kinetics of its growth, and the dependence of these parameters on the alloy composition and maraging temperature have been studied by many workers using the Mössbauer effect. The effect of molybdenum on the Mössbauer pattern of iron was shown in Fig. 1 and first studied by Marcus *et al.* (1967). As the effect of cobalt and nickel on the Mössbauer pattern of iron is to modify the average hyperfine field and broaden the resonance lines (Johnson *et al.*, 1963), the principle effects seen during maraging treatments are associated with the redistribution of molybdenum in the sample. These effects were first seen in the study of precipitation in the Fe–Mo binary system, studied by Marcus *et al.* (1967).

The spectrum for Fe–6 at. % Mo is shown in Fig. 4. This spectrum was interpreted as the superposition of three six-line spectra associated with iron atoms with zero, one, or two nn or nnn molybdenum atoms. The decomposition of the alloy upon heat treatment may be understood by reference to the equilibrium phase diagram. At 910°C, the alloy is homogeneous, single phase, but upon quenching and subsequent annealing at lower temperatures,

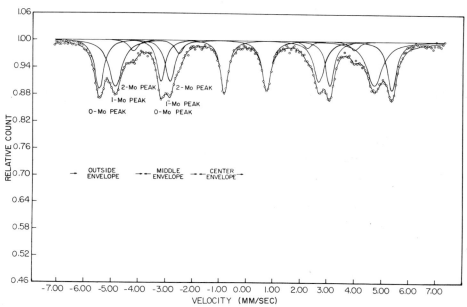

**Fig. 4.** Mössbauer spectrum of Fe–6 at. % Mo alloy quenched from 1050°C. (Marcus *et al.*, 1967.)

a precipitate of composition $Fe_2Mo$ or $Fe_3Mo_2$ is expected, depending on aging temperature, with a matrix depleted of molybdenum. The hexagonal precipitates are paramagnetic and were shown to produce quadrupole split doublets near zero velocity (referenced to pure iron). In Fig. 5, the aging sequence is shown for decomposition at 650°C. The development of the paramagnetic precipitate and the decrease in molybdenum in the matrix are evident. By curve fitting, the fraction of iron atoms in the precipitate was determined. Decomposition was also followed at 550°C, and by application of the standard kinetic treatment of Zener–Wert–Avrami (see Fine, 1964), the development of the precipitate was characterized as the thickening of large cylindrical shapes with an activation energy of 59 kcal/gm mole in agreement with values of 60 kcal/gm mole known for self diffusion of molybdenum in iron. The main advantage of the Mössbauer technique for kinetic analysis is its ability to determine the fraction transformed directly rather than by assuming some empirical and possibly incorrect relationship between some physical property, such as resistivity, and the fraction transformed.

An early stage of decomposition at 550°C was observed in which the local atomic order in the alloy changed (evidenced by changes in the intensities of subspectra in the matrix spectrum) before the development of any

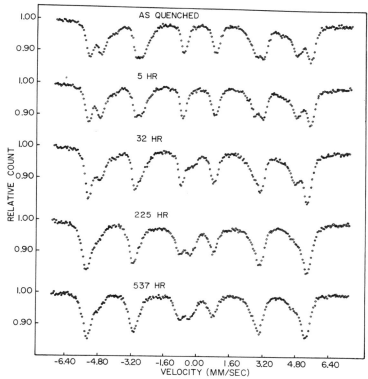

**Fig. 5.**  Mössbauer spectra for quenched Fe–6 at. % Mo aged 0–537 hr at 650°C. Note the disappearance of structure in the six-line spectrum and the development of a paramagnetic peak near zero velocity as $Fe_2Mo$ precipitates. (Marcus *et al.*, 1967.)

precipitate. This change in local order due to clustering was confirmed by the observation of the development of x-ray diffuse scattering near Bragg reflections.

These results on the binary Fe–Mo decomposition were used by Marcus *et al.* (1966) in their study of the Fe–18Ni–8Co–5Mo maraging steel (composition in weight percent). Following standard heat treating procedures, the steel was air cooled from 820°C and annealed 3 hr at 480°C. This treatment, known to produce maximum hardness, was accompanied by depletion of molybdenum and nickel from solution (decrease in the molybdenum substructure and increase in average hyperfine field was seen), but no iron-bearing precipitate was detected. Subsequent annealing for 41 hr at 480°C did produce an iron-bearing precipitate with isomer shift corresponding to $Fe_2Mo$. The conclusion drawn by these workers was that the hardening was due to a nickel-bearing precipitate, and the $Fe_2Mo$ appeared only after maximum hardness was achieved.

This subject has since received the attention of many workers who confirmed many of the observations of Marcus *et al.*, but also added to the understanding of the decomposition in more complex alloys. Gruzin *et al.* (1972a) studied Fe–16Ni–5Mo(N16M5) and Fe–16Ni–15Co–5Mo(N16C15M5). Their interpretation in the N16M5 alloy followed that of Marcus *et al.*, but they observed a two-stage hardening increase, the first associated with local atomic order (clustering of molybdenum), the second with the development of the $Fe_2Mo$ precipitate. Their results for the N16C15M5 were identical with those of Marcus *et al.*, with a depletion of molybdenum from solution within 30 min at 500°C, but no significant development of $Fe_2Mo$ until after 10 hr at 500°C. In a subsequent paper, Gruzin *et al.* (1972b) showed that in alloys without cobalt, the $Fe_2Mo$ phase appears in a considerably shorter aging time compared to similar alloys with cobalt. Similar results were obtained comparing Fe–Ni–W and Fe–Ni–Co–W alloys. These results were attributed to dominant short-range order effects as found in an Fe–12Co alloy. Gruzin *et al.* suggest that the tendency to local Fe–Co ordering retards the ultimate development of the $Fe_2Mo$ precipitate.

Genin *et al.* (1974) studied Fe–12.2Ni–6.2Mo and Fe–17.3Ni–8.4Co–2.9Mo (atomic percents) alloys, extending the work described above and their own electron microscopic observations (Bourgeot *et al.*, 1973). Their Mössbauer data for the former alloy was analyzed in a manner similar to that of Marcus *et al.* and confirmed the microscopic evidence of decomposition of $Fe_2Mo$ along dislocations in the martensite phase, resulting in kinetics characteristic of cylindrical geometry. The alloy containing cobalt had been shown by microscopy to nucleate not at dislocations, but instead randomly throughout the matrix. The Mössbauer results showed depletion of molybdenum, causing changes in the matrix spectrum, but no iron-bearing precipitate even after 1024 hr at 350°C (an advanced stage of homogeneous precipitation).

Combining quantitative analysis of the Mössbauer spectra with lattice parameter data from the electron microscopy, Genin *et al.* were able to show that the precipitate in the Fe–Ni–Mo alloy was almost totally composed of iron and molybdenum in the ratio $(Fe_{0.93}Ni_{0.07})_2Mo$ in disagreement with the formula (Fe,Ni)Mo obtained by chemical analysis and electron probe microanalysis on extracted precipitates. This result is consistent with those of Gruzin *et al.*, suggesting that the early hardening is due not to a nickel-bearing precipitate as suggested by Marcus *et al.*, but rather to molybdenum clustering.

In a study of maraging of Fe–10Mn–8Co–6Mo by Jones and Kapoor reported by Jones (1973), a paramagnetic peak corresponding to $Fe_2Mo$ was found, implying that manganese behaves similarly to nickel in these alloys.

*b. Vicalloy Magnetic Steels.* Permanent magnets of the vicalloy type (Fe–Co–V) are used extensively throughout the world. Although a great deal of effort has been expended to understand the origin of the high coercivity in these alloys, many questions remained unanswered until recently. The history, technological importance and scientific study of vicalloys was reviewed by Oron *et al.* (1969). In this paper, these workers describe the application of Mössbauer spectroscopy to the problem. Depending on heat treatment, early studies had indicated the presence of three phases in these alloys: $\alpha$, bcc disordered and ferromagnetic; $\alpha'$, bcc ordered and ferromagnetic; and $\gamma$, fcc paramagnetic. In a preliminary study of an alloy containing (wt. %) 52Co, 10V, 38Fe, Gorodetsky and Shtrikman (1967) studied samples quenched from 950°C and then cold-rolled, followed by heat treatment to increase $H_c$ from 30 Oe to 150 Oe. Their results are shown in Fig. 6. The

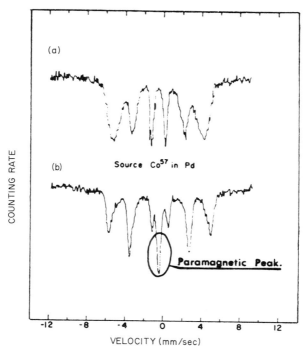

**Fig. 6.** Mössbauer spectra of Vicalloy at room temperature in different phases—(a) rolled after quenching from ~950°C; (b) after heat treatment for permanent magnet properties. The main changes observed in the heat-treated state (b) as compared with as-quenched (a) are the occurrence of a paramagnetic absorption peak, a higher interval field, and a narrowing of the absorption lines. The source was [57]Co in palladium. (Gorodetsky and Shtrikman, 1967.)

heat treatment results in narrowing the matrix spectral lines corresponding to ordering of the FeCo $\alpha$ to form $\alpha'$, and the simultaneous precipitation of the paramagnetic $\gamma$ phase. The presence of this paramagnetic phase in the vicalloy heat treated for maximum energy product was thus unequivocally confirmed.

Oron *et al.* (1969) confirmed these results of Gorodetsky and Shtrikman in several commercial vicalloys and carried the studies further. A sample quenched from 1000°C to room temperature was found to be completely paramagnetic, but upon quenching in liquid nitrogen, partially transformed to the $\alpha$ phase, which did not disappear upon warming to room temperature. This observation implied the irreversible formation of a ferromagnetic phase in the cryogenic range, rather than a magnetic transition or superparamagnetic effects as had been suggested by earlier workers. Detailed analysis of the relative intensities of the six lines in the ferromagnetic spectrum was interpreted as a change in the easy axis magnetization direction from perpendicular to the foil surface in cold rolled samples to parallel to the foil surface after heat treatment in accord with the preferred crystal orientation found in these samples using electron diffraction. These results are in general agreement with data on easy magnetization axes obtained employing a magnetometer.

In summary, cold-rolled vicalloy was found to consist entirely of the $\alpha$ phase. Proper heat treatment caused formation of a limited amount of paramagnetic $\gamma$, which considerably raised the coercive force without greatly lowering the magnetization. Overheating leads to a larger proportion of the paramagnetic phase lowering the magnetization still further. The roll of crystalline anisotropy was emphasized by observation of the change in preferred spin orientation observed.

*c. Iron–Nickel–Tin Invar Alloys.* Eliezer *et al.* (1973a) studied alloys of $(Fe_{64}Ni_{36})_{1-x}Sn_x$, $0 < x < 4.8$, aged for various times at 700°C. The addition of tin to this Invar alloy was intended to improve the mechanical properties by precipitation hardening. In this rather unique system, considerable information could be obtained from the Mössbauer effect by examining both the $^{57}Fe$ and the $^{119}Sn$ resonances. The investigators found that the solubility of tin at 1130°C of at least 4.73 at. % is reduced to 1.8 at. % at 700°C. Precipitation hardening was achieved by the formation of an intermetallic compound identified as $Ni_3Sn_2$ using the $^{119}Sn$ Mössbauer effect and later confirmed by x-ray diffraction. This precipitate increases the hardness of the alloy by $\approx 60\%$, but the loss of nickel from solution produces an undesirable increase in thermal expansion coefficient. By adding sufficient nickel to keep the final matrix composition near that of Fe–Ni Invar, a desirable combination of mechanical hardness and low thermal coefficient of expansion was achieved.

In a subsequent study, Eliezer *et al.* (1973b) examined the effects of plastic deformation on the Fe–Ni–Sn alloys described above. As-cast alloys containing 4.73 at. % tin show $^{57}$Fe patterns characteristic of the fcc Fe–Ni phase and $^{119}$Sn patterns characteristic of $Ni_3Sn_2$, which precipitates during the cooling of the casting. Plastic deformation either by rolling or filing does not destroy the $Ni_3Sn_2$ precipitates, but results in an $^{57}$Fe pattern that includes both that expected for the fcc Fe–Ni and the superimposed spectrum from $\alpha$-iron. This mixed fcc and bcc structure was confirmed by x-ray diffraction. Eliezer *et al.* suggest that due to rapid cooling of the melt in the water-cooled arc-melting crucible, regions of the matrix adjacent to the $Ni_3Sn_2$ are richer in iron than the average for the Fe–Ni matrix. These regions contribute a paramagnetic component to the spectrum for the as-quenched alloys as expected from the early Mössbauer work on Fe–Ni alloys of Johnson *et al.* (1963). Subsequent plastic deformation may then transform these low-nickel regions to bcc by a martensitic transformation (Klosterman, 1969). To test these conclusions, Eliezer *et al.*, carried out a computer simulation, adding Mössbauer spectra of cold-rolled $\alpha$-Fe, cold-rolled 64Fe–36Ni and cold-rolled, solution-treated, and quenched 60.97Fe–34.3Ni–4.73Sn. These three spectra, shown in Figs. 7a–c were weighted by trial and error to produce Fig. 7d, which is in excellent agreement with the experimental spectrum for cold-rolled 60.97Fe–34.3Ni–4.73Sn shown as Fig. 7e.

*d. Iron–Copper Alloys.* While the yield stress of solution-treated Fe–Cu alloys increases rapidly with aging, a precipitate is observed only in overaged samples. This precipitate is essentially pure fcc copper. Hornbogen (1964) suggested that the hardening is due to clustering of copper atoms in a bcc iron matrix, but was unable to obtain direct confirmation for such bcc clusters by either x-ray diffraction or electron microscopy. Lahiri *et al.* (1969) combined measurements of the elastic modulus with Mössbauer studies to confirm Hornbogen's suggestion. Dissolved copper broadens the Mössbauer lines and produces a small change in isomer shift. The lowest energy line of the six-line spectrum was broadened $\sim 25\%$ for an alloy of Fe–1.67 at. % Cu solution treated at 845°C or 1000°C; however, aging for 30 min at 475°C, corresponding to nearly maximum yield stress increase, resulted in a reduction of the Mössbauer linewidth to only 10% greater than that for pure iron, implying a significant reduction of copper in solution. Since little change in elastic modulus occurs for this aging condition, there is probably no structural change, supporting Hornbogen's postulate that bcc copper clusters are formed. After extended aging, the Mössbauer pattern is indistinguishable from that of pure iron, showing that nearly all of the copper has precipitated from solution. The elastic moduli for such extended anneals has changed significantly as expected for the presence of clusters of fcc copper.

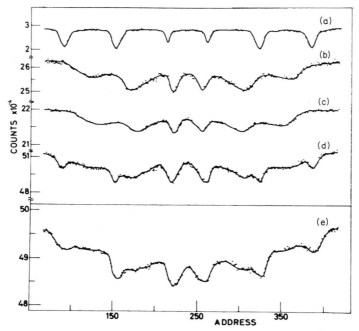

**Fig. 7.** Computer simulation of an experimental room temperature $^{57}$Fe Mössbauer spectrum: (a) α-Fe, cold-rolled; (b) 36Ni–64Fe, cold-rolled; (c) 60.97Fe–34.30Ni–4.73Sn cold-rolled, solution-treated, and quenched; (d) computer summing of (a), (b), and (c); (e) experimental spectrum of 60.97Fe–34.30Ni–4.73Sn in the cold-rolled condition. (Eliezer *et al.*, 1973b.)

## C. Spinodal Decomposition

### 1. Metallurgical Theory and Expected Effects

The thermodynamic theory of spinodal decomposition has been developed primarily by J. W. Cahn and J. E. Hilliard and is reviewed by these workers in two comprehensive articles (Cahn, 1968; Hilliard, 1970). When alloys are quenched into the region below the coherent spinodal shown in Fig. 3, no activation barrier for decomposition exists, and any fluctuation in composition from the mean value, say $c_0{}'$, is unstable and grows. In this case, the decomposition is governed solely by diffusion. Cahn and Hilliard have developed a generalized diffusion equation for this decomposition, taking into account the effect on the driving force of both the interfacial energy associated with composition variation and the elastic strain energy due to variation of lattice constant with composition. They have solved a linearized version of this equation, applicable to the early stage decomposition, and in a series of experiments carried out by Hilliard and his associates, verified the essential correctness of the theory. Analytical extensions of the theory to later aging stages is not possible, but recently, numerical solutions have been presented, notably by Langer (1973). The reader is referred to the cited papers for details of the thermodynamic treatment, as only the implications for Mössbauer effect studies will be sketched here.

The solution of the linearized diffusion equation gives the composition $c(\mathbf{r}, t)$ at any position $\mathbf{r}$ in the sample and time $t$ as

$$c(\mathbf{r}, t) - c_0' = \int A(\boldsymbol{\beta}, t)\exp(i\boldsymbol{\beta} \cdot \mathbf{r})\, d\boldsymbol{\beta}, \qquad (9)$$

in which $c_0'$ is the mean composition and $A(\boldsymbol{\beta}, t)$ is the amplitude at time $t$ of the composition fluctuation of wave vector $\boldsymbol{\beta}$ ($|\boldsymbol{\beta}| = 2\pi/\lambda$, where $\lambda$ is the wavelength of this Fourier component). The development of $A(\boldsymbol{\beta}, t)$ with time depends exponentially on a generalized diffusion coefficient in such a way as to enhance the fluctuations having wavelength very near to some value $\lambda_m$. Thus, after a very short aging time, the composition profile is well approximated by three mutually orthogonal cosine waves of some amplitude $A(\lambda_m, t)$ and wavelength $\lambda_m$. Regardless of the details of the composition fluctuation, one expects that the early stages of decomposition via this mechanism will result in an alloy containing a range of local compositions surrounding the initial mean value $c_0'$, with no evidence of precipitates of the equilibrium compositions $c_1$ and $c_2$ (see Fig. 3) appearing until late in the aging sequence. In a ferromagnetic alloy in which the mean hyperfine magnetic field $H_{\text{eff}}$ varies strongly with composition, one expects an initial broadening of the Mössbauer spectrum, followed later by evidence of a well-defined two-phase structure in which each phase has a Mössbauer spectrum characteristic of the equilibrium concentrations $c_1$ and $c_2$.

## 2. Experimental Results

*a. Cu–Ni–Fe.* In an early attempt to see evidence of spinodal decomposition using the Mössbauer effect, Nagarajan and Flinn (1967) studied the alloy Cu–Ni–Fe, long believed by other workers to be a classic example in which spinodal decomposition occurs. Their results, however, were not consistent with the expectations for spinodal decomposition, but rather suggested nucleation and growth of a second phase of near-equilibrium composition. This discrepancy is not yet cleared up, but it is perhaps noteworthy that in their magnetic and electron microscopic studies of several Cu–Ni–Fe alloys, Butler and Thomas (1970) gave evidence of a rapid decomposition during the quench, so that even the as-quenched alloys in the

Mössbauer study may have been well beyond the early stages of decomposition. One must look to other, more sluggish systems for evidence of the early stages.

   *b. Fe–Cr.* The low-temperature portion of the Fe–Cr phase diagram is very similar to the idealized diagram shown as Fig. 3. Development of the intermetallic FeCr σ-phase may be completely suppressed by rapid quenching. The rapid variation of $H_{eff}$ with concentration observed by Johnson *et al.* (1963) makes this an ideal system in which to observe spinodal decomposition using the Mössbauer effect. Indeed, the early stages of decomposition in this system cannot be observed by conventional x-ray or electron microscopic techniques due to the near identity of atomic number and size of the iron and chromium.

   Yamamoto (1964) examined alloys in the composition range 20–50 at. % Cr, as quenched and as annealed for 150 hr at 500°C. The Mössbauer patterns of the near-equilibrium-annealed specimens confirmed the presence of the miscibility gap, but revealed nothing about the mode of decomposition. Subsequent Mössbauer studies by Chandra and Schwartz (1971a, b) and De Nys and Gielen (1971) have confirmed the spinodal mode of decomposition for alloys near equiatomic concentration aged at 475°C. Similar observations were made by Ettwig and Pepperhoff (1970) and Gruzin *et al.* (1971) but not specifically identified by them as due to spinodal decomposition. The results of these studies may be characterized by reference to one.

   Chandra and Schwartz (1971a, b) carried out extensive studies of decomposition in Fe–60Cr at 475°C. The Mössbauer spectrum of the as-quenched alloy is shown in Fig. 8. The solid curve is a fit based on Eq. (1) and empirical parameters a and b as determined for dilute chromium in iron by Wertheim

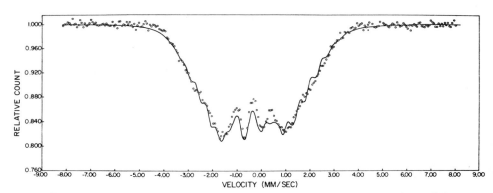

**Fig. 8.** Mössbauer spectrum for as-quenched Fe–60 at. % Cr: (○) experimental data; (——) calculated using Eq. (1) and assuming random solid solution. (Chandra and Schwartz, 1970.)

*et al.* (1964). Although this model neglects anisotropic hyperfine interactions, the fit is sufficiently good that it lends confidence to the assumption that the as-quenched alloy is essentially random. Room temperature spectra taken for a series of aging times at 475°C are shown in Fig. 9. Note that in the

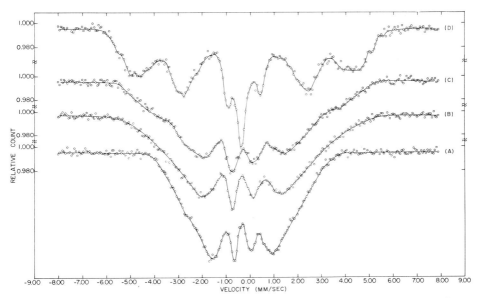

**Fig. 9.** Mössbauer spectra of Fe–60 at. % Cr measured at room temperature: (A) the as-quenched alloy; (B, C, D) after 30, 80, and 1300 hr, respectively, at 475°C. The solid curves are hand drawn through the experimental points. (Chandra and Schwartz, 1970.)

near-equilibrium two-phase mixture in (D), a strong paramagnetic peak from the chromium-rich phase appears near zero velocity, and the large hyperfine field of the iron-rich phase gives rise to considerable resonant absorption near ±5 mm/sec. By contrast, the spectrum for a 30 hr age (B) shows no paramagnetic peak, no significant absorption at ±5 mm/sec, but rather a general broadening as expected for the range of compositions produced by spinodal decomposition. These effects are more dramatically in evidence in Fig. 10, which shows spectra taken at 200°C, above the Curie point of the as-quenched alloy. The paramagnetic peak of the as-quenched alloy (A) broadens after 30 hr aging at 475°C, with no evidence of the near equilibrium iron-rich phase, which appears in (D) for the alloy aged 1059 hr.

In a related study, the miscibility gap in the ternary Fe–Cr–V has been determined using the Mössbauer effect. Mima and Yamaguchi (1970) studied iron-rich ternary alloys. The shape of the miscibility gap was determined by

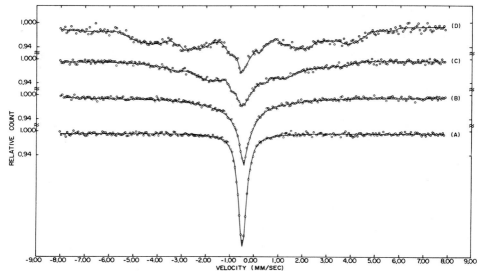

**Fig. 10.** Mössbauer spectra of Fe–60 at. % Cr measured at 200°C: (A) the as-quenched alloy; (B, C, D) after 30, 114.5, and 1059 hr, respectively, at 475°C. The solid curves are hand drawn through the experimental points. (Chandra and Schwartz, 1970.)

differential thermal analysis. After extensive aging at 480°C, the hyperfine fields of the iron-rich phase were used to establish the tie line in the two-phase region. This analysis depends on the somewhat questionable and untested assumption that the rate of change of $H_{eff}$ with concentration of one solute is independent of the concentration of the other solute—a sort of Mathiesson's rule for average hyperfine fields.

   *c. Alnico Alloys.*   The hard permanent magnets known as the alnicos are subjected to a complex thermomagnetic treatment to produce optimum magnetic properties. The detailed atomic rearrangement associated with this treatment is complicated, involving the separation of the initial bcc alloy into two ordered, body-centered, tetragonal phases, with their long axes in the direction of the magnetic field applied during the heat treatment. The decomposition into two phases is probably via a spinodal mechanism (Cahn, 1963) with the magnetic field as the driving force resulting in elongated alternating grains of the two phases of average size about $300 \times 300 \times 4000$ Å³. The magnetic state of the two phases and the state of order in each have received considerable attention; however, the presence of iron, nickel, cobalt, aluminum, copper and titanium in the alloy limits the information available from standard x-ray and microscopic techniques. The Mössbauer effect technique has been applied to various aspects of this problem by a number of workers (Shtrikman and Treves, 1966; Wierengen and Rensen, 1966; Makarov *et al.*, 1967; Povitsky *et al.*, 1970; Belozerskiy *et al.*, 1971; and Makarov *et al.*, 1972. By far, the most complete study of a single alloy, alnico 8, has been made by the Moscow group, and that study will be sketched below.

Makarov *et al.* (1967) studied an alnico 8 alloy of composition 33.2Fe–31.5Co–12.6Ni–13.8Al–3.3Cu–5.6Ti. Four samples were subjected to various heat treatments, and the Mössbauer spectra were taken at room temperature as shown in Fig. 11. Spectrum (a) corresponds to solution treatment at 1240°C for 30 min followed by quenching in water; the presence of

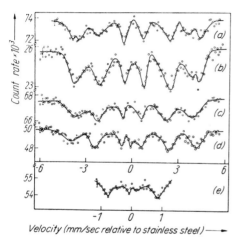

**Fig. 11.** Mössbauer spectra of the Alnico 8 alloy after different heat treatments: (a) tempering at 1240°C for 30 min and water quench; (b) tempering at 800°C for 30 min; (c) tempering at 820°C for 30 min; (d) and (e) magnetic field heat-treatment cycle. (Makarov *et al.*, 1967.)

at least two distinct hyperfine fields for iron is seen in the splitting of the first and sixth peaks and attributed to iron in the two sites corresponding to FeAl atomic ordering. This ordering tendency was confirmed by examination of the x-ray diffraction spectrum. In addition, weak x-ray satellites were seen, suggesting the onset of spinodal decomposition during the quench. Analysis of the intensity of the two subspectra of the Mössbauer pattern led to the assignment of 70% of the iron to body-centered (D) sites, and 30% to cube corner (A) sites. The magnitudes of the observed hyperfine fields suggest that the cobalt and nickel are also predominantly on D sites, and the aluminum, titanium, and copper occupy the A sites.

The spectra in Figs. 11b and c correspond to aging at 800°C and 820°C, respectively, each for 30 min. Most noticeable is the loss of resolution of the two well-defined hyperfine fields seen in Fig. 11a. In this temperature range, the decomposition into a two-phase $\alpha + \alpha'$ structure occurs. Clearly, nothing can be said about the atomic redistribution from these spectra, but it is known from chemical analysis of extracted particles from overaged specimens that the α-phase enriches in nickel and aluminum, while the α'-phase enriches in iron and cobalt. It is not clear whether the broad Mössbauer spectra should be attributed to the concentration fluctuations of the spinodal decomposition, to a change in atomic order, or, as suggested by these authors, to changes in the hyperfine fields at A sites and D sites due to changes in their near-neighbor composition.

Spectra (d) and (e) in Fig. 11 correspond to the alloy subjected to the entire thermomagnetic heat treatment (TMT)—solution treatment at 1200°C, isothermal anneal in a magnetic field at 800°C for 12 min, aging at 650°C for 5 hr, aging at 550°C for 20 hr. The paramagnetic line near zero velocity is the most noticeable difference between these spectra and (b) and (c). A similar paramagnetic line was found by Shtrikman and Treves (1966) and by Wierengen and Rensen (1966) after TMT of alnico 5. This paramagnetic peak is attributed to the increase of aluminum in the α phase, resulting in a significant number of iron atoms with five or more aluminum atoms in the first coordination shell. As pointed out by Stearns (1966), the hyperfine field at such iron atoms in the binary Fe–Al alloys have zero hyperfine field. From the intensity of the paramagnetic line in Fig. 11d, it was estimated that about 7% of the iron atoms are in such paramagnetic surroundings.

In their second paper, Povitskiy *et al.* (1970) differentiate between the effects of the two anneals at 650 and 550°C. The former is shown to primarily influence the distribution of alloy components among the two phases, resulting in an increase in the degree of tetragonality and the appearance of the paramagnetic Mössbauer peak. The second aging primarily affects the degree of atomic order in the aluminum-rich phase, as evidenced by increase in the paramagnetic iron fraction and increase in the x-ray superstructure lines.

These experiments were extended by Makarov *et al.* (1972) by examination of the Mössbauer spectra taken at temperatures as low as 4.2°K and in the presence of a 42 KOe magnetic field. In particular, they wished to examine the suggestions made by (1) Albanese *et al.* (1970) that the entire α phase is paramagnetic—about 20% of the iron atoms are in the α phase rather than the 7% reported as paramagnetic by Makarov *et al.* (1967)—and by (2) Belova *et al.* (1969), who suggested that a correlation between the temperature dependence of magnetization and the Mössbauer data would exist only if the α phase were superparamagnetic. In this 1972 paper, Makarov *et al.*, demonstrate that the paramagnetic α-phase peak seen at room temperature does disappear at lower temperatures and is replaced at 4.2°K by a second six-line spectrum. Further examinations of field and temperature dependence of this spectrum confirmed the superparamagnetic behavior of the α phase. In addition the relative areas of the two spectra at 4.2°K correspond to 25 ± 5%

Fe in the $\alpha$ phase, in agreement with the chemical analysis of the electro-chemically separated phases as noted by Albanese *et al.* (1970). The lower value of 7% obtained from room temperature data is ascribed to a difference in the temperature dependence of the Mössbauer fractions of iron in the two phases. Thus, these experiments confirm the suggestions of Albanese *et al.*, and Belova *et al.*

While this study of alnico 8 demonstrates the extensive information about atomic rearrangement possible using the Mössbauer effect, it also serves to emphasize the importance of using other experimental techniques, such as diffraction, microscopy, magnetization, and chemical analysis, to separate out the complex features found in decomposition of multicomponent systems.

*d. Fe–Be Alloys.*  A complex sequence of phase changes accompanies aging of alloys of composition $Fe_{1-x}Be_x$, $0.20 \leqslant x \leqslant 0.25$. When aged at 400°C or below, modulated structures are formed, first followed by precipi-tates of a metastable phase. The modulated structures are themselves atom-ically ordered, and consequently, as for the alnico steels, this subject may be viewed as an example of both spinodal decomposition and order–disorder. Aging in the temperature range above 400°C results in precipitation first of a metastable phase and then of an equilibrium phase of composition $FeBe_2$.

Yagisawa (1973a) studied the decomposition of Fe–23 at. % Be aged (1) 94 hr at 400°C, (2) 24 hr at 600°C, and (3) 240 hr at 650°C. Although the metastable precipitate was shown to have the same crystal structure at 400 and 600°C by earlier electron diffraction results, the relative intensities of the two subspectra in the Mössbauer data indicate that the concentration of beryllium in the precipitate formed at 400°C after 94 hr is 0.45 compared with 0.61 for precipitation at 600°C for 24 hr. Thus, one is led to the conclusion that phase separation at 400°C may involve a continuous variation in con-centration of the precipitate. The observations tend to confirm the evidence for spinodal decomposition developed by earlier workers using x-ray dif-fraction and electron microscopy.

More detailed studies of the low-temperature aging sequence of these alloys have been carried out by Dehtiar *et al.* (1969), who studied Fe–20Be aged at 400°C and Yagisawa (1973b) who studied Fe–23Be aged at 300°C. While Dehtiar *et al.*, suggest that ordering of the FeAl type occurs prior to phase separation at 400°C, Yagisawa believes that both effects proceed simul-taneously. He analyzes his data in such a way as to obtain a measure of the beryllium concentration in the high- and low-beryllium regions of the de-composing structure ($c_H$ and $c_L$, respectively) and a measure of the long-range order in each phase (denoted $\alpha_H$ and $\alpha_L$). He finds that even in the as-quenched alloy $c_H \approx 0.3$ and $c_L \approx 0.2$ in an alloy of average composition $c = 0.23$. As aging proceeds at 300°C, $c_H$ increases slowly to $\sim 0.35$ after 100 hr, at which time $\alpha_H = 1.0$, corresponding to the maximum allowable order at that composition. On the other hand, $c_L$ decreases to $\sim 0.08$ at 100 hr, but $\alpha_L$ has already reach zero at 10 hr, at which time $c_L \approx 0.12$. One must be careful not to give too much credence to these numbers, as several assumptions in the data analysis are untested—anisotropic hyperfine interactions are neglected; a single-order parameter is assumed for each phase, neglecting nonrandom disorder (see Section IV for more on this subject); most importantly, a single unique composition is taken for each phase, whereas one expects a range of compositions based on theories of spinodal decomposition. This last oversimplification may be responsible for the fact that the estimated change in magnetic hyperfine field per beryllium neighbor in these composition-modulated alloys is found to be twice the values observed for beryllium-poor matrices of the same composition produced by classic precipitation at temperatures above 400°C. Nevertheless, the general conclusions about simultaneous development of phase separation and order–disorder are in accord with electron microscopic observations.

### D. Retained Austenite

#### 1. Background

Transformation of steels from the fcc high-temperature phase (austenite) to a low-temperature bcc or bc tetragonal phase (martensite) is one of the oldest and most common means to achieve high strength. The fraction transformed in a given steel subjected to high-temperature anneal and quenched below the $M_s$ (martensite start) temperature is influential in determining the mechanical properties and consequently of considerable interest to the metallurgical engineer. Common techniques for determining the austenite retained after the quench include magnetic measurements, optical metallographic techniques, and x-ray diffraction. Of these, x-ray diffraction is perhaps the most precise, but its absolute accuracy may be severely affected by preferred polycrystalline grain orientation in the sample. An alternative method that can achieve high absolute accuracy independent of the degree of preferred orientation would find considerable application in the steel industry. Such a technique may be available using the Mössbauer effect.

The first suggestions for using the Mössbauer effect for studies of retained austenite were made by Epstein (1963) in reference to some unpublished work of P. A. Flinn, and by Marcus et al. (1966), who studied the effect of thermal treatment on the austenite concentration in 17–7 PH stainless steel. Christ and Giles (1968) noted that such studies would be complicated by the presence of other paramagnetic phases with isomer shift near that of austenite. Their work emphasized the importance of identifying the presence of other phases by x-ray diffraction before concluding that the only contribution to the para-magnetic Mössbauer peak was due to austenite. When only two phases are present, the paramagnetic resonance peak appears near zero velocity relative to the ferromagnetic spectrum—i.e., in the region of near-zero resonant absorption by the martensite. Consequently, austenite concentrations as low as 0.5% may be detected with high precision. It should also be noted that in some high alloy steels (notably those with high nickel content) the austenite is also ferromagnetic, and a rather large austenite content ($> 5\%$) must be present to be detected, let alone quantitatively analyzed. While many of the experiments reported here have been carried out in transmission geometry, the recent development of $\gamma$-ray detectors for backscattering geometry enables one to carry out these experiments as nondestructive tests.

The accuracy of the Mössbauer technique for retained austenite may be assessed by comparison of results with those obtained by other techniques or by analysis of known standard samples. Chow and Bogner (1970) studied retained austenite in a 46100 steel using a backscattering detector described by Chow et al. (1969). Their results agreed with those determined by x-ray diffraction studies of the same samples over an austenite fraction range of 2.7–39%. Abe and Schwartz (1974) studied standard samples of Fe–27 at. % Ni in transmission geometry. Following the procedure described by Asano (1969), they prepared powder specimens of the martensitic and austenitic phases and mixed them in varying proportions to produce a set of samples of well-defined austenite concentrations as specified in column one of Table II. The spectra were analyzed by two techniques. In technique 1, the ratios of intensities of paramagnetic to ferromagnetic resonance spectra were determined and corrected for the effects of finite sample thickness, granularity (Bowman et al., 1967), and differing Mössbauer fraction in the two phases. In technique 2, the absolute intensity of only the austenite resonance line was analyzed. This latter approach required knowledge of the Mössbauer source fraction (Abe and Schwartz, 1973) and gave results indistinguishable from those of technique 1 and in excellent agreement with the known austenite fractions. These same samples have been examined using a backscattering experiment (Flinn, 1973b) with equal success. Thus, in determination of the austenite fraction of a two-phase sample, the Mössbauer effect method can compete with the x-ray technique and may be favored for samples with a high degree of preferred orientation and low austenite concentration.

## 2. *Experimental Results*

The Mössbauer effect has been used to determine the austenite/martensite ratio in several problems of metallurgical interest. Hirsch and Eliezer (1972) have studied the effect of deformation on the mechanical properties of Fe–Ni alloys. In Fig. 12, the Mössbauer spectra of these alloys of Fe–Ni are shown

**TABLE II**

Analysis of Standard Austenite Samples in
Fe−27 at. % Ni Powders[a]

| Paramagnetic fraction—Standardized (%) | Paramagnetic fraction—MES (%) | |
| :---: | :---: | :---: |
| | Technique 1 | Technique 2 |
| 5.42 | 5.95 | 5.89 |
| | 5.77 | 5.82 |
| 9.74 | 10.0 | 10.5 |
| | 10.1 | 10.1 |
| 18.4 | 19.8 | 19.8 |
| | 19.1 | 19.3 |
| 27.2 | 28.6 | 28.3 |
| | 28.7 | 28.4 |
| 44.9 | 45.5 | 44.4 |
| | 46.8 | 45.8 |
| | 44.4 | 46.0 |

[a] Abe and Schwartz (1974).

in the cold-rolled and annealed condition. As evidenced most clearly by the size of the paramagnetic austenite peak, the effects of cold rolling on transforming austenite to martensite are more pronounced as composition approaches 29 at. % Ni. The change in concentration of martensite, of course, produces dramatic effects on the magnetic properties, reducing the saturation magnetization of the 28 and 29% nickel alloys to zero in the annealed condition from 1.2 and 0.5 kG, respectively, in the cold-rolled condition as shown in Fig. 13.

Swartzendrubber and Bennett (1972) studied the effects of light surface grinding on a sample of spheroidized iron carbide in ferrite. Using backscattering geometry, they were able to determine the austenite concentration at 1−2 mil depth in the sample using both the 14.4 KeV $\gamma$-rays and 6.3 KeV x-rays and found $\sim 1\%$ austenite in the bulk of the sample. The surface could be examined using the weakly penetrating internal conversion electrons, and from these measurements, they identified 14% austenite in the surface layers of depth $\sim 0.05$ $\mu$m. The combined use of weakly and strongly penetrating radiation thus allows one an exciting probe for the examination of phase transitions confined to the surface of the sample.

Swartzendrubber et al. (1974) have extended their use of the scattering technique to the study of stainless steel welds and castings. They discuss the

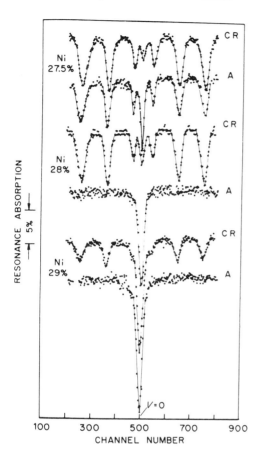

**Fig. 12.** Mössbauer patterns of cold-rolled (CR) and annealed (A) samples of nickel–iron alloys. The nickel percentage is given near each pattern. (Hirsch and Eliezer, 1972.)

**Fig. 13.** Saturation magnetization of cold-rolled and annealed nickel–iron alloys as a function of nickel percentage; $f = 2$ Hz. (Hirsch and Eliezer, 1972.)

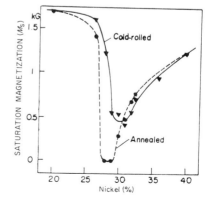

need for absolute determination of ferrite (martensite) concentration as expressed by the Welding Research Council (WRC) and propose the Mössbauer effect as the technique for these determinations. In these studies of cast and welded structures, the phase transformation are not necessarily diffusionless, and consequently, small corrections to the measured iron ratios in the ferrite and austenite phases were made to obtain appropriate volume fraction ratios of the two phases. They compare their results of studies of several cast and welded samples with the so-called "ferrite number" (volume fraction of ferrite) as determined by a magnetic method approved by the WRC. The agreement is satisfactory for samples with less than 20% ferrite, but the Mössbauer data are systematically lower than the ferrite number for samples with more than 20% ferrite. They suggest that the origin of this discrepancy is in the WRC method; however, their analytical treatment of the data does not correctly account for the attenuation of the scattered beam in the case of a magnetically split spectrum (Flinn, 1973b). This problem has not yet been treated correctly and is particularly complex for the case of alloys in which considerable overlap of the resonance lines occurs.

Sharma *et al.* (1974) examined the effects of plastic deformation on martensitic 1045 steel using the Mössbauer effect in transmission geometry. Although no retained austenite was detected by x-ray diffraction in as-quenched samples, the Mössbauer effect disclosed the presence of $2.9 \pm 0.3\%$ austenite. After deformation of 5% by swaging, some of the martensite transformed, leaving only $0.6 \pm 0.1\%$ austenite. The precision of these measurements was checked by comparisons of the calculated density change associated with such a transformation $(-0.107 \pm 0.016\%)$ and the experimentally observed density change $(-0.061 \pm 0.002\%)$. While this agreement may be satisfactory considering the small amounts austenite, there is a possibility that the Mössbauer data are even better than implied by the density differences. The effect of deformation on the martensite may result in disordering of carbon, producing a more dense cubic martensite and partially cancelling the decrease in sample density attributable to transformation of retained austenite.

## IV. Order–Disorder

### A. Measurement Theory

Atomic arrangements in alloys are most completely determined using diffraction techniques, and so it is natural to find descriptions of atomic order given in terms of parameters amenable to determination by diffraction. In general such a diffraction study leads to the measurement of the conditional probabilities $p^{A/B}(\mathbf{r}_i)$, the probabilities of finding a B atom at the end of vectors $\mathbf{r}_i$ when an A atom is at the origin. These quantities are often abbreviated $p_i^{A/B}$, where the subscript $i$ refers to the vector $\mathbf{r}_i$ and interpreted as the probabilities of finding a B atom at a site in the $i$th neighbor shells of an A atom. The diffraction results are usually expressed as a set of Warren short-range order parameters (Cowley, 1950) $\alpha_i$ defined by the relation

$$p_i^{A/B} = c(1 - \alpha_i). \tag{10}$$

It is important to emphasize that $p_i^{A/B}$ is a conditional or pair probability (the discrete analogue of the pair density function in liquids), and describes only the number of pairs of AB atoms separated by the vector $\mathbf{r}_i$. Thus $p_1^{Fe/Mo}$ gives the probability of finding a molybdenum atom at a particular site in the nn shell of an iron atom, but it tells nothing about the occupation of the remaining seven sites in the nn shell. It is not correct to assign the probability $p_1^{Fe/Mo}$ to each of these sites simultaneously as suggested by Heilmann and Zinn (1967) and Brümmer $et$ $al.$ (1972) in their Mössbauer studies of ordered alloys.

A somewhat oversimplified classification of atomic ordering systems may be made based on the qualitative behavior of the $\alpha_i$ values. When $\alpha_i \to 0$ as $\mathbf{r}_i \to \infty$, the system possesses local order only. Such systems are usually further divided into those possessing "short-range order" ($\alpha_1 < 0$ and the parameters usually oscillate between positive and negative values, becoming smaller as $|\mathbf{r}_i|$ increases), in which a preference for unlike near neighbors is predominant, and those possessing "clustering" ($\alpha_1 > 0$), in which a preference for like near neighbors is predominant. However, these definitions are merely suggestive, and in many real systems the distinction between short-range order and clustering is blurred. In both of these examples, the atomic order is evidenced by diffuse scattering in the diffraction pattern. At the other extreme, when $\alpha_i \to$ constant as $\mathbf{r}_i \to \infty$, the system exhibits long-range order and the diffraction pattern contains superlattice reflections. In this case, it is more convenient to divide the crystal lattice into sublattices and inquire about the atomic occupancy of the various sublattices. For example, in a binary AB alloy in which a division into equivalent sublattices a and b is possible, with the A atoms mostly on the a sites, and the B atoms mostly on the b sites, we may define a long-range order parameter $S$, as was first done by Bragg and Williams (1934), by

$$S = (r_a - c)/c, \qquad c \leqslant \tfrac{1}{2}. \tag{11}$$

Here $r_a$ is the fraction of a sites that are occupied by the proper A atoms. When order is perfect, all A atoms will be on a sites, $r_a = 1$, and $S = 1$. On the other hand, when there is no correlation between the occupancy of distant sites, both sublattices will be occupied by atoms in the ratio of their concentrations, $r_a = c$, and $S = 0$. When long-range order exists, $p^{A/B}(\mathbf{r}_i)$ for large $\mathbf{r}_i$ does not depend on $\mathbf{r}_i$ except in that the two sites may be on the same sublattice or on different sublattices. Thus, for the AB alloy,

$$\alpha_i \to \alpha_i{}^0 S^2 = \begin{cases} 4c(1 - c)S^2, & i \text{ even} \\ -4c(1 - c)S^2, & i \text{ odd} \end{cases} \tag{12}$$

(Cowley, 1950). In the present context, it is important to emphasize that equations of the type of Eq. (12) are correct only for large values of $\mathbf{r}_i$. In Mössbauer determinations of order in alloys, sensitivity is usually limited to the first several neighbor shells, and in general for those shells, $\alpha_i \neq \alpha_i{}^0 S^2$. In the $Cu_3Au$ system with $S = 0.8$, a detailed x-ray diffuse scattering study by Schwartz and Cohen (1965) showed that $\alpha_1 = 1.25\alpha_1{}^0 S^2$; i.e., the disorder was not random over short distances. In all the studies of long-range order described below, it has been assumed that $\alpha_i = \alpha_i{}^0 S^2$, and consequently, an unknown error is inherent in the quoted values of $S$.

Early examples of Mössbauer studies of long-range atomic order include the estimates of the degree of order in Fe–Rh alloys by Shirane et al. (1963) and the prediction of CsCl type ordering in Fe–V by Preston et al. (1966). More recent attempts to extract the Warren short-range order parameters $\alpha_i$ from the Mössbauer spectra of ferromagnetic iron alloys can be traced to the paper of Heilmann and Zinn (1967). Schwartz and Asano (1974) have recently pointed out that while approximately valid in the case of nearly complete order, the analytical approach of Heilmann and Zinn contains serious errors that preclude its application to alloys with partial long-range or short-range order, as was done in the case of Fe–12.3 at. % Al considered by Brümmer et al. (1972). Heilmann and Zinn have assumed that a binomial distribution may be used to calculate the atomic configuration. That is, if $w(i, j)$ is the fraction of iron atoms with $i$ and $j$ nn and nnn solute atoms, they write

$$w(i, j) = \binom{z_1}{i}(P_1)^i(1 - P_1)^{z_1 - i}\binom{z_2}{j}(P_2)^j(1 - P_2)^{z_2 - j} \tag{13}$$

where $z_1$ and $z_2$ are the numbers of nn and nnn sites, respectively. As Schwartz and Asano point out in the derivation of Eq. (13) it is assumed that the quantity $P_1$ (or $P_2$) is the probability of finding a solute atom at a specific site in the nn (or nnn) shell and that this probability is the same for each nn (or nnn) site. Thus, what is required is a site probability, not the conditional probability $p_1^{A/B}$ (or $p_2^{A/B}$) as suggested by Heilmann and Zinn. In fact, for an alloy with no long-range order, the site probability is the atomic fraction $c$ regardless of the degree of local order. Clearly with $P_1 = P_2 = c$, Eq. (13) does give a correct evaluation of $w(i, j)$ for a random atomic arrangement. When the alloy has extensive long-range order, and one assumes that $\alpha_i =$

$\alpha_i{}^0 S^2$ even for nn and nnn sites, it is approximately correct to set $P_1 = p_1^{A/B}$ and $P_2 = p_2^{A/B}$ and to use Eq. (13) to calculate $w(i, j)$.

As pointed out in Section II of this article, it is sometimes possible to use single crystal absorbers and unambiguously measure $w(i, j)$ for a limited set of $(i, j)$. While these measured $w(i, j)$ do contain information about the local order in the alloy, there is no known analytical way to derive the $w(i, j)$ from the Warren short-range order parameters. The reverse process would be possible if the entire set of $w(i, j)$ could be determined, since, for example, for the nn shell $p_1^{A/B} = \bar{n}_1/c_1$, where $\bar{n}_1 = \sum_{i,j} iw(i, j)$. However, the large shifts in hyperfine fields for $i + j \gtrsim 3$ preclude the separation of these subspectra from other lines in the pattern, and even though those values of $w(i, j)$ may be small, they are strongly weighted in the calculation of $\bar{n}$.

Fortunately, as pointed out by Schwartz and Asano (1974), there are numerical techniques for obtaining the $w(i, j)$ from a known set of $\alpha_i$. Gehlen and Cohen (1965) have developed a technique for the computer simulation of an atomic arrangement consistent with the measured concentration $c$, long-range order parameter $S$, and short-range order parameters $\alpha_i$. This program has been extended and discussed extensively by Gragg (1970). No energetic model is assumed. The "atoms" are merely rearranged in the computer to satisfy the geometrical requirements of the measured order parameters. From such a simulation, one may obtain pictures of the atomic arrangements, or for the present problem, obtain the desired $w(i, j)$ by a simple counting procedure. An alternative procedure for obtaining this information is the statistical mechanical approach of Clapp (1971). In this approach, a pairwise interaction relationship between the short-range order parameters and the configurational energy is assumed, and the most probable distribution of multisite configurations is numerically calculated. In this analysis, computational difficulties have limited the available information to $w(i) = \sum_j w(i, j)$, but for systems in which the hyperfine fields are primarily influenced by nn solute atoms, this should suffice.

The Gehlen–Cohen–Gragg program allows one to see the magnitude of the error associated with calculating the $w(i, j)$ from Eq. (13) with $P_i = p_i^{A/B}$, as suggested by Heilmann and Zinn. The case of Fe–6.1 at. % Mo will be considered using the measured short-range order parameters $\alpha_1 = 0.069$, $\alpha_2 = 0.042$, $\alpha_3 = 0.024$ (Ericsson *et al.*, 1970). Using Eq. (13), one obtains the results listed in the first line of Table III for the fraction of iron atoms with zero, one, two, and more than two molybdenum atoms in the first two neighbor shells. By contrast, when the Gehlen–Cohen–Gragg program is used with these measured $\alpha_i$, the results given in the second line of Table III are obtained. Note, for example, that an error of 44% is made in the ratio $w(1)/w(0)$ by using the incorrect analysis in this alloy, which exhibited only relatively weak clustering.

**TABLE III**

Calculated Values of $w'(l)^a$ for Fe−6.1 at. % Mo[b]

|  | $w'(0)$ | $w'(1)$ | $w'(2)$ | $w'(l > 2)$ |
|---|---|---|---|---|
| Calculated from Eq. (13) with $P_1$ from Eq. (10), $\alpha_1 = 0.069$ | 0.436 | 0.373 | 0.148 | 0.043 |
| Calculated from Gehlen–Cohen–Gragg Program (Gehlen and Cohen, 1965 and Gragg, 1970) | 0.498 | 0.296 | 0.133 | 0.073 |

[a] $w'(l) = \sum\limits_{\substack{i,j \\ i+j=l}} w(i,j)$ = fraction of iron atoms with $l$ molybdenum atoms in the nn and nnn shells.

[b] Schwartz and Asano (1974).


## B. Experimental Results

### 1. FeNi₃

The atomic order in FeNi₃ is known to be of the AuCu₃ type, with perfect order corresponding to iron at corners and nickel at faces of an fcc lattice. Studies with diffraction are difficult due to the similar x-ray and neutron scattering amplitudes of iron and nickel, and this system has been studied extensively using the Mössbauer effect. Since the hyperfine field of iron is only weakly sensitive to local environment, the Mössbauer spectra from alloys with different states of order differ primarily in the linewidths of the six lines observed. The problem reduces to fitting observed line profiles with calculated curves in which the hyperfine field changes due to nn and nnn iron atoms, $\Delta H_1$, and $\Delta H_2$, the hyperfine field for the perfectly ordered state $H(0, 6)$, and the long-range order parameter $S$ (or the short-range order parameter $\alpha_1$) are variables.

The first study of this system was made by Heilmann and Zinn (1967), whose analytical procedure has been discussed above. Typical results of the fitting are shown in Fig. 14 for samples quenched from 700°C (disordered) and annealed at 370°C for 1551 hr ($S \approx 0.9$) and compared to a spectrum from pure iron. The Mössbauer determinations of $S$ for alloys annealed below 500°C and quenched are in general agreement with, but somewhat smaller than, those obtained by previous diffraction studies; however, achieving equilibrium in these alloys at such low temperatures is unlikely, and different thermal histories may account for the observed differences in $S$. Samples quenched from above 500°C could be fit with short-range order parameters $\alpha_1 = 0$; however, the error limits on this fit are quoted as $\pm 0.2$, which allows for considerable short-range order. Quoted values of $H(0, 6)$,

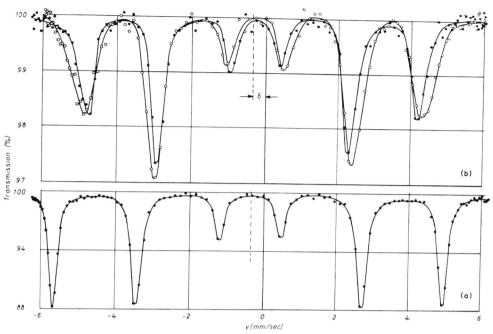

**Fig. 14.** Mössbauer spectra of (a) iron and (b) FeNi$_3$ (Ni–25 at. % Fe) in two states of order—(●) after annealing for 1551 hr at 370°C and (○) quenched from 700°C. (Heilmann and Zinn, 1967.)

$\Delta H_1$, and $\Delta H_2$ are 279 KOe, 10.25 KOe/Fe atom, and 3.0 KOe/Fe atom, respectively.

Subsequent studies of FeNi$_3$ have been made by Lutts and Gielen (1970), who qualitatively confirmed the observations of Heilmann and Zinn, and by Polakova *et al.* (1973). In this latter study, five alloys ranging in composition from 58 to 85 at. % nickel were examined after various metallurgical treatments. The data were fit with several superimposed sets of six-line patterns, and from the relative intensities and hyperfine fields associated with these the near neighbor atomic distributions and magnetic moments were inferred. No systematic study of atomic order was made, and no order parameters are quoted.

The most complete study of order in FeNi$_3$ was made by Drijver *et al.* (1974), who studied samples heat-treated for various times at temperatures between 361 and 490°C. They found maximum attainable values of $S = 0.90 \pm 0.02$ at 361°C, and $S = 0.84$ at 490°C. Their values of $H(0, 6)$, $\Delta H_1$, and $\Delta H_2$ were 276.6 KOe, 11.1 KOe/Fe atom, and 2.7 KOe/Fe atom, in excellent agreement with the results of Heilmann and Zinn.

### 2. FeNi

Atomic ordering in FeNi is exceedingly sluggish below the critical temperature of 320°C but may be accelerated by electron or neutron irradiation or by enhanced diffusion due to excess vacancies trapped in a rapid quench from elevated temperatures. Using irradiation, Gros and Pauleve (1970) have produced extensive order in single-crystal foils of Fe–50Ni and studied these specimens using the Mössbauer effect. The ordering is AuCu type, a tetragonal structure consisting of alternating [100] planes of iron and nickel in a fully ordered structure. In such a system, all iron sites would be expected to be equivalent. Instead, as the atomic order was increased, two distinct hyperfine fields were observed, interpreted by the authors as due to two types of ordered domains with local fields differently oriented to the local symmetry axis. The data were analyzed following the procedures outlined by Heilmann and Zinn with the additional complication that the hyperfine fields have a large anisotropic component in this tetragonal alloy, and that the relative amounts of the two types of ordered domains depend on the annealing conditions. When, as a result of these complications, long-range order was assumed perfect in each type of domain, it was found necessary to introduce a third "disordered" spectrum to fit the data. At best, one must consider this an imprecise description of the atomic arrangement in these alloys.

Ferro *et al.* (1971) studied single crystal foils of Fe–50Ni that were quenched from 1000°C to trap vacancies and subsequently annealed at low temperatures. While their Mössbauer spectra showed signs of splitting into two superimposed spectra as seen by Gros and Pauleve, no attempt was made to estimate the degree of order.

### 3. Iron–Aluminum

The iron–aluminum system has been extensively studied, with interest devoted to its rich variety of magnetic and structural properties. Early Mössbauer studies include those of Ono *et al.* (1962), Friedman and Nicholson (1963), and the earlier cited papers of Stearns (1964, 1966), all of which emphasized studies of hyperfine fields. Huffman and Fisher (1967) examined cold-worked and annealed specimens of compositions ranging from 30 to 50 at. % aluminum. They discussed the presence of magnetic behavior in the cold-worked powders and its absence in fully ordered alloys in terms of the local atomic configurations that develop at antiphase domain boundaries introduced by the deformation.

Most workers who have studied the iron–aluminum system have concentrated on alloys at or near the stoichiometric $Fe_3Al$ composition. The complex phase diagram in this composition range has been the source of much confusion and debate in the metallurgical literature, and unfortunately, the Mössbauer studies described below were all carried out without benefit of the most recent determinations of the phase diagram (Swann et al., 1969, 1972; Okamoto and Beck, 1971). Although these recent phase diagrams differ on several minor points, they are in accord in their disagreement with previous diagrams. Thus, the precise metallurgical state of the samples studied by the Mössbauer workers is in some doubt. Furthermore, although the details of ordering in $Fe_3Al$ had been studied in some detail with x-ray diffraction (Rudmann, 1960), only the recent papers of Kuz'min and Losiyevskaya (1970), Losiyevskaya and Kuz'min (1972), and Losiyevskaya (1973) make use of this structural information. The low-temperature structure of $Fe_3Al$ may be described as four interpenetrating face-centered cubic sublattices, on two of which (the $\alpha$ sublattices) the number of Fe(Al) are similar during the ordering, while on the other two, $\beta$ and $\gamma$, the numbers of iron and aluminum atoms depend on the aluminum concentration and the degree of long-range order. Losiyevskaya (1973) has emphasized the necessity of considering the two independent long-range order parameters for the $\beta$ and $\gamma$ sites, as well as local correlations between these sublattices. At temperatures above $\sim 600°C$ the $Fe_3Al$ type, order is replaced by an ordered arrangement based on the FeAl superlattice. At stoichiometry, ordered FeAl is composed of two interpenetrating simple cubic sublattices, with iron on one and aluminum on the other, to form a CsCl type order.

In several papers, Cser et al. (1967, 1969, 1970) describe studies of alloys in the composition range 24.3–26.4 at. % aluminum (Fig. 15). Mössbauer spectra were made at temperature after allowing 2 hr for the alloy to come to equilibrium. As the temperature is decreased, a discontinuous isomer shift between 850 and 800°C is evidence of the first-order transition to FeAl-type ordering. Below 650°C, magnetic ordering sets in and two superimposed spectra are seen, one paramagnetic, the other ferromagnetic, presumably due to iron with iron-poor and iron-rich local environments, respectively. Further changes in isomer shift and hyperfine fields at $\sim 600°C$ signal the onset of $Fe_3Al$-type ordering. A complex spectrum is observed as the temperature decreases, until at the lowest ordering temperatures, two superimposed ferromagnetic spectra are seen and attributed to iron with eight or four near neighbor iron atoms. Ordering kinetics were examined in a semiquantitative way by measuring the time-dependent change of various Mössbauer parameters for an alloy held at 580°C for various times. No attempt was made to extract long-range order parameters.

Lesoille and Gielen (1970) studied ordering in alloys ranging from 11.5 to 25.6 at. % aluminum. They also examined the spectrum from ordered and disordered $Fe_3Al(C)$. They note that even the 0.2% carbon present in the Armco iron used to make samples was sufficient to stabilize a significant

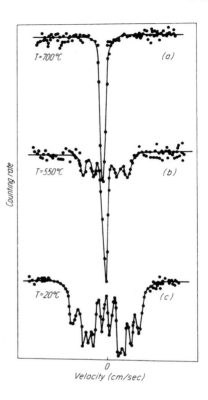

**Fig. 15.** Mössbauer spectra in Fe–24.3 at. % Al measured at different temperatures (velocity scale is sinusoidal). Samples were equilibrated for 2 hr at each temperature before taking the Mössbauer spectrum. (Cser *et al.*, 1967a.)

amount of $Fe_3Al(C)$ and further confuse the already complex spectra of iron–aluminum alloys. Although "ordered" and disordered spectra were observed, no long-range order parameters were determined.

Kuz'min and Losiyevskaya (1970) studied Fe–Al alloys with concentrations from 20 to 31 at. % aluminum, slowly cooled to 24°C and quenched from 400, 500, 600, and 900°C. In addition to Mössbauer spectra, x-ray diffraction was employed to identify the phases present. The Mössbauer spectra were analyzed by a procedure equivalent to that described earlier by Heilmann and Zinn (1967), assuming only a single average long-range order parameter. The larger hyperfine field shifts, $\Delta H_1$ and $\Delta H_2$ in iron–aluminum alloys (compared to those in Fe–Ni alloys), allowed fitting of the spectra with several superimposed six-line patterns rather than looking principally at linewidths, as was necessary with the $FeNi_3$ alloy studied by Heilmann and Zinn. The $Fe_3Al$ alloy quenched from 400°C corresponded to an order parameter of $S = 0.88$. The presence of iron with a field corresponding to five iron neighbors in alloys with less than 25% aluminum had been noted earlier by Huffman and Fisher (1967) and Friedman and

Nicholson (1963) and attributed by these workers to the presence of an $Fe_{13}Al_3$ phase. Kuz'min and Losiyevskaya studied a 22 at. % alloy slowly cooled to room temperature, and found a good fit to the observed spectrum with $S = 0.91$ in an $Fe_3Al$ structure. There was no need to suppose the existence of an $Fe_{13}Al_3$-type phase, in agreement with the more recent phase diagram determinations. More detailed studies of the long-range order were carried out by Losiyevskaya and Kuz'min (1972), allowing for the two long-range order parameters for the $\beta$ and $\gamma$ sublattices. No consideration of local atomic correlation (nonrandom disorder) was made; however, this problem was later treated in an approximate manner by Losiyevskaya (1973) using a statistical thermodynamic analysis. When the relative intensities of the lines in the Mössbauer spectra of alloys of stoichiometric composition were compared to the calculated values, it was shown that in the slowly cooled alloys, the aluminum atoms replace iron atoms preferentially on the $\beta$ sublattices. The complexity of the ordering in this system might yield to a combined x-ray diffraction and Mössbauer study, but no such experiments have yet been carried out.

The Mössbauer effect has been used in the study of short-range order in Fe–12.3 at. % Al by Brümmer *et al.* (1972). As pointed out in Section IV,A, this study suffers from the unwarranted assumption that Eq. (13) may be used to describe $w(i, j)$, the fraction of iron atoms with $i$ and $j$ nn and nnn, respectively. The experimental Mössbauer spectra do show significant change when annealed at 300°C for 10 hr compared to quenching from 800°C or cold rolled, and short-range order is certainly in evidence. The values of $\alpha_1 = -0.1$ and $\alpha_2 = -0.04$ give good fits to the annealed spectrum, while the assumption of complete randomness results in equally satisfactory fits to both the quenched and cold-rolled spectra. While the magnitude of the short-range order parameters obtained should not be given too much credence, the negative sign for $\alpha_1$ is consistent with the short-range order expected in a system that exhibits long-range order at higher concentrations.

### 4. Iron–Silicon

Cranshaw *et al.* (1966) demonstrated the power of single-crystal specimens in their study of the hyperfine fields of Fe–Si alloys with silicon concentrations between 4.9 and 8.6 at. %. The presence of anisotropic hyperfine field effects allowed them to identify the contribution to the spectrum from nn and nnn sites unambiguously, and to conclude from the intensity of the subspectra that silicon atoms are not randomly distributed. They suggest that Si–Si nearest neighbor pairs are forbidden, but make no attempt to describe their results in terms of short-range order parameters. The nonrandom nature of these alloys was confirmed by Gustin *et al.* (1970), who showed qualitative evidence of atomic rearrangement in an Fe–3.3 at. % Si alloy. Differences in the Mössbauer spectra of two foils, one quenched from 1200°C, the other held at 570°C for 3 hr, were attributed to local atomic ordering.

Papadimitriou and Genin (1972) studied an alloy containing 7 at. % silicon slow cooled from 750°C. They analyzed the Mössbauer data as the superposition of three six-line spectra and associated these with iron with 0, 1, or 2 nn silicon atoms. The relative intensities of these lines are not consistent with random atomic arrangement, and the authors suggest the presence of $Fe_{15}Si$-type ordering. The presence of such a phase was earlier suggested by Fallot (1936) as an explanation for his magnetic data but never confirmed by diffraction techniques. Furthermore, no such phase has been found in the recent phase diagram determination by Warlimont (1968). It may be that the combined effects of finite sample thickness distortion and anisotropic magnetic hyperfine fields discussed in Section II are the cause of the deviations of the observed $w(i)$ from those expected for $Fe_3Si$-type ordering. In any case, the presence of $Fe_{15}Si$-type ordering should not be assumed confirmed until such corrections to the Mössbauer data are made.

A detailed study of spectra from alloys containing 0.85–24.9 at. % silicon was made by Häggström et al. (1973). The data were analyzed in a manner similar to that described by Heilmann and Zinn (1967). As was true for Fe–Al alloys, the larger shifts in hyperfine field, $\Delta H_1$, allowed for a more detailed experimental determination of the $w(i)$ (only near neighbor effects were considered). No long-range order parameters were calculated, but explanation of the experimental data for the low-silicon alloys required the assumption of a less than random number of Si–Si near neighbors, in accord with the results of Cranshaw et al. (1966). These workers also suggest the possibility of an $Fe_{15}Si$-type order as explanation. Alloys in the composition range 11.0–18.4 at. % silicon gave complex spectra, suggesting the presence of two phases, in accord with the phase diagram of Warlimont (1968); however, x-ray diffraction data on these samples were not presented. The alloys of 23.8 and 24.9 at. % silicon gave relative intensities of the subspectra in general agreement with the expected $Fe_3Si$ ordering, but no long-range order parameters were determined.

### 5. Iron–Gallium

Newkirk and Tsuei (1971) have presented a comprehensive study of metastable alloys of iron with up to 25 at. % gallium. The samples were prepared by rapid quenching from the melt using a piston and anvil "splat-cooling" apparatus, and studied with x-ray diffraction and Mössbauer techniques. Spectra were fit assuming nn and nnn effects on hyperfine fields, and the relative iron fractions $w(i, j)$ were determined. For concentrations less than 20 at. %, the alloys appeared random, as might be expected for rapid quench from the liquid. However, for the 20 and 25% alloys, the spectra showed considerable evidence of atomic order. The complexity of the spectrum is evidenced in Fig. 16 for the $Fe_3Ga$ sample, showing the computer fit results allowing for twelve different atomic configurations. In Fig. 17, the probability of occurrence of the various configurations is shown for the first

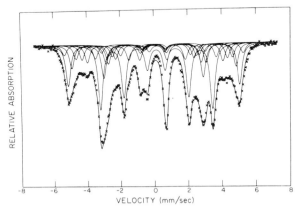

**Fig. 16.**  Absorption spectrum of the metastable alloy $Fe_{75}Ga_{25}$ including the peaks of the various $(i, j)$ components. (Newkirk and Tsuei, 1971.)

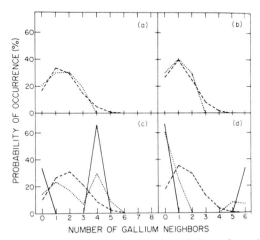

**Fig. 17.**  Probability of the occurrence of various $(i, j)$ configurations for the alloys $Fe_{1-x}Ga_x$ based on: (---) random atomic arrangement; (——) perfect $Fe_3Al$-type ordering; (···) experimental results. (a) First shell, $x = 0.20$; (b) second shell, $x = 0.20$; (c) first shell, $x = 0.25$; (d) second shell, $x = 0.25$. (Newkirk and Tsuei, 1971.)

and second neighbor shells of the 20 and 25% alloys. Agreement between experimental (dotted lines) and random (dashed lines, calculated from Eq. (13) assuming $P_1 = P_2 = c = 0.25$) indicates little if any long-range order in this alloy. On the other hand, the random assumption does not work for the 25% alloy, which shows a definite tendency toward ordering of the $BiF_3$ ($Fe_3Al$) type. The solid curves in Fig. 17 are drawn for perfect $BiF_3$ ordering, and no estimate of the observed long-range order parameter is made from the data. It appears from these observations that the metastable FeGa phase diagram is very similar to the equilibrium FeAl phase diagram.

## 6. $Fe_{1-x}Mo_x$

Schwartz and Asano (1974) have studied the local atomic order in Fe–3.2 at. % Mo and in an unpublished study (Asano, 1974) have also examined alloys with 1.8 and 3.8 at. % molybdenum. The Mössbauer data were corrected for finite sample thickness distortion, and since single crystal specimens were used, the effects of the anisotropic hyperfine magnetic field were included. The results of this analysis have been presented earlier in Fig. 2 and in Table I as a set of measured $w(i, j)$. Asano and Schwartz (Asano, 1974) also measured the x-ray diffuse scattering from these samples and determined the short-range order parameters $\alpha_i$. Using the Gehlen–Cohen–Gragg program described in Section IV,A, a set of $w(i, j)$ were obtained from the x-ray data. Although the x-ray data indicated a weak tendency toward clustering and the Mössbauer data indicated a weak tendency toward short-range order, both agreed within the estimated error with the assumption of random atomic distribution. These results demonstrate the possibilities in the uses of the Mössbauer effect for study of local arrangements, but they also emphasize the care needed to obtain meaningful results.

### Acknowledgments

The author is indebted to many people for his education in this field, but to none more than his former students. It is a pleasure to acknowledge the stimulating contributions over the years of Drs. H. L. Marcus, D. Chandra, N. Abe, and A. Asano. The support of the National Science Foundation through the Northwestern University Materials Research Center is gratefully noted.

### References

Abe, N. (1972). Ph.D. Thesis, Northwestern Univ., Evanston, Illinois.
Abe, N., and Schwartz, L. H. (1973). *In* "Mössbauer Effect Methodology" (I. J. Gruverman, ed.), Vol. 8, p. 249. Plenum Press, New York.
Abe, N., and Schwartz, L. H. (1974). *Mater. Sci. Eng.* **14**, 239.
Albanese, G., Asti, G., and Criscuolli, R. (1970). *IEEE Trans. Magn.* **6**, 567.

Asano, A. (1974). Ph.D. Thesis, Northwestern Univ., Evanston, Illinois.

Asano, A., and Schwartz, L. H. (1973). *Proc. Conf. Magn. Magn. Mater.*, *Boston, Massachusetts, Nov. 1973*, (C. D. Graham, Jr. and J. J. Rhyne, eds.), AIP Conf. Proc. #18, p. 262.

Asano, H. (1969). *J. Phys. Soc. Japan* **27**, 542.

Belova, V. M., Nikolaev, V. I., Spephanovich, S. Yu., and Yakimov, S. S. (1969). *Fiz. Tverd. Tela* **11**, 3662.

Belozerskiy, G. N., Grinblat, Yu. N., and Shapiro, A. I. (1971). *Fiz. Metal. Metalloved.* **32** (2), 301.

Bokshtein, B. S., Voitkovskii, Yu. B., and Shvartsman, A. B. (1968). Translated from *Zavod. Lab.* **34**, No. 1, 34.

Bourgeot, J., Maitrepierre, P. L., Manenc, J., and Thomas, B. J. (1973). *Mém. Sci. Rev. Met.* **70** (2), 125.

Bowman, J. D., Kankeleit, E., Kaufman, E. N., and Persson, B. (1967). *Nucl. Instrum. Methods* **50**, 13.

Bragg, W. L., and Williams, E. J. (1934). *Proc. Roy. Soc. A* **145**, 699.

Brümmer, O., Dräger, G., and Mistol, I. (1972). *Ann. Phys.* **28**, 135.

Butler, E. P., and Thomas, G. (1970). *Acta Met.* **18**, 347.

ByKov, G. A., and Hien, P. Z. (1963). *Sov. Phys.-JETP* **16**, 646.

Cahn, J. W. (1963). *J. Appl. Phys.* **34**, 3581.

Cahn, J. W. (1968). *Trans. AIME* **242**, 166.

Chandra, D., and Schwartz, L. H. (1971a). *In* "Mössbauer Effect Methodology" (I. J. Gruverman, ed.), Vol. 6, p. 79. Plenum Press, New York.

Chandra, D., and Schwartz, L. H. (1971b). *Met. Trans.* **2**, 511.

Chow, H. K., and Bogner, R. L. (1970). *Develop. Appl. Spectrosc.* **8**, 229.

Chow, H. K., Weise, R. F., and Flinn, P. A. (1969). Rep. to USAEC Contract No. AT-(30-1)-4023.

Christ, B. W., and Giles, P. M. (1968). *Trans. AIME* **242**, 1915.

Clapp, P. C. (1971). *Phys. Rev. B* **4**, 255.

Cowley, J. M. (1950). *Phys. Rev.* **77**, 669.

Cranshaw, T. E. (1964). *Rev. Mod. Phys.* **36**, 395.

Cranshaw, T. E. (1972). *J. Phys. F* **2**, 615.

Cranshaw, T. E. (1974). *J. Phys. E* **7**, 1.

Cranshaw, T. E., Johnson, C. E., Ridout, M. S., and Murray, G. A. (1966). *Phys. Lett.* **21**, 481.

Cser, L., Ostanevich, J., and Pál, L. (1967). *Phys. Status Solidi* **20**, 581, 591.

Cser, L., Konczos, G., Nagy, D. L., Ostanevich, Yu. M., and Pál. L. (1969). *Proc. Conf. Appl. Mössbauer Effect, Tihany, 1969.*

Cser, L., Ostanevich, Yu. M., and Pál, L. (1970). *Phys. Status Solidi* **42**, K147.

Dehtiar, I. Ya., Nizin, P. S., Nishchenko, M. M., and Chuistov, K. V. (1969). *Proc. Conf. Appl. Mössbauer Effect, Tihany, 1969* p. 435.

De Nys, T., and Gielen, P. M. (1971). *Metall. Trans.* **2**, 1423.

Dibar-Ure, M. C., and Flinn, P. A. (1971). *In* "Mössbauer Effect Methodology" (I. J. Gruverman, ed.), Vol. 7, p. 245. Plenum Press, New York.

Drijver, J. W., DeGroot, K., and van der Woude, F. (1974). *Int. Conf. Appl. Mössbauer Effect, Bendor, France, Sept. 1974, J. Phys.* (*Suppl. Colloq.* **C–6**) **35**, 465.

Eliezer, Z., Weiss, B. Z., Ron, M., and Nadiv, S. (1973a). *J. Appl. Phys.* **44**, 419.

Eliezer, Z., Nadiv (Niedzwiedz), S., Ron, M., and Weiss, B. Z. (1973b). *Mater. Sci. Eng.* **11**, 269.

Epstein, L. M. (1963). *Proc. Symp. Phys. Non-Destructive Testing, 4th* p. 166. S.W. Res. Inst., San Antonio, Texas.

Ericsson, T., Mourikis, S., and Cohen, J. B. (1970). *J. Mater. Sci.* **5**, 901.

Ettwig, H., and Pepperhoff, W. (1970). *Arch. Eisenhuettenw.* **41**, 471.

Fallot, M. (1936). *Ann. Phys.* **6**, 305.

Ferro, A., Griffa, G., and Ortalli, I. (1971). *Nuovo Cimento II. B* **3**, 269.

Fine, M. E. (1964). "Phase Transformations in Condensed Systems." Macmillan, New York.

Flinn, P. A. (1967a). *Advan. Chem. Ser.* Oct., 21.

Flinn, P. A. (1967b). *In* "Experimental Methods of Materials Research" (H. Herman, ed.), p. 163. Wiley (Interscience), New York.

Flinn, P. A. (1971). *In* "Applications of Low Energy X- and γ-Rays" (C. A. Ziegler, ed.), p. 123. Gordon and Breach, New York.

Flinn, P. A. (1973a). *Proc. Conf. Mössbauer Spectrosc. Bratislava, Czech.*

Flinn, P. A. (1973b). Private communication.

Friedman, E. A., and Nicholson, W. J. (1963). *J. Appl. Phys.* **34**, 1048.

Gehlen, P. C., and Cohen, J. B. (1965). *Phys. Rev.* **139**, A844.

Genin, J. M., Le Caer, G., Maitrepierre, Ph., and Thomas, B. J. (1974). *Scripta Met.* **8**, 15.

Gonser, U. (1971). *In* "Introduction to Mössbauer Spectroscopy" (L. May, ed.), Chapter 8. Plenum Press, New York.

Gorodetsky, G., and Shtrikman, S. (1967). *J. Appl. Phys.* **38**, 3981.

Gragg, J. E., Jr. (1970). Ph.D. Thesis, Northwestern Univ., Evanston, Illinois.

Gros, Y., and Pauleve, J. (1970). *J. Phys.* **31**, 459.

Gruzin, P. L., Sheshin, V. I., and Alekseev, G. (1971). *Metod. Izotop. Indikatorov Nauch Issled Prom Proizvod.* 134.

Gruzin, P. L., Rodionov, Yu. L., Zharov, Yu. D., Mkrtchyan, V. S., Edneral, A. F., and Perkas, M. D. (1972a). *Sov. Phys.-Dokl.* **17**, 64.

Gruzin, P. L., Rodionov, Yu. L., Mkrtchyan, V. S., and Li, Yu. A. (1972b). *Sov. Phys.-Dokl.* **17**, 475.

Gustin, C., Gielen, P. M., and de Croylaan, G. (1970). *Scripta Met.* **4**, 925.

Häggström, L., Grånäs, L., Wäppling, R., and Devanarayanan, S. (1973). *Physica Scripta* **7**, 125.

Heilmann, A., and Zinn, W. (1967). *Z. Metallk.* **58**, 113.

Hilliard, J. E. (1970). *In* "Phase Transformations" (H. I. Aaronson, ed.), p. 497. Amer. Soc. Met., Metals Park, Ohio.

Hirsch, A. A., and Eliezer, Z. (1972). *IEEE Trans. Magn.* Sept. 690.

Hornbogen, E. (1964). *Trans. ASM* **57**, 120.

Huffman, G. P., and Fisher, R. M. (1967). *J. Appl. Phys.* **38**, 735.

Johnson, C. E., Ridout, M. S., and Cranshaw, T. E. (1963). *Proc. Phys. Soc.* **81**, 1079.

Johnson, D. P., and Dash, J. G. (1968). *Phys. Rev.* **172**, 983.

Jones, R. D. (1973). *Iron and Steel* **46**, 33, 137.

Klostermann, J. A. (1969). *Symp. Mechanism Phase Trans. Cryst. Solids, Manchester, 1969* 143.

Kuz'min, R. N., and Losiyevskaya, S. A. (1970). *Phys. Metal. Metalloved.* **29**, 569.

Lahiri, S. K., Chandra, D., Schwartz, L. H., and Fine, M. E. (1969). *Trans. AIME* **245**, 1865.

Langer, J. S. (1973). *Acta Met.* **21**, 1649.

Lesoille, M. R., and Gielen, P. M. (1970). *Phys. Status Solidi* **37**, 127.

Lin, Tsung-Ming, and Preston, R. S. (1974). *In* "Mössbauer Effect Methodology" (I. J. Gruverman, C. W. Seidel, and D. K. Dieterly, eds.) vol. 9, p. 205. Plenum, New York.

Losiyevskaya, S. A. (1973). *Phys. Status Solidi (a)* **16**, 647.

Losiyevskaya, S. A., and Kuz'min, R. N. (1972). *Russ. Met.* **3**, 141.

Lutts, A., and Gielen, P. M. (1970). *Phys. Status Solidi* **41**, K81.

Makarov, E. F., Povitskii, V. A., Granovskii, E. B., and Fridman, A. A. (1967). *Phys. Status Solidi* **24**, 45.

Makarov, V. A., Granovskii, E. B., Makarov, E. F., and Povitskii, V. A. (1972). *Phys. Status Solidi* **14**, 331.

Marcus, H. L., Schwartz, L. H., and Fine, M. E. (1966). *Trans. ASM* **59**, 468.

Marcus, H. L., Fine, M. E., and Schwartz, L. H. (1967). *J. Appl. Phys.* **38**, 4750.

Margulies, S., and Ehrman, J. R. (1961). *Nucl. Instrum. Methods* **12**, 131.

Mima, G., and Yamaguchi, M. (1970). *Trans. Jap. Inst. Met.* **11**, 239.

Nagarajan, A., and Flinn, P. A. (1967). *Appl. Phys. Lett.* **11**, 120.

Newkirk, L. R., and Tsuei, C. C. (1971). *J. Appl. Phys.* **42**, 5250.

Okamoto, H., and Beck, P. A. (1971). *Met. Trans.* **2**, 569.

Ono, K., Ishikawa, Y., and Ito, A. (1962). *J. Phys. Soc. Japan* **17**, 1747.

Oron, M. Shtrikman, S., and Treves, D. (1969). *J. Mater. Sci.* **4**, 581.

Papadimitriou, G., and Genin, J. M. (1972). *Phys. Status Solidi (a)* **9**, K19.

Polakova, J., Lauermannova, J., and Zemcik, T. (1973). *Proc. Conf. Mössbauer Spectrosc., Bratislava, Czech.*

Povitsky, V., Granovsky, E., Fridman, A., Makarov, E., and Pashkov, P. P. (1970). *IEEE Trans. Magn.* **6**, 215.

Preston, R. S., Lam, D. J., Nevitt, M. V., Van Ostenburg, D. O., and Kimball, C. W. (1966). *Phys. Rev.* **149**, 440.

Ron, M. (1973). Int. Atomic Energy Agency Sm-159/32, Vienna.

Rosencwaig, A., Ron, M., Kidron, A., and Shechter, H. (1969). *J. Phys. Chem. Solids* **30**, 359.

Rudmann, P. S. (1960). *Acta Met.* **8**, 321.

Schwartz, L. H. (1970). *Int. J. Nondestruct. Testing* **1**, 353.

Schwartz, L. H., and Asano, A. (1974). *Proc. Int. Conf. Appl. Mössbauer Effect, Bendor, France, Sept. 1974 J. Phys. (Suppl. Colloq.* **C–6**) **35**, 453.

Schwartz, L. H., and Cohen, J. B. (1965). *J. Appl. Phys.* **36**, 598.

Sharma, V. K., Breyer, N. N., Abe, N., and Schwartz, L. H. (1974). *Scripta Met.* **8**, 699.

Shenoy, G. K. (1974). *In* "Mössbauer Effect Methodology" (I. J. Gruverman, C. W. Seidel, and D. K. Dieterly, eds.) vol. 9, p. 277. Plenum, New York.

Shirane, G., Chen, C. W., Flinn, P. A., and Nathans, R. (1963). *Phys. Rev.* **131**, 183.

Shtrikman, S., and Treves, D. (1966). *J. Appl. Phys.* **37**, 1103.

Stearns, M. B. (1964). *J. Appl. Phys.* **35**, 1095.

Stearns, M. B. (1966). *Phys. Rev.* **147**, 439.

Stearns, M. B. (1971). *Phys. Rev.* **4**, 4069.

Stearns, M. B. (1974). *Proc. Conf. Magn. Magn. Mater., Boston, Massachusetts, Nov. 1973* (C. D. Graham, Jr. and J. J. Rhyne, eds.), AIP Conf. Proc. #18 p. 257.

Swann, P. R., Duff, W. R., and Fisher, R. M. (1969). *Trans. TMS-AIME* **245**, 851.

Swann, P. R., Duff, W. R., and Fisher, R. M. (1972). *Met. Trans.* **3**, 409.

Swartzendruber, L. J., and Bennett, L. H. (1972). *Scripta Met.* **6**, 737.

Swartzendruber, L. J., Bennett, L. H., Schoefer, E. A., DeLong, W. T., and Campbell, H. C. (1974). *J. Welding* **53**, 1–S.

Vincze, I., and Aldred, A. T. (1974). *Phys. Rev. B* **9**, 3845.

Vincze, I., and Campbell, I. A. (1973). *J. Phys. F* **3**, 647.

Warlimont, H. (1968). *Z. Metallk.* **59**, 595.

Wertheim, G. K., Jaccarino, V., Wernick. J. H., and Buchanan, D. N. E. (1964). *Phys. Rev. Lett.* **12**, 24.

Wierengen, J. S., and Rensen, J. G. (1966). *Z. Angew. Phys.* **21**, 69.

Yagisawa, K. (1973a). *Phys. Status Solidi (a)* **16**, 291.

Yagisawa, K. (1973b). *Phys. Status Solidi (a)* **18**, 589.

Yamamoto, H. (1964). *Jap. J. Appl. Phys.* **3**, 745.

# COLLOID AND
# INTERFACE CHEMISTRY

# 3

# *Corrosion and Interfacial Reactions*

**Gary W. Simmons**
**Henry Leidheiser, Jr.**

Center for Surface and Coatings Research
Lehigh University
Bethlehem, Pennsylvania

## I. Introduction

The purpose of this chapter is to describe the applications of Mössbauer spectroscopy to the fundamental and applied aspects of corrosion phenomena. Although Mössbauer spectroscopy is applicable to only a limited number of metals, the technique nevertheless has provided and will continue to provide a tool for practical as well as fundamental corrosion studies primarily of iron, tin, cobalt, and alloys containing these elements. The considerable information available in the literature on the characterization by Mössbauer spectroscopy of iron and tin compounds provides a strong basis for applying $\gamma$-ray resonance spectroscopy to corrosion studies. The isomer shift, quadrupole splitting, and magnetic hyperfine splitting have been measured for a large number of organic and inorganic compounds formed during corrosion of these metals. In particular, the oxides and hydroxides of iron have been studied extensively. Mössbauer spectroscopy is, therefore, readily applicable to qualitative analysis of corrosion products. In many cases, direct quantitative measurements can be made of the corrosion product(s) that consist either of a single phase or of a complex mixture of corrosion species. A particular advantage of the Mössbauer technique is that analysis can be made of either amorphous or crystalline corrosion products. In addition to the analytical applications, it is possible to determine some of the chemical and physical properties of oxides that are basic to the understanding of corrosion phenomena. For example, the relationship between the stoichiometry of $Fe_3O_4$ and its conductivity has been determined by Mössbauer spectroscopy. Experimental techniques have been developed that permit studies of corrosion films with thickness ranging from tenths of a nanometer to several micrometers. Furthermore, it is possible in many cases to conduct studies *in situ*. Since the various Mössbauer spectroscopic techniques used in corrosion studies have been described in a previous paper (Leidheiser *et al.*, 1973a), only a brief description of the experimental methods will be presented. Emphasis will be placed on the application of Mössbauer spectroscopy to phenomena important to understanding corrosion processes. The subjects that are covered include qualitative and quantitative analysis of corrosion products, particle size of corrosion products, solvated corrosion products, defect structure of corrosion products, kinetic studies, passivity and corrosion inhibition, corrosion beneath a coating, stress corrosion cracking and hydrogen embrittlement, and diffusion. Selected papers are discussed to illustrate the application of Mössbauer spectroscopy to these subjects.

## II. Experimental Methods

Mössbauer spectra can be obtained by transmission, scattering, and emission techniques. The details of each of these spectroscopic methods have been described elsewhere in this volume. Nevertheless, since each experimental method has features that makes it unique to particular types of corrosion studies, it is appropriate to describe briefly each method in terms of its applicability.

## A. Transmission Techniques

Transmission Mössbauer spectroscopy is directly applicable to corrosion studies of thin metal foils, and the major application has been in studies of relatively thick corrosion products ($10^3$–$10^5$ nm). *In situ* studies of the corrosion of thin films and foils are possible with suitably designed reaction cells with windows that are essentially transparent to the Mössbauer $\gamma$ rays. The corrosion products on thick materials can be studied by transmission after removal from the substrate. The relative concentrations for any number of phases 1, 2, 3, . . . , $n$ can be determined from the relationship

$$N_n \bigg/ \sum_{j=1}^{n} N_j = (A_n/f_n) \bigg/ \sum_{j=1}^{n} (A_j/f_j), \tag{1}$$

where for each phase, $N$ is the number of atoms per cubic centimeter of the Mössbauer element, $A$ is the area under the resonance peak(s), and $f$ is the recoil-free fraction. Absolute quantitative analysis is possible, but is more difficult, since it is necessary to determine accurately the background counts and the fundamental parameters such as recoil-free fraction, linewidth of the source, and the resonance cross-sections (see pp. 23 and 41–43).

## B. Reflection Techniques

Gamma rays, x rays, or electrons produced after resonance absorption may be used for reflection spectroscopy. The energies and yields for the resonantly scattered radiation in the case of iron will now be given. After resonance absorption of the 14.4 keV $\gamma$ rays by $^{57}$Fe, approximately 90% of the excited nuclei revert to the ground state by conversion electron emission, of which 80% are K electrons with energies of 7.3 keV. The other 10% of the nuclei decay to the ground state by re-emission of 14.4 keV $\gamma$ rays. Approximately 60% of the atoms that are ionized as a result of K conversion subsequently release energy by the emission of KLL Auger electrons with 5.4 keV energy, and the remaining 40% undergo de-excitation by emission of 6.3 keV $K_\alpha$ x rays. The resonantly scattered $\gamma$ rays and x rays are applicable for studying advanced stages of corrosion, and the conversion and Auger electrons are more suitable for investigating thin layers of corrosion product. The major advantage of reflection spectroscopy is that bulk specimens can be studied while the corrosion products are still attached to the substrate. Another significant advantage of scattering techniques is the relatively high signal-to-noise ratio. In transmission experiments, the background count rate (or nonresonance count rate) is largely due to the $\gamma$ rays that have not been absorbed by the specimen, consequently, the total background counts do not differ appreciably from the number of incident $\gamma$ rays. On the other hand, in scattering geometry, the detector is shielded from the direct irradiation by the source, and most of the background, therefore, originates from nonresonant scattering of radiation in the absorber. The cross sections for the nonresonant phenomena in scattering experiments are a few orders of magnitude lower than for the resonance effects; consequently, higher signal-to-noise is possible for reflection than for transmission spectroscopy. Because of the high internal conversion coefficients, most reflection Mössbauer spectroscopy studies have utilized the x rays and electrons associated with internal conversion rather than the resonantly scattered $\gamma$ rays. The main features of the x ray and electron scattering techniques, in terms of applicability to corrosion studies, will be discussed separately.

There is a significant difference in surface sensitivity between conversion x-ray and conversion electron reflection Mössbauer spectroscopy owing to differences in depths from which the two types of measured radiation originate within a specimen. The applicability of the x-ray reflection method to corrosion studies will be discussed first. Swanson and Spijkerman (1970) have determined experimentally that 78% of backscattered x-ray signal comes from the first $5 \times 10^{-4}$ cm of surface depth in iron and that 93% of the signal originates in the first $1.2 \times 10^{-3}$ cm. Quantitative analysis in x-ray scattering experiments requires a detailed description of both the attenuation of resonant $\gamma$ rays as they enter the absorber and the attenuation of resonantly produced x rays that are emitted from specific depths within the scatterer. Terrell and Spijkerman (1968) have derived theoretical expressions for the intensity of the resonantly scattered x-ray signal as a function of resonator thickness. The determination of thickness for a single oxide phase on iron is possible with these expressions by using the appropriate resonant absorption cross section for the 14.4 keV $\gamma$ ray and mass absorption coefficient for the 6.3 keV x ray. Thickness determinations are, however, difficult for multiphase corrosion layers, since the attenuation of resonant $\gamma$ rays and scattered x rays would have to be considered separately for each phase.

Mössbauer resonant scattering detected via conversion electrons offers the opportunity for obtaining qualitative and quantitative information about the chemical and physical properties of thin corrosion layers. The relatively high yield of resonantly produced electrons of moderately low energies makes possible Mössbauer spectroscopy with a high surface sensitivity, since the escape depth of electrons originating within an absorber is limited by a high probability for inelastic scattering. Swanson and Spijkerman (1970) have determined experimentally the escape depth of resonantly produced electrons in natural iron by taking Mössbauer spectra of iron films of known thickness deposited onto a stainless steel substrate. From the areas under the resonance peaks for iron and stainless steel, they calculated that 65% of the signal originates within the first 60 nm of the surface. Simmons *et al.* (1973) have demonstrated that oxide films of approximately 5 nm in thickness can be detected on $^{57}$Fe-enriched iron surfaces. For backscattered K conversion electrons, Krakowski and Miller (1972) have derived expressions for the area under resonance peaks and for the electron signal intensity at maximum resonance. The integrals describing the electron signal are expressed in terms of the reduced resonator thickness $\mu_K t$ and the ratio $\mu_R/\mu_K$, where $\mu_K$ and $\mu_R$ are the linear absorption coefficients for conversion electrons and for resonant $\gamma$ rays, respectively. Quantitative analysis, in this case, is possible by using the appropriate values of $\mu_K$ and $\mu_R$ for the specimen under study. Single-phase corrosion films are amenable to quantitative analysis, but unfortunately such analysis is not easily accomplished for multiphase and/or multilayered films. The major shortcoming of the resonant electron scattering technique is that the specimens must be placed inside a flow counter. Since changes in the composition of the flow gas strongly influence the counting efficiency, it is not possible to conduct experiments either under *in situ* conditions or at low temperatures. Furthermore, some corrosion products may change composition by dehydration in the relatively dry environment of the flow counter.

Mössbauer spectra of corrosion films as a function of depth can be obtained by energy analysis of K conversion electrons, since the electron energy losses are a function of escape depth. A schematic diagram of the experimental arrangement used for energy analysis of resonantly produced electrons is shown in Fig. 1. Krakowski and Miller (1972) have determined

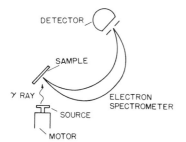

**Fig. 1.** Arrangement for Mössbauer scattering experiments with energy analysis of conversion electrons. (Bonchev *et al.*, 1969.)

theoretically the factors that limit the depth resolution of electron-scattering Mössbauer experiments. They concluded that for scattered electrons with energies greater than approximately one-half of the initial conversion electron energy, the depth resolution is sufficient to permit Mössbauer spectra to be obtained from selected regions near a specimen surface. Furthermore, for good depth resolution, sufficient energy separation between the conversion and Auger electrons is required. The latter criterion is satisfied in the case of tin, since there is a 16.6 keV difference between the energies of the L conversion and LMM Auger electrons. Prospects for the application of this spectroscopic technique with good depth resolution are not as favorable in the case of iron, since the energy separation of the K conversion and KLL Auger electrons is only 1.9 keV.

## C. Emission Techniques

Emission spectra of Mössbauer nuclei are obtained by doping a specimen with a source isotope and performing a conventional transmission experiment with a single-line resonant absorber. In the case of $^{57}$Co-doped specimens, for example, the emission Mössbauer spectra arises from the $^{57}$Fe probe atoms, and the observed isomer shift, quadrupole splitting, and magnetic hyperfine splitting provide chemical and physical information about the host matrix. For corrosion studies, this technique has the important advantage that investigations can be readily carried out with possible surface sensitivity of less than one atomic layer. Surface sensitivity is obtained simply by controlling the thickness of the doped layer on the specimen surface. Studies of corrosion in aqueous environments are possible, since the low scattering cross section of $\gamma$ rays allows measurements to be made *in situ* while corrosion is occurring.

Some potential problems arise with the application of the emission technique when the source isotope decays by electron capture. In the case of $^{57}$Co, for example, electron capture produces an ionized K level in $^{57}$Fe daughter atoms, and the de-excitation that follows by Auger cascade can produce charge states as high as $7+$ in the valence levels of $^{57}$Fe (Pollak, 1962). If these nonequilibrium charges on the daughter $^{57}$Fe atoms have a lifetime of the order of the nuclear excited state ($\sim 10^{-7}$ s), then the emission spectrum will show resonance lines that do not represent the intrinsic properties of the parent $^{57}$Co atoms. The stability of charges produced by Auger aftereffects depends upon the chemical and physical environment of the $^{57}$Fe nucleus in the host. Normal charge states are observed with emission spectroscopy for metals and alloys that are doped with $^{57}$Co, since charges on the $^{57}$Fe atoms are rapidly equilibrated by highly mobile conduction electrons. The charge equilibration is slower for insulating materials, and nonequilibrium charges as high as $3+$ have been observed (Wickman and Wertheim, 1968). Stability of these charge states in insulators has been attributed in some cases—CoO, for example—to localized lattice defects and nonstoichiometry. Lattice energy, crystal field effects, and electron exchange with neighboring cations, however, may also contribute in some compounds to the formation or stabilization of ionic charges different from the parent ionic state. The possible formation of charged $^{57}$Fe species different from the original charge on $^{57}$Co imposes a complication in the interpretation of spectra, nevertheless emission spectroscopy can be useful for obtaining important information of chemical and physical interest to corrosion studies.

## III. Analytical Applications

There are essentially four kinds of data that can be utilized to identify a specific chemical compound by Mössbauer spectroscopy—the isomer shift, the magnitude of quadrupole splitting, magnetic hyperfine splitting, and changes in the magnitude of these parameters as a function of temperature. Table I lists the values of the nuclear hyperfine interactions for eleven oxides and hydroxides of iron that might be expected to be formed on iron or its alloys under different conditions of corrosion. It is apparent from the tabulation that it is relatively easy to distinguish between $Fe(OH)_2$, FeOOH, FeO, $Fe_2O_3$, and $Fe_3O_4$, but there are difficulties in distinguishing between the various forms of FeOOH. Part of this difficulty arises from the fact that the reference spectra are not reliable and that the preparations from which spectra were obtained probably were not totally one form of the hydroxyoxide. Some examples of the analytical applications of Mössbauer spectroscopy to the characterization of corrosion products will be given. No attempt has been made to cite every reference in the literature on the analytical applications.

The formation of rust on iron in the presence of saturated water vapor at room temperature and by periodic immersion in calcium chloride solution was studied by Dezsi *et al.* (1964). The rust formed in water vapor consisted of 50% by weight $\alpha$-$Fe_2O_3$, $\sim$40% $\gamma$-FeOOH, and $\sim$10% $\beta$-FeOOH, and corrosion in the salt solution produced equal amounts of $\beta$-FeOOH and $\gamma$-FeOOH. The conclusions derived from spectroscopy were supported by thermoanalytical measurements.

Bancroft *et al.* (1971) removed the air-formed oxide and the oxide formed upon immersion of iron in chromate solution by stripping the films from the iron substrate in $CH_3OH$—Br medium. The stripped films were collected on filter paper, and the ferrous bromide was removed by heating the samples at 300°C in dry nitrogen. The product was broken up, sandwiched between two pieces of adhesive tape, and the Mössbauer spectra taken by the transmission method. The air-formed film consisted of $Fe_3O_4$ and $\gamma$-$Fe_2O_3$. Mössbauer spectra of specimens exposed to chromate solution showed resonance lines associated with $Fe_3O_4$ and $\gamma$-$Fe_2O_3$ and additional resonance peaks that were attributed to the mixed oxide $Fe_2O_3\cdot3Cr_2O_3$. Information derived from corrosion products stripped from the surface by chemical methods must, however, be viewed with some reservation, since the stripping process may significantly alter the corrosion products.

**TABLE I**

Mössbauer Parameters for Iron Oxides, Hydroxides, and Hydroxyoxides

| Compound | Method of Preparation | Site | Temp. (°K) | H (kOe) | Q.S. (mm/sec) | I.S. (mm/sec) | I.S. vs. Fe (mm/sec) | Magnetic Trans. Temp. (°K) | Ref.[a] |
|---|---|---|---|---|---|---|---|---|---|
| α-FeOOH | Natural | — | 0 | 510 | — | — | — | 393.3 | 1 |
| | | | 90 | 500 | — | — | — | — | 1 |
| | Ppt. from Cl⁻ and NO₃⁻ solns. | | 300 | 389 | — | +0.26 (Pt) | +0.61 | — | 1 |
| | | — | 80 | 481 | 0.17 | +0.20 (Cu) | +0.43 | — | 2 |
| β-FeOOH | Ppt. from Cl⁻ soln. using ammonia or urea | — | 80 | 445 | 0.12 | +0.06 (Cu) | +0.29 | 295 (d) | 2 |
| | | | | | 0.20 | +0.14 (Cu) | +0.37 | | |
| β-FeOOH small particles, spindle-shaped | Ppt. from Cl⁻ soln. | — | 80 | 485 | 0.64 | +0.64 (NaNP) | +0.38 | — | 3 |
| | | | RT$^b$ | — | 0.62 | +0.65 (NaNP) | +0.39 | — | 3 |
| δ-FeOOH | Ppt. from sulfate soln. and H₂O₂, dried at 45°C | — | 80 | 505 | — | — | — | — | — |
| | | | | 525 | | | | | — |
| γ-FeOOH | Oxidn. of Fe(OH)₂ | — | 300 | — | 1.1 | +0.50 (Cr) | +0.35 | — | 4 |
| | | — | 295 | — | 0.60 | +0.64 (NaNP) | +0.38 | — | 5, 6 |
| Fe(OH)₂ | Ppt. from sulfate soln. under anaerobic cond. | — | 95 | — | 3.13 | +1.62 (NaNP) | +1.36 | — | 5 |
| | | | 195 | — | 3.06 | +1.55 (NaNP) | +1.29 | — | — |
| | | | 298 | — | 2.92 | +1.44 (NaNP) | +1.18 | — | — |
| | Not given | — | 4.2 | 200 | 3.06 | +1.48 (SS) | +1.22 | — | 7 |
| Fe₀.₉O | Heat in CO₂–CO | Fe²⁺ (A) | 297 | — | 0.46 | +0.68 (Cu) | +0.91 | — | 8 |
| | | Fe²⁺ (B) | 297 | — | 0.78 | +0.63 (Cu) | +0.86 | — | — |
| | | Fe²⁺ (A) | 203 | — | 0.58 | +0.73 (Cu) | +0.96 | — | — |
| | | Fe²⁺ (B) | 203 | — | 1.00 | +0.70 (Cu) | +0.93 | — | — |

| | | | RT[b] | | | | | | |
|---|---|---|---|---|---|---|---|---|---|
| FeO | — | — | | — | 0.8 | +0.93 (Fe) | +0.93 | 198 | 9 |
| $Fe_3O_4$ | Natural | $Fe^{3+}$ (A) | 298 | 493 | 0 | +0.25 (Fe) | +0.25 | — | 10 |
| | | $Fe^{3+}$ (A + B) | 77 | 514 | 0 | +0.44 (Fe) | +0.44 | — | — |
| | | $Fe^{3+}$ (A + B) | 4.2 | 516 | −0.06 | +0.42 (Fe) | +0.42 | — | — |
| | | $Fe^{2+,3+}$ (B) | 298 | 460 | 0 | +0.65 (Fe) | +0.65 | — | — |
| | | $Fe^{2+}$ (B) | 77 | 499 | −0.76 | +0.94 (Fe) | +0.94 | — | — |
| | | $Fe^{2+}$ (B) | 4.2 | 510 | −0.89 | +0.99 (Fe) | +0.99 | — | — |
| $Fe_{2.83}O_4$ | Reference 11 | $Fe^{3+}$ (A) | 298 | 490 | 0 | +0.23 (Fe) | +0.23 | — | 10 |
| | | $Fe^{3+}$ (A + B) | 77 | 517 | 0 | +0.42 (Fe) | +0.42 | — | — |
| | | $Fe^{3+}$ (A + B) | 4.2 | 518 | 0 | +0.44 (Fe) | +0.44 | — | — |
| | | $Fe^{2+,3+}$ (B) | 298 | 459 | 0 | +0.62 (Fe) | +0.62 | — | — |
| | | $Fe^{2+}$ (B) | 77 | 503 | −1.50 | +0.78 (Fe) | +0.78 | — | — |
| | | $Fe^{2+}$ (B) | 4.2 | 509 | −1.66 | +0.74 (Fe) | +0.74 | — | — |
| $\alpha$-$Fe_2O_3$[c] | Heated in air at 1500°C | | 296 | 518 | +0.42 | +0.39 (Fe) | +0.39 | 260° and 956° | 12 |
| | | | 83 | 542 | −0.69 | — | — | — | — |
| $\gamma$-$Fe_2O_3$ | Not given | A | 300 | 502 | — | +0.33 (Cr) | — | — | 13 |
| | | B | 300 | 503 | — | +0.55 (Cr) | — | — | — |

[a] *Key to references.* (1) van der Woude and Dekker (1966), (2) Moreira *et al.* (1973), (3) Voznyuk and Dubinin (1973), (4) Dézsi *et al.* (1967), (5) Pritchard and Mould (1971), (6) Terrell and Spijkerman (1963), (7) Miyamoto *et al.* (1967), (8) Johnson (1969), (9) Vaughan and Drickamer (1967), (10) Evans and Hafner (1969), (11) Kullerud *et al.* (1968), (12) Kündig *et al.* (1966), (13) Khalafalla and Morrish (1972).*

[b] RT = room temperature.

[c] See van der Woude (1966) for data up to 900°K in graphical form.

Pritchard *et al.* (1971) determined the Mössbauer spectra of corrosion products removed from a mild steel high-temperature water circuit in a nuclear reactor and compared these with the spectrum of the product obtained from hydrothermal decomposition of $Fe(OH)_2$. The similarity of the spectra suggested that the decomposition of $Fe(OH)_2$ to $Fe_3O_4$ is an important reaction in determining the nature of the products formed during start-up, and that the relatively large $Fe_3O_4$ particles were in part responsible for the subsequent protection of the mild steel from further corrosion.

Gonser *et al.* (1966) used transmission Mössbauer spectroscopy to follow the internal oxidation and reduction of iron in a copper–iron (2.2 at. % Fe) alloy. At low partial pressures of oxygen ($\sim 3 \times 10^{-4}$ Torr) and at 850°C, incomplete oxidation of Fe to FeO was observed. Hydrogen annealing at 900°C completely reduced the FeO. The large fraction of $\alpha$-iron observed after reduction indicated that the incoherent $\gamma$-precipitates that were formed during the hydrogen anneal transformed readily on cooling without plastic deformation. Reaction of the copper–iron alloy at higher partial pressure of oxygen (0.2 Torr) produced complete oxidation to $Fe_3O_4$. The stoichiometry of the magnetite corresponded to $Fe_{2.87}O_4$ as deduced from the intensities of the $Fe^{3+}$ and $Fe^{2+}$ resonance lines. Attempts to oxidize the specimens further to $\gamma$-$Fe_2O_3$ resulted in the formation of $CuFeO_2$ (delafossite).

Terrell and Spijkerman (1968) used the x-ray reflection technique to determine the corrosion product formed when a steel plate was exposed to HCl vapor. The spectrum was interpreted to be from $\beta$-FeOOH, and the film thickness was estimated to be $2 \times 10^{-3}$ cm.

Simmons *et al.* (1973) used the K conversion electron reflection method to study oxide films formed by the thermal oxidation of $^{57}$Fe enriched iron specimens at 225, 350, and 450°C. Mössbauer spectra were obtained for oxide thicknesses ranging from approximately 5 to several tens of nm. Figure 2 shows the spectra of iron after oxidation at 225°C for various times. The $Fe_3O_4$ phase shown in Fig. 2e was estimated to be approximately 12 nm thick. The equal intensities of the A site and B site resonance lines indicate that the oxide formed at 225°C is either nonstoichiometric $Fe_3O_4$ corresponding to $Fe_{2.91}O_4$ or is a mixture of the stoichiometric oxides $Fe_3O_4$ and $\gamma$-$Fe_2O_3$. In either case, it can be concluded that oxidation at 225°C produces a cation-deficient $Fe_{3-v}O_4$ oxide phase. Oxidation at 350°C for short times produced a duplex film consisting of $Fe_3O_4$ and $\alpha$-$Fe_2O_3$ as shown in Fig. 3. The spectrum obtained after oxidation at 450°C indicated that a single oxide, nearly stoichiometric $Fe_3O_4$, was formed. The absence of $\alpha$-$Fe_2O_3$ after the short-time oxidation at 450°C was attributed to an increased cation flux at the higher temperature.

Bonchev *et al.* (1969) have obtained Mössbauer spectra of corrosion films on tin as a function of depth by energy analysis of $^{119}$Sn L conversion

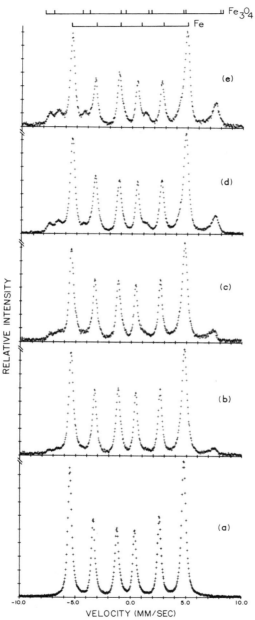

**Fig. 2.**   Conversion electron Mössbauer spectra of iron oxidized at 225°C for specific times (a) before oxidation, (b) 5 min, (c) 15 min, (d) 120 min, and (e) 1000 min. (Simmons *et al.*, 1973.)

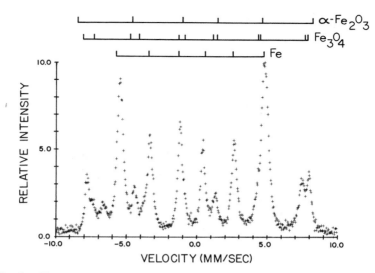

**Fig. 3.**   Electron conversion Mössbauer spectrum of iron after 5 min oxidation at 350°C. (Simmons *et al.*, 1973.)

electrons. An electron spectrometer was used to focus conversion electrons of a definite energy onto a scintillation detector (see Fig. 1). A $^{119}Sn$-enriched tin specimen was exposed to bromine vapor for approximately 10 sec and the Mössbauer spectra were obtained as a function of the magnetic focusing field of the spectrometer. Figure 4 shows the Mössbauer spectra obtained from electrons originating at different depths. For electrons with lowest measured energy, a Mössbauer spectrum of the $\beta$-tin substrate was produced, and for electrons of higher energies, corrosion layers of $SnBr_2$ and $SnBr_4$ were detected. A similar experiment was performed with tin exposed to fuming nitric acid vapors. The relative intensities of the $\beta$-tin and $SnO_2$ Mössbauer resonance lines observed in this case were measured as a function of electron energy. The relatively small dependence that was found for the $\beta$-$Sn/SnO_2$ intensity ratio as a function of electron energy was attributed to a highly inhomogeneous oxide film.

Knowledge of the composition of a solid corrosion product in relation to the composition of the alloy from which it is formed is often of interest. Although Mössbauer spectroscopy may generally not be the analytical tool chosen for application, quantitative analysis can be made in some systems as shown by Pella and DeVoe (1970). The concentration of tin in copper–tin alloys was obtained with an error of the order of 3% as determined by comparison with National Bureau of Standards certified values. The method involves conversion of the material to $SnO_2$, mixing with an $Al_2O_3$ diluent,

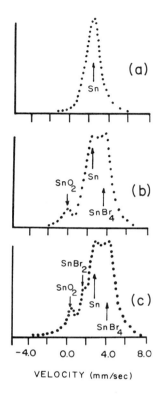

**Fig. 4.** Conversion electron spectra of tin treated with bromine vapors. The energy of the electrons detected increase from (a) to (c). (Adapted from Bonchev *et al.*, 1969.)

and comparing the Mössbauer spectrum with an internal tin reference and calibration curves. A full description of the technique has been given (Pella and DeVoe 1971).

## IV. Particle Size of Corrosion Products

In order to identify or to characterize fully oxides and corrosion products, it is necessary to consider how the magnetic properties of finely divided specimens affect their Mössbauer spectra. When the particle size of a magnetic compound approaches approximately 10 nm, the magnetic hyperfine splitting is not observed, and the spectra exhibit only a singlet or doublet resonant peak(s). Studies of fine particles of $\alpha\text{-Fe}_2\text{O}_3$ are presented as an example (Kündig *et al.*, 1966). Figure 5 shows the spectra of $\alpha\text{-Fe}_2\text{O}_3$ as a function of average particle size ranging from $<10$ nm to bulk oxide. The most widely advanced interpretation of this behavior is known as super-paramagnetism. Each individual particle is considered to be a single magnetic domain, and the magnetization vector is held in an easy direction by

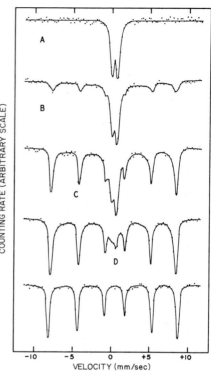

**Fig. 5.** Mössbauer spectra of α-$Fe_2O_3$ as a function of the average particle size: (A) < 10 nm, (B) 13.5 ± 1 nm, (C) 15 ± 1 nm, and (D) 18 ± 1 nm. (Kündig *et al.*, 1966.)

the crystalline field. The energy required for the magnetization vector to go from one easy direction to another is denoted as $2K$, and the probability that this vector will spontaneously change its direction is proportional to $\exp(-2Kv/kT)$, where $v$ is the particle volume. When the particle volume is small, the relaxation time for this change in magnetization direction approaches the lifetime of the nuclear excited state ($\sim 10^{-7}$ sec). The magnetic hyperfine structure consequently begins to collapse when the particles reach a critical size as shown in Fig. 5. Since the relaxation time is proportional to $\exp(-2Kv/kT)$, it is possible to obtain a magnetic hyperfine split spectrum for fine particles by lowering the temperature. Figure 6 shows the Mössbauer spectra for α-$Fe_2O_3$ with an average particle diameter of 13.5 nm as a function of temperature. Note that an increasing degree of magnetic hyperfine splitting is observed as the relaxation time for the fluctuation of magnetization direction is decreased by the decreasing temperature. It is necessary to consider the possibility that the specimens may behave as small superparamagnetic particles when corrosion products are examined by Mössbauer spectroscopy. The resolution of the hyperfine splitting in finely

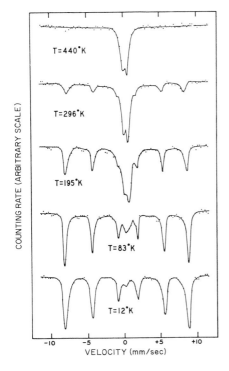

**Fig. 6.** Mössbauer spectra of α-Fe$_2$O$_3$ with average particle size 13.5 ± 1 nm at various temperatures. (Kündig et al., 1966.)

divided corrosion products at low temperatures is invaluable to the identification of the oxide phase(s). Furthermore, the superparamagnetic behavior can be useful in the actual determination of average particle size.

## V. Solvated Corrosion Products

The hydrolysis of ferrous and ferric ions is a function of pH, ionic strength, and the electrolyte. Details of hydrolysis reactions are important in the determination of mechanisms of corrosion phenomena that take place in solutions.

Vértes et al. (1970a) studied the formation of α-FeOOH and α-Fe$_2$O$_3$ from suspensions of Fe(OH)$_3$ by Mössbauer spectroscopy with the purpose of differentiating between the two proposed mechanisms given below:

$$2Fe(OH)_3 + OH^- \to 2Fe(OH)_4^- \nearrow Fe_2O_3 + 3H_2O + 2OH^- \atop \searrow 2FeOOH + 2H_2O + 2OH^- \tag{2}$$

$$2Fe(OH)_3 \to Fe_2O_3 + 3H_2O$$
$$2Fe(OH)_3 + OH^- \to 2Fe(OH)_4^- \to 2FeOOH + 2H_2O + 2OH^- \tag{3}$$

The mole fraction $FeOOH/(FeOOH + Fe_2O_3)$ as a function of concentration of $Fe(OH)_3$ suspension is shown in Fig. 7. Since the mole fraction of FeOOH is a function of the suspension concentration, it is apparent that mechanism (3) described the formation of $Fe_2O_3$ and FeOOH. If mechanism (2) were operative, one would have expected the relative amounts of the two oxides to be independent of suspension concentration.

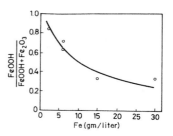

**Fig. 7.** Change in mole fraction of $\alpha$-FeOOH as a function of suspension concentration. (Vértes *et al.*, 1970.)

The hydrolysis of $Fe^{3+}$ ions in perchloric acid solutions has been investigated by means of Mössbauer spectroscopy (Dezsi *et al.*, 1968). Spectra of frozen solutions were recorded after specimens were prepared by quenching solutions to liquid nitrogen temperatures. Spectra of frozen solutions of ferric perchlorate as a function of pH are shown in Fig. 8. The recoil-free fraction of $Fe(H_2O)_6^{3+}$ ions was too low for spectra to be obtained in frozen solutions of pH 0.1. Spectra of $Fe(H_2O)_6^{2+}$, on the other hand, were observed for solutions at this pH. The lower f factor of the hexaquo ferric ion complex was attributed to the smaller ionic radius and higher positive charge of the ferric ion as compared with the ferrous ion. The $Fe^{3+}$ ions are consequently not tightly bound in frozen solutions because of a decrease in the strength of the hydrogen bonds that bind the water molecules to the second coordination sphere around the central ion.

The quadrupole split lines shown in Fig. 8 at pH 1.0–1.6 were attributed to the hydrolysis species $[FeOH(H_2O)_5]^{2+}$. The relatively high quadrupole splitting (1.4 mm/sec) is a result of the low symmetry of this hydroxo complex. The center line observed for solutions of the same pH (1.0–1.6) was associated with $[Fe(OH)_2(H_2O)_4]^+$. The width of this center line is a consequence of the small quadrupole splitting that is expected from the noncubic environment of the $Fe^{3+}$ ion in the $[Fe(OH)_2(H_2O)_4]^+$ complex. The spectra from solutions of pH higher than 1.6 were from the dimeric ions, $[Fe(OH)(H_2O)_4]_2^{4+}$. The increase in the isomer shift to higher velocities for this complex indicates a higher ionic character of the ferric ion in the dimer than in the other two hydroxo complexes. The results of the Mössbauer studies were consistent with spectrophometric studies of ferric ion hydrolysis.

The hydrolysis of $Fe^{3+}$ in sulfuric acid solutions over the same range of

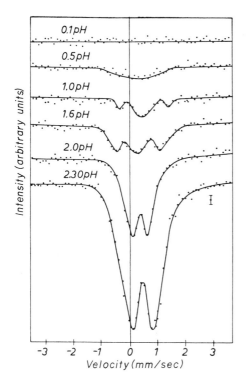

**Fig. 8.**   Mössbauer spectra of frozen aqueous solutions of ferric perchlorate. Source is $^{57}$Co in stainless steel. (Dézsi *et al.*, 1968.)

pH was different compared with the perchloric acid system. Vértes *et al.* (1970) have concluded that in sulfuric acid none of the hydroxo complexes are formed prior to complete hydrolysis. This conclusion was further substantiated by the fact that the electron exchange in the $Fe^{2+}$–$Fe^{3+}$ sulfuric acid system was independent of pH, whereas in perchloric acid, the rate of electron exchange dropped with increasing concentrations of $[Fe(OH)_2(H_2O)_4]^+$ and $[Fe(OH)(H_2O)_4]_2^{4+}$ complexes. Vértes *et al.* (1970) unfortunately do not provide a clear interpretation of the single resonance line observed in the $Fe^{3+}$–sulfuric acid system. The appearance of additional resonance lines at pH 2.7, however, was attributed to the formation of colloidal $Fe(OH)_3$.

Vértes *et al.* (1973) have extended their hydrolysis studies to include the effects of $Cl^-$ and $NO_3^-$ anions. The electron exchange reaction between $Fe^{2+}$ and $Fe^{3+}$ was also studied as a function of pH in the presence of different anions. Additional Mössbauer studies of frozen solutions that give information about equilibria in the hydrolysis of ferrous and ferric ions are warranted because of the importance of these equilibria in occluded corrosion cells such as pits and crevices (Brown, 1970) and in the industrial production of ferric oxide pigments by corrosion.

## VI. Defect Structures of Corrosion Products

It is well known that the defect structure of an oxide film is related to the mechanism of oxidation of metals at high temperatures. The type and concentration of defects in the oxide determine the electron and ionic transport properties that control the rate of film growth. Mössbauer spectroscopy has been successfully applied to the determination of defect structures in FeO and $Fe_3O_4$. The results of these studies are described in this section.

### A. FeO

The phase diagram of $Fe_{1-x}O$ indicates that the $Fe_{1-x}O$ phase is thermodynamically stable only at temperatures above 843°K, and below this temperature $Fe_{1-x}O$ disproportionates to the stable products Fe and $Fe_3O_4$ (Fender and Riley, 1969). It is possible, however, to obtain a series of the metastable phases by rapid quenching from the equilibrium state to temperatures at which diffusion is slow. The advantage of using Mössbauer spectroscopy in this case is that the sensitivity of the Mössbauer parameters to local defect arrangements permits detection of defect aggregation at an earlier stage than possible by other experimental techniques such as x-ray spectroscopy. Greenwood and Howe (1972a) have extensively studied the defect structure of $Fe_{1-x}O$ phases by Mössbauer spectroscopy.

The room-temperature spectra of $Fe_{1-x}O$ samples quenched from 1520°K into water are shown in Fig. 9. Although the spectra in Fig. 9 were obtained under optimum quenching conditions, it is evident that some disproportionation to $Fe_3O_4$ occurred for compositions $Fe_{0.88}O$ to $Fe_{0.86}O$. The Mössbauer spectra can be related to the defect structure through the quadrupole splitting of $Fe^{2+}$. A quadrupole splitting is observed for an $Fe^{2+}$ ion that is adjacent to a cation vacancy owing to the electric-field gradient produced at the nucleus by the unequal electron population of the nondegenerate $T_{2g}$ levels. Further distortions from cubic symmetry caused by more than one adjacent vacancy or by adjacent $Fe^{3+}$ cations further split the $T_{2g}$ levels and cause larger quadrupole splittings.

For $Fe_{0.947}O$, an acceptable computer fit was obtained with five Lorentzian peaks. The peak labeled A was attributed to one component of a quadrupole split $Fe^{3+}$ doublet; the other component is unresolved and combined with peak C. Peaks B and C, as well as peaks D and E, were attributed to two different quadrupole doublets of $Fe^{2+}$. Acceptable computer analysis of the $Fe_{0.900}O$ spectrum was, however, not obtained. It was found in this case that many more peaks overlap to produce a complex spectrum that could not be resolved into a sum of individual Lorentzian

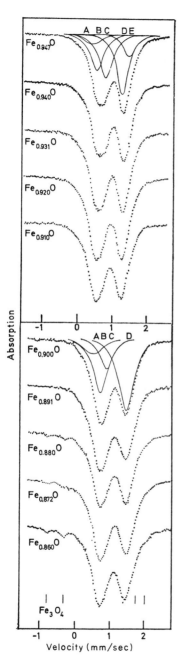

**Fig. 9.** Room-temperature Mössbauer spectra of $Fe_{1-x}O$ quenched from $1520°K$ into water, showing the progressive change of the resonance profile as a function of composition. (Greenwood and Howe, 1972a.)

peaks. A detailed interpretation in terms of resolvable components could, therefore, be offered only for small values of $x$.

The $Fe_{1-x}O$ data were interpreted on the basis of cluster structures that have been established by diffraction experiments (Koch and Cohen, 1969; Cheetham *et al.*, 1971). The smallest cluster was taken to be four vacant cation sites situated around one tetrahedral $Fe^{3+}$ in the NaCl cubic type lattice, $['Fe^{3+}(\square_+)_4]^{5-}$. The largest cluster was taken to be a fourfold cluster represented as $['Fe_4^{3+}(\square_+)_{13}]^{14-}$. The twelve nearest-neighbor cations and the six next-nearest-neighbor cation sites have the most influence on the quadrupole splitting of $Fe^{2+}$. There are, therefore, thirty-six such $Fe^{2+}$ sites surrounding the single cluster of four vacancies. Five $Fe^{3+}$ ions are distributed among these thirty-six peripheral sites to obtain a neutral charge on the vacancy cluster. For the fourfold cluster, fifty-four $Fe^{2+}$ sites are affected by the thirteen vacancies in this cluster. Fourteen of these sites are, however, occupied by $Fe^{3+}$ ions to obtain the necessary charge balance. Based on this model, it is possible to construct histograms of the proportion of $Fe^{2+}$ ions which have from 0 to 9 defects, either $Fe^{3+}$ or vacant cation sites, for different compositions. Histograms for the compositions $Fe_{0.918}O$ and $Fe_{0.947}O$ are shown in Fig. 10. The histogram of the single-cluster model for $Fe_{0.907}$

**Fig. 10.** Histograms showing the proportion $P(n)$ of $Fe^{2+}$ ions having from 0 to $n$ neighboring defect ($Fe^{3+}$ or cation vacancies) calculated for the single-cluster model (dashed line) and the fourfold-cluster model (solid line) for two typical compositions. The effect of disproportionation on defect distribution is shown in the bottom histogram. The single-cluster and fourfold-cluster structures are pictured. (Greenwood and Howe, 1972a.)

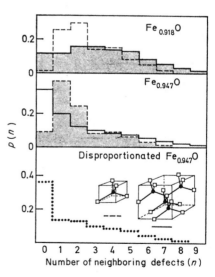

predicts that most of the $Fe^{2+}$ ions have mainly two neighboring defects. The spectrum of $Fe_{0.947}$ shown in Fig. 9 is therefore consistent with the presence of single clusters. The $Fe_{0.918}O$ histogram unfortunately does not predict significantly different quadrupole splitting distributions between the two models. This composition, however, is known to contain fourfold clusters, and the Mössbauer spectra of samples of similar composition were claimed to be consistent with this model. The appropriate distributions of quadrupole splittings for the two compositions are given by the shaded areas in Fig. 10. The change from single clusters in $Fe_{0.947}O$ to fourfold clusters in $Fe_{0.918}$ results in an increase in the number of $Fe^{2+}$ ions having more than two neighboring defects, a fact that is reflected in the Mössbauer spectra. The range of $Fe^{2+}$ environments increases with increases in $x$, and consequently the range of quadrupole splittings made it difficult to obtain computer fits for spectra from $Fe_{1-x}O$ for high values of $x$.

## B. $Fe_3O_4$

Magnetite, $Fe_3O_4$, has the inverse spinel structure, and the unit cell consists of thirty-two oxygen ions in a fcc arrangement. In this cell, there are sixty-four interstitial sites with tetrahedral symmetry (A sites) and thirty-two interstitial sites with octahedral symmetry (B sites). Magnetite compositions range from that of stoichiometric $Fe_3O_4$ with eight $Fe^{3+}$ ions in tetrahedral and eight $Fe^{2+}$ plus eight $Fe^{3+}$ ions in octahedral sites, to that of $\gamma$-$Fe_2O_3$ with eight $Fe^{3+}$ ions in tetrahedral and thirteen and one-third $Fe^{3+}$ ions in octahedral sites of the unit cell. Mössbauer spectroscopy has been successfully employed to determine the defect structure of the magnetite series $Fe_3O_4$–$\gamma$-$Fe_2O_3$ and to relate the defect structures to the electrical conductivity of the oxides.

The distribution of vacancies in magnetite of the composition $Fe_{3-v}O_4$ may be described by three different idealized systems (Annersten and Hafner, 1973):

$$[Fe^{3+}_{1-v}\square_v]_{tet}[Fe^{2+}_{1-3v}Fe^{3+}_{1+3v}]_{oct}O_4 \tag{4}$$

$$[Fe^{3+}_{1-\tau}\square_\tau]_{tet}[Fe^{2+}_{1-3v}Fe^{3+}_{1+3v-\omega}\square_\omega]_{oct}O_4 \tag{5}$$

$$[Fe^{3+}]_{tet}[Fe^{2+}_{1-3v}Fe^{3+}_{1+2v}\square_v]_{oct}O_4 \tag{6}$$

Here, $v$ is the total number of vacancies in the spinel, $\tau$ is the vacant fraction of cations at the octahedral positions, and $\tau + \omega = v$. In systems (4) and (6), the vacancies are restricted to tetrahedral and octahedral sites, respectively, whereas, in system (5) the vacancies are distributed over both types of sites. Experimental data in support of system (4) have not been reported. The character of the Mössbauer spectra of magnetite in terms of system (5) and (6) very likely depends upon the method of preparation. The general application of Mössbauer spectroscopy to studies of $Fe_{3-v}O_4$ is illustrated in the following discussion derived from recent studies of this oxide system.

Mössbauer spectra of a series of $Fe_{3-v}O_4$ specimens at 296°K are shown in Fig. 11 (Coey et al., 1971). The spectra of stoichiometric magnetite, $v = 0$, shows only two sets of six magnetic hyperfine resonance lines. The set of

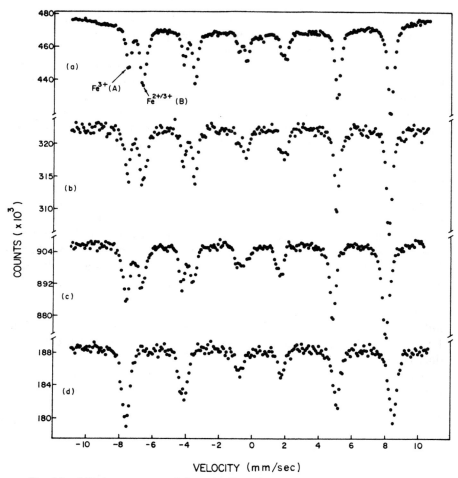

**Fig. 11.** Mössbauer spectra of the series $Fe_{3-v}O_4$ at 296°K; (a) $v = 0.00$, (b) $v = 0.03$, (c) $v = 0.09$, and (d) $v = 0.33$. (Coey *et al.*, 1971.)

lines with the highest $H_{eff}$ is attributed to the $Fe^{3+}$ ions on tetrahedral sites, and the other set to the $Fe^{2+}$ and $Fe^{3+}$ ions on the octahedral sites. The indistinguishability in the spectra of the $Fe^{3+}$ and $Fe^{2+}$ octahedral sites is because of the rapid electron exchange between these ions at temperatures above 120°K. Below this temperature, where the electron exchange is slower, the Mössbauer spectra from each of the valence states on the octahedral ions is resolved. The onset of electron exchange at 120°K and the commensurate increase in electrical conductivity is known as the Verwey transition (Verwey *et al.*, 1947).

When the value of $v$ is increased, the intensity of the B site spectrum decreases relative to the intensity of the A site spectrum, as shown in Fig. 11. This result has been interpreted to indicate that the cation vacancies are formed primarily on the octahedral sites according to system (6). For $v = 0.33$ the oxide phase is $\gamma\text{-}Fe_2O_3$ as described earlier, with $Fe^{3+}$ ions on both the tetrahedral and octahedral sites. In this oxide an externally applied magnetic field is necessary to resolve the ferric ions on the A and B sites. The spectra of the same series of oxides as given in Fig. 11 are shown in Fig. 12 with a 47 kOe field applied parallel to the $\gamma$-ray propagation direction.

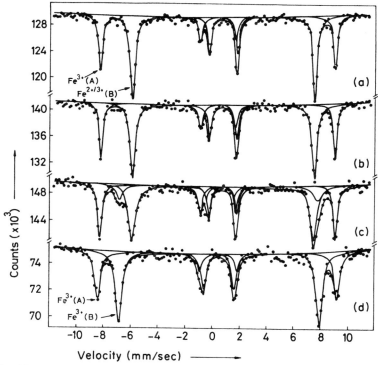

**Fig. 12.**  Mössbauer spectra of the same series shown in Fig. 11 in a 47 k Oe field applied parallel to the $\gamma$-ray propagation direction. (Coey *et al.*, 1971.)

Each six-line pattern is reduced to four lines in the presence of the magnetic field because of the alignment of the ferrimagnetically coupled spins by the external field. A comparison of Fig. 11d with Fig. 12d clearly demonstrates that the A and B $Fe^{3+}$ ions in $\gamma\text{-}Fe_2O_3$ are readily resolved by the external field.

Daniels and Rosencwaig (1969) have concluded that the rapid electron exchange in magnetite above the Verwey transition is essentially a localized phenomenon, occurring principally within available $(Fe^{2+}-Fe^{3+})$ octahedral pairs. The degree of localization increases with increasing nonstoichiometry. In other words, any nonpaired $Fe^{3+}$ octahedral cations do not participate in this exchange process. The appearance of a separate spectrum in Fig. 11c for octahedrally coordinated $Fe^{3+}$ was used to support this model. The ferric ions in excess of the number of $Fe^{2+}-Fe^{3+}$ pairs would be expected to produce a separate characteristic spectrum. Daniels and Rosencwaig (1969) suggested that the delocalization of electron exchange to $(Fe^{2+}-Fe^{3+})$ octahedral pairs accounts for the observed critical dependence of the conductivity of magnetite on its composition.

The results of most Mössbauer studies have indicated that the defect structure of $Fe_{3-v}O_4$ is best represented by system (6). Annersten and Hafner (1973), on the other hand, have studied $Fe_{3-v}O_4$ over a range of stoichiometry and found evidence for defect structure (5) for compositions close to $\gamma$-$Fe_2O_3$. Their results were inconclusive as to the possibility of tetrahedral vacancies for compositions closer to $Fe_3O_4$, however.

## VII. Kinetics of Corrosion Processes

Kinetic studies of corrosion are generally undertaken to elucidate corrosion mechanism, to estimate the service life of a material, or to evaluate the degree of protection offered by a coating. Mössbauer spectroscopy has been applied to studies of corrosion kinetics, and some typical examples are discussed.

Pritchard and Dobson (1969) studied the corrosion kinetics of iron which was heated in deoxygenated water at temperatures between 180 and 290°C. The surfaces of the specimens were enriched in $^{57}Fe$ to increase the sensitivity of the experiments. Magnetite was the only oxide phase observed, and the rate of film growth was described by $dy/dt = ky^3$. Reaction rates at different temperatures gave an Arrhenius activation energy of 156 kcal/mole. This activation energy was associated with the decomposition of the ferrous hydroxide formed at the surface of the film.

Simmons et al. (1973) studied the kinetics of air oxidation of $^{57}Fe$-enriched iron surfaces at 225°C (see Fig. 2). The conversion electron Mössbauer spectroscopic technique permitted the kinetic data to be obtained for oxide film thickness as low as 4–5 nm. Nonstoichiometric $Fe_3O_4$ was formed under the oxidation conditions, and the increase in thickness followed a logarithmic time dependence.

Channing and Graham (1972) have extensively studied by Mössbauer spectroscopy the growth behavior of $Fe_3O_4$ and $\alpha$-$Fe_2O_3$ formed on iron at 450–550°C. The transmission technique was employed, and the specimens were metal foils ($\sim 10$ $\mu$m) that were oxidized in 1 atm of oxygen. Changes in the relative amounts of the phases Fe, $Fe_2O_3$, and $Fe_3O_4$ as a function of time for different temperatures are summarized in Fig. 13. At all temperatures, it can be seen that $Fe_3O_4$ is formed initially and after the iron foil is nearly completely oxidized the $\alpha$-$Fe_2O_3$ phase begins to form at the expense of the $Fe_3O_4$ phase. The rate of oxidation of the magnetite to hematite followed a parabolic rate law. The oxidation was, therefore, controlled by ionic diffusion through the $\alpha$-$Fe_2O_3$ layer. It was not possible, however, to identify whether the diffusing species were cations or anions.

An Arrhenius plot for the growth of hematite from magnetite over the temperature range 475–550°C produced an activation energy of 49 kcal mole$^{-1}$. Since this activation energy for the oxidation process was low compared to values determined by diffusion studies (78 kcal mole$^{-1}$ for oxygen and 112–100 kcal mole$^{-1}$ for iron), Channing and Graham (1972) suggested that short-circuit diffusion of ions via grain boundaries predominated over lattice diffusion.

## VIII. Passivity and Corrosion Inhibition

The progress in research on passivity and corrosion inhibition has been limited by the lack of physical methods for elucidating under *in situ* conditions the structure and chemical properties of the relatively thin protective film ($< 10$ nm). Recent attempts have been reported on the application of Mössbauer spectroscopy to studies of protective films. These efforts have, with varying degrees of success, provided needed data for understanding passivation phenomena.

Onodera *et al.*, (1972) have studied by means of conversion electron Mössbauer spectroscopy the films formed on $^{57}$Fe-enriched iron surfaces in $NaNO_2$ solution at pH 8. Although this technique has the sensitivity for studying thin surface films, *in situ* studies in solutions are, unfortunately, not possible. After exposure to the solution, two sets of magnetic hyperfine lines were observed in addition to the resonance lines from the iron substrate. The strongest set of additional lines, with $H_{eff} = 325$ kOe and an isomer shift of 0.75 mm/sec with respect to iron, was attributed to $\alpha$-FeOOH, although the reported value of $H_{eff}$ for this oxyhydroxide is 389 kOe. The weakest set of resonance lines with a hyperfine field of 475 kOe and isomer shift of 0 mm/sec was considered to be from $\gamma$-$Fe_2O_3$, but in this case

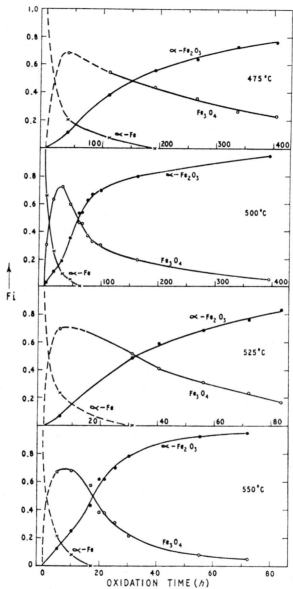

**Fig. 13.**   The fraction of total $^{57}$Fe contained within each component of an iron foil as a function of heating time in 1 atm of oxygen at the temperatures specified. Starting foil thickness 9.6 $\mu$m. (Channing and Graham, 1972.)

the isomer shift is not in agreement with previous reports. The authors attributed the discrepancies in the Mössbauer parameters to the possibility that the thin surface layers exist in an amorphous state. This interpretation, however, appears to be inconsistent with the fact that magnetic hyperfine splitting was observed in the films at room temperature.

O'Grady and Bockris (1973) have reported results on the *in situ* examination of passive films formed on iron at pH 6.8. Unfortunately, the experimental details were not described. Mössbauer spectra of specimens taken *in situ* were compared with spectra obtained after the samples were dried at room temperature. The isomer shift and quadrupole splitting were functions of drying time, but the internal field of both *in situ* and dry specimens was $470 \pm 10$ kOe. Since the temperature at which this internal field was measured was not given and the isomer shifts were not reported with respect to a specified reference, it is not possible to compare the results of O'Grady and Bockris with those obtained by Onodera *et al.* (1972). O'Grady and Bockris (1973) concluded from their results that the composition of the dried films was $\gamma$-$Fe_2O_3$. As in the case of the results reported by Onodera *et al.* (1972), O'Grady and Bockris (1973) found that the Mössbauer parameters of the *in situ* passive film did not match those of any known stoichiometric iron oxide, including hydrates. The latter authors, however, reached a different conclusion from that of the former. O'Grady and Bockris (1973) suggested that the Mössbauer data can be interpreted in terms of a structure that contains dioxy and dihydroxy bridging bonds between the iron atoms. This film was described as a polymeric film of hydrated iron oxide. The changes observed in the Mössbauer spectra of the passive layer upon drying emphasizes the necessity for *in situ* structure characterization.

Pritchard and Dobson (1973) have studied the passive film formed on iron in chromate solution by emission spectroscopy using trace amounts of [57]Co electrodeposited onto iron surfaces. Doping iron with [57]Co adds undue complication to the interpretation of spectra from the protective films on iron–cobalt specimens. Owing to differences in the chemical potentials of iron and cobalt, the cobalt may not necessarily *a priori* be incorporated as an isolated probe atom in the passive film. The significance of the chemical potential differences has been clearly demonstrated by experiments conducted in our laboratory. [57]Co-doped iron surfaces were oxidized at 250°C in air. The Mössbauer emission patterns from these specimens were characteristic of elemental metal, and no oxide phase was evident in the spectra, even though a substantially thick oxide was present on the surface. Under the conditions of these experiments, the parent [57]Co atoms were not incorporated into the oxide phase, and the Mössbauer spectra were, therefore, characteristic of [57]Fe daughter atoms in the elemental state. Because of thermodynamic and kinetic factors, the behavior of cobalt in iron during "dry oxidation" would not necessarily be the same as during corrosion and/or passivation in electrolytes. The interpretation from emission spectra of the structure of passive films on iron–cobalt is, nevertheless, likely to be ambiguous. Pritchard and Dobson (1973), therefore, did not offer an interpretation for the surface compound(s). Changes in the spectra from specimens after treatment with chromate were attributed to the presence of $Fe^{3+}$ at the surface, and no evidence for the presence of $Fe^{4+}$ or $Fe^{6+}$ was obtained, in contrast to suggestions made by Nagayama and Cohen (1962). The observed decrease in f factor for the surface film formed after chromate treatment was explained on the basis of a low degree of crystallinity.

Leidheiser et al. (1973a) have reported preliminary studies of the in situ study of the passivity of cobalt by emission Mössbauer spectroscopy. $^{57}Co$-doped cobalt surfaces were chosen for study to avoid the complication described above for the $Fe$–$^{57}Co$ system. When the $Co$–$^{57}Co$ surface is passivated, the $^{57}Co$ parent atoms behave chemically as cobalt until they decay by electron capture to $^{57}Fe$. These conditions allow the $^{57}Fe$ atoms to be incorporated as isolated Mössbauer probe nuclei into the cobalt compound that constitutes the protective surface film. The interpretation of the spectra in this case is still subject to the possible complications described in the section on experimental methods. These studies have been continued in our laboratory (Simmons et al., 1974), and the experimental details and results are described below.

The polarization of cobalt in deaerated buffered borate solution (pH 8.5) produced a classic potential versus current relationship. An active-to-passive transition occurred between $-500$ mV and $-300$ mV (versus S.C.E.). The cobalt remained passive at higher anodic potentials up to $+500$ mV. Above $+500$ mV, thick anodic film formation and oxygen evolution were commensurate with an increase in current. To characterize the surface chemical species present during polarization, emission Mössbauer spectra were obtained at specific applied cathodic and anodic potentials. Surface sensitivity was obtained by electroplating a thin layer ($<5$ nm) of cobalt doped with $^{57}Co$ onto a fresh cobalt electrodeposit. Mössbauer spectra of the $^{57}Fe$ probe atoms were obtained by resonance absorption of the emitted 14.4 keV $\gamma$ ray by a $^{57}Fe$-enriched stainless steel foil. The in situ studies were carried out with the experimental arrangement shown schematically in Fig. 14. The reaction cell was designed to minimize the amount of solution between the specimen surface and a thin Mylar window. The growth rate and stability of the surface films that formed during anodic polarization were determined from measurements of current and radioactivity as a function of time as shown in Fig. 15 for polarization at $+200$ mV and higher. At anodic potentials higher than $+200$ mV, the corrosion rate was sufficiently low that

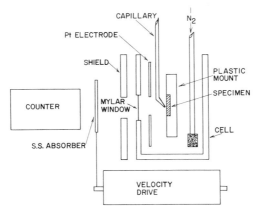

**Fig. 14.**   Experimental arrangement adopted for *in situ* studies of the passivation of cobalt surfaces using emission Mössbauer spectroscopy.

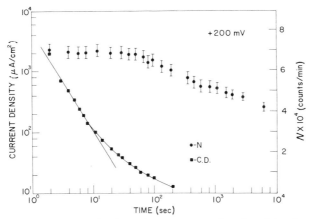

**Fig. 15.**   Changes in current density and count rate as a function of time for a $^{57}$Co-doped surface during anodic polarization in borate buffer pH 8.5.

emission spectra could be obtained before the $^{57}$Co-active-surface region of the specimen was lost.

Emission spectra of specimens during cathodic polarization at $-1100$ mV (Fig. 16) showed only the six lines expected from the magnetic hyperfine splitting of $^{57}$Fe in metallic cobalt. This result indicated that atoms at the surface of cobalt were essentially in a zero valence state during cathodic polarization and that any "oxide" formed during specimen preparation and transfer to the electrolytic cell was readily reduced by the cathodic treatment. Two additional resonance lines of equal intensity were observed in the

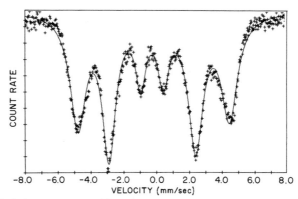

**Fig. 16.** Emission spectrum of $^{57}$Fe at the surface of a cobalt specimen during cathodic polarization ($-1100$ mV versus S.C.E.) in borate buffer pH 8.5.

emission spectra during anodic polarization at $+200$ and $+500$ mV. Figure 17 shows the spectra from a specimen polarized at $+500$ mV, and Fig. 18 was obtained by subtracting the unreacted cobalt contribution from the spectra. The isomer shift and quadrupole splitting were characteristic of high-spin $Fe^{3+}$.

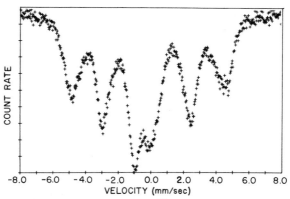

**Fig. 17.** Emission Mössbauer spectrum of $^{57}$Fe at the surface of cobalt during anodic polarization ($+500$ mV versus S.C.E.) in buffered borate pH 8.5.

In view of the possible influence of Auger aftereffects and the possibility that the charge on the $^{57}$Fe probe atom may not represent the charge of the $^{57}$Co parent, emission spectra were obtained from the cobalt oxides, hydroxides and oxyhydroxides to serve as a basis for the interpretation of the spectra from the passivated surface. $^{57}$Co-doped cobalt compounds were

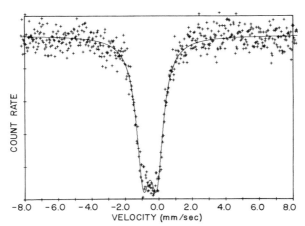

**Fig. 18.**   Emission Mössbauer spectrum of $^{57}Fe$ obtained from Fig. 17 after subtracting the spectrum of unreacted metal.

prepared according to the following reactions:

$$Co(NO_3)_2 + 2KOH \rightarrow \beta\text{-}Co(OH)_2 + 2KNO_3 \tag{7}$$

$$Co(OH)_2 \xrightarrow{1000^\circ C} CoO + H_2O \tag{8}$$

$$3Co(OH)_2 + \tfrac{1}{2}O_2 \xrightarrow{400^\circ C} C_3O_4 + 3H_2O \tag{9}$$

$$Co(OH)_2 + \tfrac{1}{2}O_2 \xrightarrow{KOH} CoOOH + H_2O \tag{10}$$

Cathodic deposition of $\alpha\text{-}Co(OH)_2$ from $Co(NO_3)_2$ solution \hfill (11)

Anodic oxidation of $\alpha\text{-}Co(OH)_2 \xrightarrow{pH\,8.5} CoOOH$ \hfill (12)

Assuming that the passive film consists of essentially a single oxide component, the emission spectra for the product of reaction (12) shown in Fig. 19 gave the best fit to the emission spectra of the passive film formed on cobalt at the two anodic potentials stated above. Although it is likely that the product of reaction (12) is a hydrated oxyhydroxide of cobalt, the structure of this compound unfortunately has not been reported in the literature. The elucidation of the structure for this oxyhydroxide may provide a key to understanding the passivation phemonena of cobalt.

Many intermetallic compounds are less active chemically in corrosive environments than the component elements. For example, the iron–tin alloy $FeSn_2$ in all but the strongest oxidizing environments is more resistant to attack than either iron or tin (Covert and Uhlig, 1957; Luner and Murray, 1963; Leidheiser, 1971). In the manufacture of tinplate, $FeSn_2$ forms at the interface between the steel substrate and the tin coating. There is strong evidence both from laboratory tests and service experience that the rate of

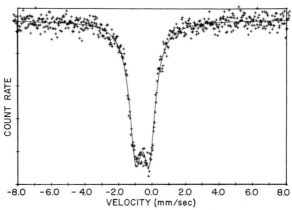

**Fig. 19.** Emission Mössbauer spectrum of $^{57}Fe$ from $^{57}Co$-doped cobalt "hydroxyoxide." The "hydroxyoxide" was obtained by anodic oxidation of $\alpha$-$Co(OH)_2$.

corrosion of tin plate in fruit acids is a function of the continuity and thoroughness of coverage of the steel substrate by the $FeSn_2$ at the interface (Gabe and Mort, 1965). Intermetallic compounds are also suspected to be responsible for the reduced efficiency of hydrogen evolution on the platinum metals under certain conditions (Buck and Leidheiser, 1964). The formation of intermetallic compounds has been proposed as the explanation for the reduced rate of corrosion of iron in fruit acids in the presence of tin ions (Buck et al., 1964) and the large amounts of tin and lead present in gold when electrically coupled to tin or lead in acid solution (Leidheiser, 1962). In the majority of cases where intermetallic compounds are suspected to be present, conventional diffraction techniques can be applied. However, when the intermetallic compound is present to the extent of a few atom layers an *in situ* analytical technique is necessary since oxidation of the surface may occur during the transfer of the sample to the diffraction chamber. Although Mössbauer studies have not been reported, the emission technique is potentially useful in this regard as already indicated by the *in situ* studies of the passive film on cobalt (Simmons et al., 1974). Three candidate systems are the cobalt–tin, iron–tin, and gold–tin systems. In the cobalt–tin system, emission spectroscopy can be carried out using a cobalt surface on which $^{57}Co$ has been electrodeposited. Under such an experimental plan, it is essential that the cobalt corrode at a very low rate in the medium under investigation. In all three systems, emission spectroscopy can be applied using $^{119m}Sn$ present in the aqueous phase at the start of the experiment. Studies of intermetallic compound formation in the iron–tin and gold–tin systems is facilitated by the fact that reference spectra on the intermetallic compounds exist. Charlton et al. (1970) have studied the $^{119}Sn$ isotope and

Charlton and Harris (1973) the $^{197}$Au isotope in the gold–tin system, and Nikolaev *et al.* (1963) have studied the $^{119}$Sn and $^{57}$Fe spectra in the iron–tin system.

## IX. Corrosion and Interfacial Reactions beneath a Coating

Leidheiser *et al.* (1973b) have demonstrated that emission Mössbauer spectroscopy may be applied to nondestructive studies of the rate of corrosion at the metal–coating interface of a polymer-coated cobalt surface. Cobalt doped with $^{57}$Co was electrodeposited onto a cobalt substrate. The mass deposited was equivalent to 5 nm, assuming uniform deposition. The specimen was then coated to a thickness of 0.001 cm with polybutadiene and cured at 200°C in air for 30 min. Figure 20 shows the emission spectrum

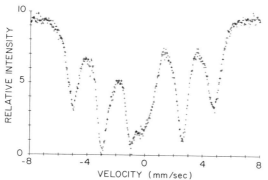

**Fig. 20.**   Emission Mössbauer spectrum after coating specimen with polybutadiene to a thickness of 0.001 cm and curing in air at 200°C for 30 min. (Leidheiser *et al.*, 1973b.)

of the freshly prepared specimen, and Fig. 21 is a spectrum after the specimen had been exposed to 3% NaCl solution for 84 hr. In addition to the six lines from the unreacted metal, a central line(s) originating from ionic cobalt was observed. In the case of the freshly prepared sample, the ionic cobalt contribution is from the thin oxide that was present on the surface prior to the application of the coating. The increase in the intensity of the center line after exposure to the salt solution was attributed to corrosion that had taken place beneath the polymer film. Changes in the ratio of resonance areas from the ionic and metallic cobalt as a function of exposure time to the salt solution provided a means of evaluating the protective properties of the coating. No attempt, however, was made to identify the corrosion product(s) in this study.

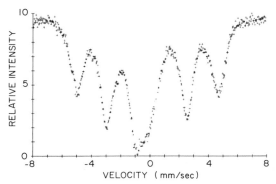

**Fig. 21.** Emission Mössbauer spectrum of coated specimen after exposure to 3% NaCl solution for 84 hr. Note that the center peak characteristic of ionic cobalt has increased in intensity relative to that shown in Fig. 20. (Leidheiser *et al.*, 1973b.)

Protective metallic coatings of zinc and aluminum on steel react with the substrate to form intermetallic compounds. These compounds consume the coating and reduce its long-term galvanic protection capabilities as well as alter the mechanical properties of the substrate–coating system. Nondestructive means for determining the amount of intermetallic compound formation at the substrate–coating interface are useful in choosing the most satisfactory operating parameters during preparation of the coated steel.

Jones and Denner (1974) used Mössbauer spectroscopy to follow the rate of intermetallic compound formation at the steel–zinc interface as a function of the time of exposure of mild steel foil to molten zinc at 480°C. The formation rate $\Delta W$ in gm/cm$^2$/min at 480°C was given by

$$\Delta W = 15.8 \times 10^{-4} t^{0.33} \tag{13}$$

Characteristic spectra of the $\gamma$, $\delta_1$, and $\zeta$ phases were determined, but the individual phases could not be identified in the transmission spectra of the galvanized steel foils.

Mössbauer spectroscopy has also been used to show that $Fe_4Al_{13}$ forms at the interface between iron and aluminum. This compound serves as a barrier limiting the rate at high temperature at which iron dissolves into the solid phase (Preston, 1972).

## X. Stress Corrosion Cracking and Hydrogen Embrittlement

Before stress corrosion and corrosion fatigue phenomena can be fully understood, the possible influence of hydrogen on crack growth must be established. It is generally accepted that conditions of pH and variations in composition and/or strain in the vicinity of the crack tip are favorable for cathodic production of hydrogen. It is possible for this hydrogen to enter the plastic zone ahead of the crack and induce embrittlement which consequently enhances crack growth. It is possible with Mössbauer spectroscopy to study *in situ* the electrolytic charging of metals and alloys. Important data on the conditions for charging, rate and activation energy of hydrogen entry, and the changes in the physical and chemical properties of the metal can be obtained.

Wertheim and Buchanan (1967) have studied nickel, $\alpha$-iron, and 310 stainless steel containing electrolytically introduced hydrogen. Figure 22 shows the spectra of [57]Fe in nickel as a function of charging time. The spectra were obtained by the emission method after [57]Co had been diffused into a nickel foil. Most of the data were obtained after quenching the specimens to liquid nitrogen temperature immediately following charging. Some studies were conducted in an electrolytic cell that made it possible to keep the specimen saturated with hydrogen at room temperature. Only two distinct spectra were observed, one characterized by the magnetic hyperfine lines of [57]Fe in pure nickel and the other exhibiting a single broad line, which was interpreted as nickel hydride. The hyperfine field of the magnetic spectrum was independent of the hydrogen content of the sample. Wertheim and Buchanan (1967), therefore, proposed that as nickel is saturated to varying degrees with hydrogen, the specimens consist of two separate phases— essentially pure nickel and nickel hydride. Only the single line from the hydride was observed after 28.5 hr of charging, and the amount of hydrogen in the specimen was determined to be 0.56 hydrogen atom per nickel atom.

The broadening and asymmetry of the nickel hydride line was attributed to the fact that, on the average, 60% of the six octahedral sites surrounding the iron atom are occupied by hydrogen. The lack of cubic symmetry gives rise to a quadrupole splitting, and the isomer shift also depends upon the number of hydrogen neighbors. The nonmagnetic nature of the hydride and the isomer shift were attributed to filling of the 3d band of nickel by electrons from hydrogen.

Hydrogen in type 310 nonmagnetic fcc austenitic stainless steel and $\alpha$-iron were also studied by Wertheim and Buchanan (1967). In these cases, the transmission technique was used with specimens quenched to liquid nitrogen temperature. The amount of hydrogen in stainless steel was equivalent to 0.3 hydrogen atoms per alloy atom. The introduction of hydrogen into the steel produced both broadening and a line shift in the same direction as that produced in nickel. The spectra of $\alpha$-iron before and after cathodic treatment were identical, in agreement with the low equilibrium solubility of hydrogen in iron of $10^{-5}$ hydrogen atoms per iron atom.

The rate of desorption of hydrogen from cathodically charged dilute nickel–iron alloys (0.4–1.0 at. % iron) was investigated by Janot and Kies (1972). These investigators also found a two-phase system as described by

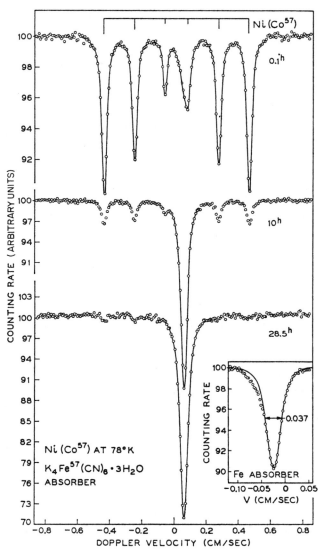

**Fig. 22.** The hyperfine structure of $^{57}$Fe in the nickel–nickel hydride system as a function of hydrogen content. The iron results from the decay of $^{57}$Co, which was diffused into nickel. The insert shows data taken with a metallic iron absorber on a threefold expanded velocity scale. (Wertheim and Buchanan, 1967.)

Wertheim and Buchanan (1967). When the hydrogen concentration approached 0.65 hydrogen atoms per metal atom, the Mössbauer data indicated that the iron had more than the average number of hydrogen neighbors. These "hydrogen clusters" surrounding the iron atoms disappeared during the early stages of desorption. Desorption then appeared to be controlled by volumetric diffusion with first-order kinetics. An activation energy of approximately 15 kcal/mole was obtained. Diffusion coefficients for dilute nickel–iron binary alloys were two to six times smaller than that for pure nickel.

## XI. Solid State Diffusion

High-temperature diffusion processes are important in understanding the kinetics of the growth of oxide, sulfide, and halide scales. Mössbauer spectroscopy offers a method of determining the average jump frequency of the resonant nuclei. Since correlation effects are not involved, the jump frequency can be more directly related to diffusion mechanisms than by conventional techniques. The nondestructive feature of the Mössbauer approach permits the measurement of temperature dependence of diffusivity with a single sample of constant composition. Ambiguities associated with other techniques, such as diffusion via grain boundaries, are avoided by this method.

The effect on the Mössbauer spectra of vibration, diffusion and jumping of resonant nuclei depend upon the time scale for each type of atomic motion. For an atom vibrating about a fixed lattice site, the recoil-free fraction depends upon the mean square amplitude $\langle x^2 \rangle$ of the oscillation in the direction of observation of the $\gamma$ ray,

$$f = \exp(-4\pi^2 \langle x^2 \rangle / \lambda^2), \tag{14}$$

where $\lambda$ is the $\gamma$-ray wavelength. Since the vibrational frequency is high compared with the inverse excited-state lifetime, the width of the Mössbauer line is essentially determined by the average lifetime of the excited state. If, on the other hand, the resonant nuclei jump instantaneously between fixed sites, random phase shifts in the $\gamma$ rays may accompany the jumps. Since the phase shifts reduce the coherency of the $\gamma$ rays, the net effect of the jumping process is a reduction in the observation time of the emission or absorption to a value less than the average excited lifetime. This reduction in time produces an increase in the uncertainty in the $\gamma$-ray emission or absorption energy, and the resonance line is consequently broadened. Singwi and Sjölander (1960) have derived the following relationship between line broadening and the average jumping frequency $\tau^{-1}$;

$$\tau^{-1} = \Delta\Gamma/2h. \tag{15}$$

The activation energy for the jumping process can be evaluated from

$$\Delta\Gamma = (\Delta\Gamma_0)\exp(-E/kT). \tag{16}$$

Recent studies by Greenwood and Howe (1972b) on line broadening caused by diffusion in cation deficient FeO provides an excellent example of the application of Mössbauer spectroscopy to studies of diffusion in an oxide system. Typical line broadening observed for FeO as a function of temperature is shown in Fig. 23. The results obtained by Greenwood and Howe (1972b) for $Fe_{1-x}O$ are summarized in Table II. The activation energies of $140 \pm 20$ kJ mole$^{-1}$ for $Fe_{0.94}O$ and $135 \pm 20$ kJ mole$^{-1}$ were calculated from the data at the two temperatures.

A mechanism for diffusion in $Fe_{1-x}O$ was derived to account for the decrease in jump frequency with the increase in cation deficiency. For an

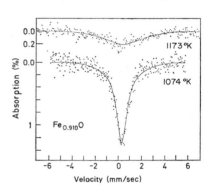

**Fig. 23.** Diffusion-broadened Mössbauer spectra of $Fe_{0.910}O$ showing the marked dependence on temperature. (Greenwood and Howe, 1972b.)

**TABLE II**

Dependence of the Jumping Frequency in $Fe_{1-x}O$ on Composition and Temperature[a]

| Composition $(1 - x)$ | $T/K$ | $\Delta\Gamma/\text{mm sec}^{-1}$ | $\tau^{-1}/\text{sec}^{-1}$ |
|---|---|---|---|
| 0.940 | 1074 | 1.18 | $4.30 \times 10^7$ |
| 0.941 | 1074 | 1.18 | $4.30 \times 10^7$ |
| 0.940 | 1074 | 1.18 | $4.30 \times 10^7$ |
| 0.934 | 1074 | 1.09 | $3.98 \times 10^7$ |
| 0.923 | 1074 | 1.19 | $4.34 \times 10^7$ |
| 0.920 | 1074 | 0.95 | $3.46 \times 10^7$ |
| 0.910 | 1074 | 0.82 | $2.99 \times 10^7$ |
| 0.910 | 1074 | 0.80 | $2.92 \times 10^7$ |
| 0.940 | 1173 | 4.03 | $1.46 \times 10^8$ |
| 0.910 | 1173 | 3.50 | $1.28 \times 10^8$ |

[a] Adapted from Greenwood and Howe (1972b).

assembly of cations, the average jump frequency is given by:

$$\tau^{-1} = \sum_n np_n\tau_0^{-1}e^{-E/kT} \tag{17}$$

where $\tau_0^{-1}$ is a constant, $p_n$ is the probability of a particular cation having $n$ adjacent vacancies, and $\sum_n np_n$ is the average number of adjacent vacancies per cation. The sums $\sum_n np_n$ were calculated as a function of composition for three different random arrays of vacancies—single vacancies, $['Fe^{3+}(\square_+)_4]^{5-}$ clusters, and $['Fe_4^{3+}(\square_+)_{13}]^{14-}$. The $\sum_n np_n$ sums calculated from these vacancy distributions predict an increase in jumping frequency for increases in the value of $x$. Since this prediction did not parallel the experimental results, Greenwood and Howe (1972a) suggested that function (17) does not adequately account for the jump frequency because of a probable increase in activation energy for a cation jumping into a vacancy associated with a cluster. The values of $\tau^{-1}$ would consequently decrease as the cluster size increases. This mechanism implies that the cluster size increases as the concentration of defects increase over the range $Fe_{0.94}O$ to $Fe_{0.91}O$.

## XII. Conclusions

The experimental methods and results that were presented in this chapter indicate that Mössbauer spectroscopy provides a unique means for studying corrosion behavior. The variety of techniques available allows for the study of corrosion from the formation of the first atomic layer to the development of corrosion layers many microns thick. Experiments can be designed for investigations *in situ* so that corrosion processes may be studied while the metal is immersed in a solution or while it is covered with a protective coating. The limitation of corrosion studies to iron, tin, and cobalt and alloys of these elements is a serious shortcoming of the Mössbauer method, but many important fundamental and applied problems can be examined.

### Acknowledgment

The author's corrosion studies reported herein were carried out through support of the Army Research Office, Durham. We express our appreciation for this support and particularly to Dr. Henry Davis for the stimulation he gave to us and to many other workers in the field of corrosion during his service with this agency.

### References

Annersten, H., and Hafner, S. S. (1973). *Z. Kristallogr. Bd.* **13**, 321.
Bancroft, G. M., Mayne, J. E. O., and Ridgway, P. (1971). *Brit. Corros. J.* **6**, 119.
Bonchev, Zw., Jordanov, A., and Ninkova, A. (1960). *Nucl. Instrum. Methods* **70**, 36.

Brown, B. F. (1970). *Corrosion* **26**, 249.
Buck, W. R., and Leidheiser, H. (1964). *Nature (London)* **204**, 107.
Buck, W. R., Heyn, A. N. J., and Leidheiser, H. (1964). *J. Electrochem. Soc.* **111**, 386.
Channing, D. A., and Graham, M. J. (1972). *Corros. Sci.* **12**, 271.
Charlton, J. S., and Harris, I. R. (1970). *Phys. Status Solidi* **39**, Kl.
Charlton, J. S., Hayes, M. C., and Harris, I. R. (1970). *J. Less Common Metals* **20**, 105.
Cheetham, A. K., Fender, B. E. F., and Taylor, R. I. (1971). *Solid State Phys.* **4**, 2160.
Coey, J. M. D., Morrish, A. H., and Sawatzky, G. A. (1971). *J. Phys.* **32**, Cl-271.
Covert, R. A., and Uhlig, H. H. (1957). *J. Electrochem. Soc.* **104**, 537.
Daniels, J. M., and Rosencwaig, A. (1969). *J. Phys. Chem. Solids* **30**, 1561.
Dézsi, I., Vértes, A., and Kiss, L. (1964). *J. Radioanal. Chem.* **2**, 183.
Dézsi, I., Keszthelyi, L., Kulgawczwk, D., Molnar, B., and Eissa, N. A. (1967). *Phys. Status Solidi* **22**, 617.
Dézsi, I., Vértes, A., and Komor, M. (1968). *Inorg. Nucl. Chem. Lett.* **4**, 649.
Evans, B. J., and Hafner, S. S. (1969). *J. Appl. Phys.* **40**, 1411.
Fender, B. E. F., and Riley, F. D. (1969). *J. Phys. Chem. Solids* **30**, 793.
Gabe, D. R., and Mort, R. J. (1965). *J. Iron Steel Inst. (London)* **203**, 64.
Gonser, U., Grant, R. W., Muir, A. H., and Wiedersich, H. (1966). *Acta Met.* **14**, 259.
Greenwood, N. N., and Howe, A. T. (1972a). *J. Chem. Soc. Dalton* **1**, 110.
Greenwood, N. N., and Howe, A. T. (1972b). *J. Chem. Soc. Dalton* **1**, 122.
Johnson, D. P. (1969). *Solid State Commun.* **7**, 1785.
Janot, Ch. and Kies, A. (1972). *In* "L'Hydrogéne dans les Métaux," p. 204. Congr. Int., Paris.
Jones, R. D., and Denner, S. G. (1974). *Scripta Met.* **8**, 175.
Khalafalla, D., and Morrish, A. H. (1972). *J. Appl. Phys.* **43**, 624.
Koch, F., and Cohen, J. B. (1969). *Acta Cryst.* **B25**, 275.
Krakowski, R. A., and Miller, R. B. (1972). *Nucl. Instrum. Methods* **100**, 93.
Kullerud, G., Donnay, G., and Donnay, J. D. H. (1968). *Z. Kristallogr.* **128**, 1.
Kündig, W., Bömmel, H., Constabaris, G., and Lindquist, R. H. (1966). *Phys. Rev.* **142**, 327.
Leidheiser, H. (1962). *Appl. Phys. Lett.* **1**, 39.
Leidheiser, H. (1971). "The Corrosion of Copper, Tin and Their Alloys," pp. 306–325. Wiley, New York.
Leidheiser, H., Simmons, G. W., and Kellerman, E. (1973a). *Croat. Chem. Acta* **45**, 257.
Leidheiser, H., Simmons, G. W., and Kellerman, E. (1973b). *J. Electrochem. Soc.* **120**, 1516.
Luner, C., and Murray, M. V. (1963). *J. Electrochem. Soc.* **110**, 176.
Miyamoto, H., Shinjo, T., Bando, Y., and Takada, T. (1967). *J. Phys. Soc. Japan* **23**, 1421.
Moreira, J. E., Knudsen, J. M., deLima, C. G., and Dufresne, A. (1973). *Anal. Chem. Acta* **63**, 295.
Nagayama, M., and Cohen, M. (1962). *J. Electrochem. Soc.* **109**, 781.
Nikolaev, V. I., Shcherbina, Yu. I., and Karchevskii, A. I. (1963). *J. Exp. Theoret. Phys. (USSR)* **44**, 775.
O'Grady, W. E., and Bockris, J. O'M. (1973). *Surface Sci.* **38**, 249.
Onodera, H., Yamamoto, H., Watanabe, H., and Ebiko, H. (1972). *Jap. J. Phys.* **11**, 1380.
Pella, P. A., and DeVoe, J. R. (1970). *Anal. Chem.* **42**, 1833.
Pella, P. A., and DeVoe, J. R. (1971). *Appl. Spectrosc.* **25**, 472.
Pollak, H. (1962). *Phys. Status Solidi* **2**, 270.
Preston, R. S. (1972). *Met. Trans.* **3**, 1831.
Pritchard, A. M., and Dobson, C. M. (1969). *Nature (London)* **224**, 1295.
Pritchard, A. M., and Dobson, C. M. (1973). *Chem. Phys. Lett.* **23**, 514.
Pritchard, A. M., and Mould, B. T. (1971). *Corros. Sci.* **11**, 1.
Pritchard, A. M., Haddon, J. R., and Walton, G. N. (1971). *Corros. Sci.* **11**, 11.

Simmons, G. W., Kellerman, E., and Leidheiser, H. (1973). *Corrosion* **29**, 227.

Simmons, G. W., Kellerman, E., and Leidheiser, H. (1974). *J. Electrochem. Soc.* **121**, 275C.

Singwi, K. S., and Sjölander, A. (1960). *Phys. Rev.* **120**, 1093.

Swanson, K. R., and Spijkerman, J. J. (1970). *J. Appl. Phys.* **41**, 3155.

Terrell, J. H., and Spijkerman, J. J. (1968). *Appl. Phys. Lett.* **13**, 11.

van der Woude, F. (1966). *Phys. Status Solidi* **17**, 417.

van der Woude, F., and Dekker, A. J. (1966). *Phys. Status Solidi* **13**, 181.

Vaughan, R. W., and Drickamer, H. G. (1967). *J. Chem. Phys.* **47**, 1530.

Vértes, A., Lázár, K., Kelemen, K., and Bognair, L. (1970a). *Radiochem. Radioanal. Lett.* **4**, 375.

Vértes, A., Suba, M. and Komor, M. (1970b). *Radiochem. Radioanal. Lett.* **4**, 365.

Vértes, A., Ranogajec-Komor, M., and Gelencsér (1973). *Acta Chim. Acad. Sci. Hung.* **77**, 55.

Verwey, E. J. W., Haayman, P. W., and Romeijn, F. C. (1947). *J. Chem. Phys.* **15**, 181.

Voznyuk, P. O., and Dubinin, V. N. (1973). *Sov. Phys.-Solid State* **15**, 1265.

Wertheim, G. K., and Buchanan, D. N. E. (1967). *J. Phys. Chem. Solids* **28**, 225.

Wickman, H. H., and Wertheim, G. K. (1968). *In* "Chemical Applications of Mössbauer Spectroscopy" (V. I. Goldanskii and R. H. Herber, eds.), pp. 604–614. Academic Press, New York.

# BIOLOGICAL STUDIES

# 4

# Experimental Techniques for Biochemical Studies

**George Lang**

Department of Physics
The Pennsylvania State University
University Park, Pennsylvania

## I. Introduction

Many of the problems encountered in the application of Mössbauer spectroscopy to the study of biological materials are common to Mössbauer spectroscopy in general. We will ignore most of these and will concentrate our attention on aspects peculiar to or particularly severe in the biological applications. We will consider only work with iron, because it represents at least 99% of Mössbauer biological applications. The most common situations require not a one-shot operation, but many measurements of different samples and sometimes the same sample many times. This is a result of the sensitivity of the material being investigated and the frequent difficulty encountered in sample preparation. In many cases, the Mössbauer measurement itself will be the best measure of success in producing a single well-characterized species. Thus, it is important to make experimental arrangements that are not only effective but are easy to use, so that there will be no impediment to carrying out cross checks and verifications of questionable or puzzling results.

## II. Intensity Problems

### A. General Problems

Because of the relatively small metal content of most iron biomolecules, the problem of achieving good signal-to-noise ratio in an experiment of reasonable duration is often severe. It will be convenient to consider a concrete case, and for this we choose hemoglobin, the "fruit fly" of biological Mössbauer work. Red cells of mammals are about 30% hemoglobin by weight, making them about 17 m$M$ in iron (four irons in a molecular weight of 68,000). A 0.5 cm thick sample of cells would have $5 \times 10^{18}$ iron atoms/cm$^2$, and $10^{17}$ atoms/cm$^2$ would be of the 2% abundant Mössbauer-active isotope. The peak resonant cross section for $\gamma$ absorption is about $2 \times 10^{-18}$ cm$^2$. The effective cross section is reduced by a factor of two by the convolution of finite source and absorber linewidths, here assumed to have the natural value. The finite recoil-free fractions of source and absorber would decrease this by another factor of two (here we assume an absorber temperature of 80°K or less). Finally, if the sample were either completely oxygenated or completely reduced, the observed intensity would be divided by quadrupole interaction between two absorption lines, yielding another factor of two. The predicted absorption dips would thus have a strength of 2.5% of the 14.4 keV $\gamma$-ray intensity, a figure that is borne out by experiment.

With a source strength of 50 mCi and typical counter and geometry, a counting rate with the above specimen of, say, 25,000/sec might be expected. If the velocity range were divided into 256 separate velocity intervals, the accumulation of 100/sec in each would then produce a fractional uncertainty (or noise), $1/\sqrt{N}$, of 0.1% or $\frac{1}{25}$ of the signal, in $10^4$ sec—a little less than 3 hr.

The rather rosy picture presented above is subjected to modifications, mostly unfavorable, when we consider the more general problem. It must be remembered that a factor $F$ lost in signal intensity requires a factor $F^2$ increase in counting time to achieve equivalent signal-to-noise ratio. The concentration of hemoglobin in red cells is extremely high—a factor 100 to 1000 times higher than appropriate for optical spectrophotometry, the range quite commonly employed for biochemical manipulations. Once hemoglobin has been removed from cells, it is difficult to achieve again the intracellular concentration. The problem is more acute for other iron proteins, most of which have lower solubilities and lower fractional iron content. Any hyperfine interaction that spreads the spectrum decreases the effective depth, since the integrated absorption area remains constant as the effective width increases. Thus, a half-reduced hemoglobin sample, with its absorption divided among four lines instead of two, would require four times as much data collection time to achieve signal-to-noise equivalent to the above. Magnetic hyperfine structure can broaden a spectrum by an effective factor of ten or more, requiring a one hundredfold or more increase in counting time. The recoil-free fraction for iron in proteins has not been accurately determined, but appears to be of order 0.6 at 4.2°K, losing 10% of this at 77°K and 40% at 195°K. At temperatures approaching the freezing point of the solution the recoil-free fraction drops precipitously.

### B. Sources

Sources are readily obtainable commercially and few experimenters prepare their own. $^{57}Co$ dissolved in chromium, palladium, and rhodium are the most popular, and linewidths closely approximating natural are obtainable with each. The last has the highest recoil-free fraction and may be obtained (from Amersham-Searle) in foils sufficiently thin that two weak sources can be stacked to make an acceptable stronger one. In all cases, of course, the quantity of interest is the rate of emission of the 14.4 keV gammas that actually escape from the source. This may be much less than the amount emitted within the source if the emitting nuclei are deep within the host metal. Shallow diffusion into a thick matrix may be acceptable, but some measure of the depth should be taken into account when the source is specified. Uniform diffusion through a thin matrix is a more clearcut arrangement, where the customer can easily calculate the loss to self-absorption. For a technical discussion of $^{57}Co$ Mössbauer sources, the reader is referred to the paper by Longworth and Window (1971).

### C. Counters

Investigation of the decay scheme of $^{57}Co$ reveals that only about 10% of the decays produce the desired 14.4 keV $\gamma$ ray, while almost all produce 122 keV radiation. After such a beam passes through a typical sample, the flux entering the counter contains twenty to forty times as many as 122 keV photons as 14.4 keV photons. The counter should be able to detect the latter at a high rate and have minimum sensitivity to the former. Scattering of the energetic photons in the sample is normally unimportant, but Compton scattering in the counter material produces output pulses distributed fairly uniformly in the range 0 to 40 keV. Some of these will lie within the acceptance window for Mössbauer gammas, providing unwanted background that adds to statistical uncertainty but not to signal in the measurement. Thus, it is desirable that the counter have low Compton cross section for the energetic gammas and good resolution at 14.4 keV so that the window may be made narrow. Energetic $\gamma$ rays that undergo photoelectric absorption give rise to counter signals far above the acceptance window, but a large number of them can saturate the electronics at high counting rates and reduce the effective efficiency and resolution. Proportional counters with noble gas fillings are the usual detectors in this energy range. Xenon is often chosen because a counter with high efficiency at 14 keV can be made small in size. Unfortunately, xenon has a K absorption edge at 35 keV, and as a result its relative efficiency at 122 keV is high, so that at high rates the large pulses tend to saturate the electronics. Krypton is better in this respect, but the presence of the K edge just under 14.4 keV gives rise to a significant escape peak. Photoabsorption of the 14.4 keV $\gamma$ in krypton often is followed by emission of $K\alpha$ radiation, and the latter has a high probability of escaping the counter. In these cases, estimated to be typically 50% of the total, the counter pulse corresponds to a $\gamma$ energy of 14.4 keV, less the $K\alpha$ energy of krypton, or about 2 keV net. Argon is a good choice as far as relative cross section is concerned, but its absolute cross section is small, requiring either a large counter or high pressure. Solid state counters, in spite of better inherent resolution, do not offer any noteworthy advantage when their Compton cross sections for the high energy gammas are considered. Resonant counters have not found significant application in biological work; their usual disadvantage is low efficiency and the difficulty of assuring efficiency independent of source velocity. Counter manufacturers sometimes specify the features of their devices under rather atypical conditions, for example, operating at high rates with a thin (but Mössbauer-thick) absorber. This is not very relevant to typical biological Mössbauer measurements.

### D. Sample Thickness

Clearly, it is advisable to obtain in the sample the highest possible concentration of the metal complex of interest. Having done this, the choice of sample thickness remains. It is easy to show that if the sample is thin in $^{57}Fe$, then the optimum thickness is independent of the $^{57}Fe$ content. The percent Mössbauer absorption will increase linearly with thickness, while counting rate decreases exponentially. The problem is modified by the presence in the 14.4 keV counting window of signals that come from other sources. Since much more penetrating $\gamma$ rays are responsible for almost all of these, this background counting rate is approximately independent of the sample thickness. We can characterize the source–counter system by $R$, the ratio, with sample absent, of true 14.4 keV signals to background counts. It can be shown that when $R = \infty$, the figure of merit (signal-to-noise ratio for fixed counting time) is optimized for a sample that reduces the 14.4 keV beam by a factor $e^{-2}$. The maximum is not sharp however, and thicknesses of one or three absorption lengths have a figure of merit reduced by only 15%. For very small $R$, the optimum thickness is one absorption length. If $R$ be decreased to 1.0 by the addition of random signals in the counting window, the optimum thickness is 1.2 absorption lengths and the figure of merit is reduced to about 45% of its value above. Again the maximum is a soft one with the figure of merit decreasing by less than 20% for a 50% change either way in sample thickness. The rule of thumb, then, is that samples that are thin in the Mössbauer isotope should have a thickness of one to two nonresonant absorption lengths.

Most samples will consist mainly of water and protein. Typical protein has a total $\gamma$ cross section at 14.4 keV of about 1.3 $cm^2/gm$, while water is about 1.7 $cm^2/gm$. The cross sections of the elements increase rapidly with atomic number, those in the sequence from oxygen to krypton being well represented by $\sigma = 0.003Z^3$ $cm^2/gm$. Thus high concentrations of salts may drastically increase the opacity of the sample; in particular, salting out is usually not a suitable method of concentrating protein for Mössbauer studies. Solvents containing sulfur or chlorine are strongly absorbing and severely restrict sample thickness.

In experiments with samples with low $^{57}Fe$ content, the exclusion of iron not in the chemical state under investigation is particularly important. Part of this problem is in the chemistry of sample preparation, which we will not discuss here. The other part, iron in counter windows, dewar windows, radiation shields, and sample holders, should not be neglected. As an example, a measurement of common household aluminum foil at room temperature yielded a 0.15% Mössbauer absorption from a single layer 0.0007 inch thick. This could be several times the signal in a particularly difficult case. It is probably best to eliminate metal from counter and dewar windows as much as possible. The radiation shields in helium dewars, which must be metallic, should be made from high-purity aluminum, rolled as the meat in a non-ferrous protective sandwich. Beryllium is often contaminated with iron as well, and again precautions should be taken.

## E. *Isotopic Enrichment*

Since $^{57}$Fe is only 2% abundant in nature, it is clear that an extremely large increase in signal is in principle possible by replacing all iron in the sample with isotopically enriched material. At a price of about $3 per milligram for 90% $^{57}$Fe, this is usually the biggest bargain available, provided reasonable efficiency can be achieved in incorporation. The problems here are either biochemical or physiological. Enrichment by supplying isotopically enriched iron to a growing organism is possible. The largest animal so enriched is the monkey (Krasney, 1975), where iron was provided by interperitoneal injection in anemic animals. Growth of microorganisms in enriched media is often more manageable; here, success depends on selection of an organism with simple tastes, so that relatively iron-free analytical grade chemicals may be used to make up the growth medium. To enrich an organism that sups on beef broth, for example, would require immense quantities of isotope either to enrich a cow or to swamp the normal iron in the broth. Exclusion of unwanted ubiquitous iron requires careful attention to detail, and often the first attempt is not successful. Direct chemical exchange of iron, where possible, involves usually negligible investment in $^{57}$Fe. Here the problems are biochemical, principally the verification that biologically active material is produced.

## III. Cryostats and Sample Temperature

In the cases examined thus far, the zero-field splittings of low-lying multiplets in biological molecules are equivalent to several degrees Kelvin or larger. Moreover, the splitting of Kramers' doublets in applied fields can be made equivalent to several degrees Kelvin with easily attainable magnetic fields. Generally, biomolecules are magnetically dilute and have no low-temperature magnetic ordering. (Ferritin is the single exception that comes to mind.) Thus, the lowest temperature required for Mössbauer measurement of biomolecules is normally that of liquid helium, 4.2°K, or at worst, the 1.3°K or so conveniently attainable by pumping on ordinary liquid helium. The range 4.2°K up to room temperature is the one of predominant interest, and the cryogenic systems used need not be highly sophisticated. A number of mundane but important requirements should be mentioned. It is often necessary to measure many different samples and to install and retrieve them without melting. Thus, top-loading cryostats are most convenient, and any system that requires warm-up or loss of vacuum in changing samples is impractical in the long run. A reasonable economy in helium consumption is also desirable, since long counting times often are involved. A related requirement, that of good counting geometry, requires that the cryostat have a small-diameter tail with large windows. Instruments used by the author satisfying these requirements are shown in figures 5 and 6 of Lang (1970); production versions of these are now made commercially and are available from the United Kingdom Atomic Energy Establishment, Harwell.

It is most convenient to make measurements at the temperatures of the usual cryogens, helium, nitrogen, and dry ice, for 4.2°K, 77°K, and 195°K respectively. Dichlorodifluoromethane (Freon-12), with a boiling point of 243°K, is sometimes useful as well. For both dry ice and freon, care should be taken that the refrigerant is in equilibrium with its gaseous phase at 1 atm. If the vaporized material is swept away by air currents, lowering of the temperature by several degrees may result. For temperatures intermediate between those of common cryogens, controlled electrical heating is usually resorted to. One type of device for doing this is shown in Fig. 1. A brass

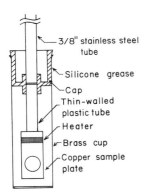

3/8" stainless steel tube

Silicone grease

Cap

Thin-walled plastic tube

Heater

Brass cup

Copper sample plate

**Fig. 1.**   Section of variable-temperature sample insert for Mössbauer measurements. The brass cup has Mylar beam windows, here obscured by the sample plate.

cup is made of thin-walled brass tubing. Holes for the $\gamma$ beam are covered with 0.005 inch Mylar foil fastened with epoxy cement. The lip of the cup is slightly flared. The cap has an outer cylindrical surface, which is a sliding fit in the cup; at the top, it is provided with slight outward taper, which serves to limit the depth to which the cap can be inserted. The cap is connected to a cryogenic stainless steel tube, which is provided with electrical feed-throughs and a pumping port at its upper end. The copper sample mount is suspended by a low-conductivity plastic support from the cap and contains a heater and thermocouple or other temperature sensor.

In use, the sample is mounted and wrapped with high-purity aluminum foil, silicone grease being used to enhance thermal contact with the plate. The cap and lip are coated with silicone grease, joined together, and the cup is plunged into liquid nitrogen. When the thermocouple indicates that the temperature of the copper plate has fallen to that of liquid nitrogen, the assembly is inserted into the helium cryostat. This freezes the air inside the device and provides good insulation. When the temperature of the sample has fallen to the desired value, power is supplied to the heater to maintain the temperature. A balance must be made between speed of cooling and helium boil-off rate from the heat leak of the plastic support. Practical values appear to require an hour or two for the cool-down from nitrogen to helium temperature. The process can be speeded by admitting some helium exchange gas, but this must then be pumped out, and it may require more time in the end to achieve stable conditions. To a considerable extent, this depends on the controller. A satisfactory control scheme involves using the heater resistance as a temperature sensor, part of a feedback bridge in a Wien bridge oscillator (Window, 1969).

The Janis Research Company's variable-temperature cryostat is being used by a number of Mössbauer groups (Fig. 2). It is made with a re-entrant

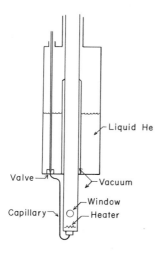

**Fig. 2.** Schematic showing principle of Janis Super Varitemp dewar. Only the helium pot is shown. Nitrogen chamber and nitrogen radiation shields, etc., are eliminated for clarity.

helium pot so that the long tailpiece containing the sample is attached above the helium level at a point where the temperature is well above that of boiling helium. At equilibrium, the sample reaches a temperature of order 100°K but can be cooled to temperatures ranging down to 4.2°K by means of a valve and capillary tube that admit helium to the sample region. For fine control, and for higher temperatures, the gas entering the sample region may be electrically heated. The system is not outstandingly efficient with respect to helium consumption, but it has proved to be convenient and very reliable.

There is usually no need for great precision in sample temperature measurements. Gold–iron thermocouples have high thermoelectric power at low temperature and nearly constant sensitivity. For measurements in high magnetic fields at temperatures below about 10°K, they are subject to field-dependent sensitivity, a situation that is complicated if the measuring and reference junctions are both in regions of high but different and not well-known field strength. For fields of order 50 kG, the glass carbon resistor sensors (Lake Shore Cryotronics) are more suitable. The capacitance sensors of the same manufacturer are almost completely unaffected by high fields, but suffer by having a maximum in the capacitance–temperature curve at about 60°K.

## IV. Magnetic Fields

Magnetic fields may be applied to biological molecule samples for a variety of purposes. In the dilute paramagnets, fields of the order of 100 g are sufficient to decouple electron and nuclear spins, since their magnetic interaction corresponds to only about 20 g at the electrons. For such measurements, a permanent magnet is suitable; these are most cheaply made by adapting a magnetron magnet. For longitudinal fields (relative to the $\gamma$ beam) at 4.2°K, persistent currents in superconducting lead washers mounted against the sample provide a convenient source. These can be charged by passing a sufficiently strong magnet over the dewar tailpiece, driving the material to its normal state, and trapping flux as it returns to superconduction. Electric solenoids or Helmholtz pairs may also be used to provide these small fields. Care should be taken to filter the power supply well and/or to protect the velocity pickup of the source driver from induced voltages in order to avoid spurious effects.

High fields are used to determine signs and asymmetry parameters in diamagnets and in high-temperature fast-relaxing paramagnets. They are also used to mix states that have zero field splittings. For the former use, 50 or 60 kG field strength is virtually always more than adequate; and fields of such strength will provide significant mixing in almost all known biological materials with zero field splittings of the ground spin multiplet. It may also be desirable to achieve the high-field limit, with Zeeman splitting much larger than zero field splitting. This is possible with 50 kG or so in the rhombic chelators, such as transferrins and mycobactins, which have splittings of the order of 1°. For the high-spin ferrous and ferric hemes, the splittings are of the order of 10°, and no available practical system could take them to the high-field limit. Thus 50 or 60 kG is a reasonable field strength, and there is for most work little point in striving for marginally higher field at the expense of other features.

In designing a superconducting magnet system, thought must be given to optimizing counting geometry. In the first instance, good counting geometry requires a small magnet. Magnet dimensions are determined by field strength and even more by homogeneity requirements. Mössbauer experiments do not require high homogeneity; a 1 or even 2% variation over the sample is normally acceptable and permits construction of a relatively compact magnet. The magnet bore will normally be required to accommodate a counter–dewar arrangement to permit sample temperature variation. A reasonable magnet bore is in the range 1.0–1.5 inches. A vertical solenoid axis is consistent with the usual methods of cryostat construction, and a choice must be made between longitudinal and transverse $\gamma$-beam directions. The transverse scheme is most convenient because it permits the source to be mounted outside the cryostat and directly on the velocity transducer, but a price is paid for this convenience. The magnet must be made in the form of a split Helmholtz-like pair with a beam hole through the gap. The field at the sample will be smaller than the maximum field in the bore, and the source-to-counter distance will be the diameter of the cryostat tail. Normally, the beam will have to traverse a helium-filled diameter of the magnet, suffering roughly a 20% absorption in a typical 5 inch diameter. The stray field at the source in the transverse setup is typically 3 kG or so and can be considerably reduced by a shield in the form of a short piece of soft iron pipe. If three of these are mounted at 120° spacing about the cryostat tail, the force on the magnet is balanced and the beam exit window is not obscured. The vertical $\gamma$-beam arrangement requires the source to be inside the cryostat and necessitates a long drive rod from a transducer mounted at the top. This delicate assembly must be removed each time a sample is changed. If a field-bucking winding is provided for the source region, close spacing makes possible a highly efficient counting geometry.

The long duration of Mössbauer measurements dictates operation of the magnet in persistent mode. It should be realized that high-field magnets have finite time constants, by virtue of small resistances in the joints, and a suitable low decay rate must be specified. A fraction of a percent per day is possible without heroic measures and is sufficient in most cases. It is common for current leads to be the primary major heat leak in a superconducting magnet setup. Their optimization for persistent mode is, of course, quite different from optimization for driven mode; the considerations can be rather involved, but the intended user should at least make sure the designer is working on the right problem.

With respect to the interpretation of data, there is no strong preference for longitudinal or transverse geometry. Although it was not recognized by the authors of some of the early papers, calculation of magnetically perturbed spectra of powder samples involves no more computing time for the transverse case than the longitudinal case, at least when isotropic recoil-free fraction is assumed.

## V. Physical State of Samples—Containment

The preponderance of biological Mössbauer work will be concerned with proteins. These exist in nature in intimate association with water, and must be regarded as different, though related, chemical species when the water is removed. Mössbauer experiments require for practical purposes that the sample be solid, however, so that the choice is generally between lyophilized material or frozen solutions. The latter clearly are preferable, because measurements indicate that well-defined iron environments can be observed, while in the dry material absorption lines are wide, suggesting the presence of a variety of conformations. The effect is most clearly seen in a comparison of paramagnetic spectra of lyophilized and frozen forms; ferricytochrome c is a good example (Lang *et al.*, 1968).

Measurements on frozen solutions bring up a number of questions, some of which have not been well investigated. The possible difference between an active enzyme in liquid water, with freedom to flex parts of its structure, and the same material locked into a solid matrix is a subject of some concern. In virtually all cases, deeply frozen biomolecules regain full biological activity when thawed, but this is of course not the same question. In addition, we are often dealing with materials in which electronic excited states have energies of a hundred or a few hundred wavenumbers. Occupation of these may be essential to the function at physiological temperatures and at the same time not observable in the temperature range required for the experiment. This is unfortunate, but in the search for complete understanding of a system, the study of its ground state is an essential and useful step.

Although frozen protein samples usually give rise to Mössbauer spectra characteristic of a single iron environment, there is at least one noteworthy exception. Many ferric hemoproteins can exist as thermal mixtures of high- and low-spin states, the relative populations depending on temperature, pH, and ionic strength of the solution. In both Mössbauer (Lang *et al.*, 1969) and susceptibility (Iizuka and Kotani, 1969) experiments the populations have been found to depend upon freezing rate as well. In the work of Iizuka, the effect of rapid freezing and subsequent low-temperature annealing has been worked out in some detail. The freezing-rate effects must result at least in part from the segregation of small solute molecules into an increasingly concentrated liquid phase as the solution freezes. However, this may not be the only operative mechanism—the differing mechanical forces on the protein afforded by the various phases of ice could also be relevant.

Although we are not aware of any systematic comparison of glassy versus crystalline frozen solution of proteins, a striking difference is known in experiments with small-model molecules. Solution in a diamagnetic solvent provides a way to achieve spin dilution and consequent low spin–spin relaxation rate in paramagnets. However, if the solvent becomes crystalline upon freezing it is likely that the solid solubility will be low, so that the paramagnetic species may well exist in regions of very high local concentration. One example of this is the case of hemin chloride in acetic acid, dimethylsulfoxide, and acetone (Moss et al., 1969), in all of which the spectrum is similar to that of solid crystalline hemin. In tetrahydrofuran solution, which forms a glass, the hemin remains dispersed, and a magnetically split spectrum characteristic of slow spin relaxation is seen (Lang et al., 1970).

Liquid samples are conveniently frozen and measured in open-topped polyethylene cuvettes, which may be molded rather cheaply once the appropriate dies are made. A drawing of one type may be seen in the article of Lang (1970). If sample holders are machined from Lucite, care should be taken to round inside corners to distribute the stress developed on freezing. Chlorinated plastics are of course to be avoided because of their $\gamma$ absorption. Materials that must be kept anaerobic can ordinarily be frozen in nitrogen in a nitrogen-filled glove box and subsequently handled in air if it is done with reasonable dispatch. Additional protection can be afforded by a layer of silicone grease on the surface. Sample cuvettes can be kept cold while being mounted for insertion in the cryostat by working on them in and over a shallow tray full of liquid nitrogen. With a bit of dexterity, this can be done without use of forceps, because the gas film formed about an immersed finger can provide momentary insulation. Picking up nitrogen-soaked cloth or cotton wool is to be avoided, however, as well as touching cold metal with wet fingers. A constant problem that arises with sealed sample holders is the accumulation of liquid nitrogen or helium inside as the result of an imperfect seal, and the subsequent destruction of the holder when it is warmed slightly.

## References

Iizuka, T., and Kotani, M. (1969). *Biochim. Biophys. Acta* **194**, 351–363.
Krasney, S. (1975). Private communication.
Lang, G. (1970). *Quart. Rev. Biophys.* **3**, 1–60.
Lang, G., Herbert, D., and Yonetani, T. (1968). *J. Chem. Phys.* **49**, 944–950.
Lang, G., Asakura, T., and Yonetani, T. (1969). *J. Phys. C* **2**, 2246–2261.
Lang, G., Asakura, T., and Yonetani, T. (1970). *Phys. Rev. Lett.* **24**, 981–983.
Longworth, G., and Window, B. (1971). *J. Phys. D* **4**, 835–839.
Moss, T. H., Bearden, A. J., and Caughey, N. S., (1969). *J. Chem. Phys.* **51**, 2624–2631.
Window, B. (1969). *J. Phys. E* **2**, 894.

# 5

# Biological Iron Transport and Storage Compounds

**W. T. Oosterhuis**[†]
**K. Spartalian**[‡]

Physics Department
Carnegie-Mellon University
Pittsburgh, Pennsylvania

† *Present address*: Division of Materials Research, National Science Foundation, Washington, D. C.
‡ *Present address*: Physics Department, The Pennsylvania State University, University Park, Pennsylvania.

## I. Introduction

### A. Importance of Iron in Biology

Iron is especially important in the life processes, as it is directly involved in the transport and storage of oxygen (hemoglobin and myoglobin), electron transport (cytochromes and ferredoxins), the metabolism of hydrogen by dehydrogenase, the utilization of peroxide (peroxidases), and several other vital metabolic functions. It is safe to say that with a few possible exceptions in the bacterial world, there would be no life without iron. Generally, the importance of iron in various life processes comes about because of the ability of iron to exist in two stable oxidation states [Fe(II) and Fe(III)] and occasionally as Fe(I) and Fe(IV). The oxidation–reduction potentials of iron that occur in various metabolic functions can vary over a wide range, depending upon the ligands attached to the iron. Nature has been very clever in utilizing molecules to fulfill various needs, and it is amazing how frequently iron is of central importance.

Iron is the most abundant element in the earth (29% by mass), although most of it is in the core (about 25%). The remaining 4% is in the earth's crust, mostly in the form of $FeO$ and $Fe_2O_3$. The relevant biological questions are, how is the iron converted from mineral to protein, and once in the protein, what is the specialized role played by the iron in the function of that protein?

### B. Why Mössbauer Spectroscopy Is Useful

Spectroscopy of all kinds has been employed in the study of biological macromolecules, but if one is interested in the spectra due to the iron alone, the contribution to the spectrum from the rest of the molecule must be filtered out. Mössbauer spectroscopy employs radiation specific for $^{57}Fe$ (2% naturally abundant) so that there are no signals (absorption peaks) due to the other atoms. This is a distinct advantage when dealing with one or two iron atoms among several thousand, since one knows that any signal obtained is due only to the iron in the sample. Mössbauer spectroscopy is especially useful in the determination of the chemical or electronic state of the iron in these molecules. The spin–orbit interaction and the ligand or crystalline electric fields determine the electronic structure of these iron complex ions, which in turn determine the hyperfine interactions between the electrons and the $^{57}$Fe nucleus. Thus, from a study of the hyperfine interactions that give rise to complicated multipeak Mössbauer spectra, one can determine the electronic state of the iron. This is important, since the electronic (chemical) states are of fundamental biological importance and one can work from these to obtain information about the molecular conformation.

## II.  Iron Transport Compounds

In this chapter, we concern ourselves with the means by which iron is obtained and secured for the living organism through the iron storage and transport proteins, and in particular how Mössbauer effect spectroscopy can help one to understand the chemical state of the iron in these molecules and how this information is related to the molecular conformation.

Iron transport compounds are found low on the evolutionary scale, in the world of microbes and bacteria where there is fierce competition for the available supply of nutrients. These organisms require iron only in small quantities, but it is nevertheless vital, primarily for growth and respiration. The compounds that serve the function of iron transport for these organisms are called siderochromes (Neilands, 1966). The siderochromes are divided into two categories according to their function—sideramines, which are growth factors, and sideromycins, which are antibiotics. Both sideramines and sideromycins have an extremely high affinity for ferric iron, which ensures efficient chelation of the metal even at low iron concentrations.

Iron transport proteins—transferrin, conalbumin, and lactoferrin (Feeney and Komatsu, 1966)—which occur in birds and mammals, have many of the same properties as the siderochromes described in the preceding paragraph, in particular, a strong affinity for ferric iron and a much weaker affinity for ferrous iron. There are other similarities in their electronic structure, which will be revealed later in this discussion. We will be primarily concerned here with the spectra in iron transport proteins where the iron is in the high-spin ferric ($^6$S) configuration with five unpaired 3d electrons, one in each of the five 3d orbital states, leaving no orbital angular momentum ($L = 0$) and having spin $S = \frac{5}{2}$.

These iron complexes used for the accumulation of iron possess many common features. Probably the most important is the very large binding constant ($pK \cong 28$–$32$) for ferric iron and the relatively weak attraction for ferrous iron. The production of these very efficient iron chelators is stimulated by the growth of the organism in an iron-deficient medium. The local environment at the iron site consists of three bidentate ligands, which grip the iron usually with two oxygen atoms. There are two possible arrangements as shown in Fig. 1. The iron site in one has orthorhombic symmetry and in the other it has a threefold symmetry axis or trigonal symmetry. If only the near neighbor environment of the oxygen were considered, then the iron would sit in a site of octahedral symmetry. The siderochromes are red-brown in color with an absorption maximum near 4500 Å. They are all in the high-spin ferric state and all exhibit a strong, nearly isotropic ESR resonance at or near $g = 4.3$.

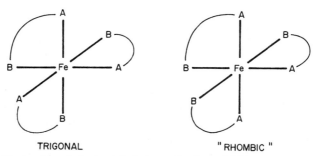

TRIGONAL          " RHOMBIC "

**Fig. 1.**   The two inequivalent octahedral coordinations of the iron atom to three bidentate ligands. In actual cases, the octahedra are distorted.

The molecular weights range from nearly 55,000 in transferrin (with two iron sites per molecule) to about 726 in enterobactin. The solubility is also wide ranging. Enterobactin is quite soluble in water, methanol, and ethanol, whereas mycobactin $P$ is soluble only in chloroform, slightly soluble in methanol, and insoluble in water. Most siderochromes are thought to exist in the growth medium as soluble entities, but recent studies have shown that mycobactin $P$ is probably localized in the cell membrane, attracting ferric iron through the outer surface of the cell and releasing it as ferrous iron to the inner cytoplasm.

It is hoped that Mössbauer effect studies on these iron-scavenging complexes will provide insight to the molecular conformation. It is also possible that these results may be useful in the identification of the molecular species.

## III. Magnetic and Electric Hyperfine Interactions

The electronic configuration and the local environment can be studied through the magnetic and electric hyperfine interactions of the $^{57}$Fe nucleus with its electrons. These interactions determine the features of the Mössbauer spectrum.

The isomer shift (see Chapter 1) reflects the effect of the electronic charge density on the mean values of the nuclear ground- and excited-state energy levels. The isomer shift can thus be used to group iron compounds according to their oxidation and spin state. This is due to the shielding effects of the valence electrons on the s electrons at the nucleus.

The magnetic hyperfine interaction between the atomic nucleus with spin $\mathbf{I}$ and its $n$ electrons is

$$\mathcal{H}_M = 2g_N\beta_N\beta_e \sum_{i=1}^{n} \left\{ \frac{\mathbf{l}_i \cdot \mathbf{I}}{r_i^3} + \frac{3(\mathbf{I} \cdot \hat{r}_i)(\mathbf{s}_i \cdot \hat{r}_i) - \mathbf{s}_i \cdot \mathbf{I}}{r_i^3} - \frac{8\pi}{3} \delta(\hat{r}_i)\mathbf{s}_i \cdot \mathbf{I} \right\} \quad (1)$$

The sum is over the $n$ electrons.

The first term is the interaction of the nuclear magnetic moment (corresponding to spin $\mathbf{I}$) with the current loops made by the orbiting electrons with angular momentum $\mathbf{l}_i$. The second term is the dipole–dipole interaction between the nuclear ($\mathbf{I}$) and electronic ($\mathbf{s}_i$) spin magnetic moments. The last term is the Fermi contact term, which describes the interaction of the nuclear moment with the s electrons, which are polarized by the unpaired electrons of the ion with spin $\mathbf{s}_i$. It is sometimes useful to think of the electrons as providing an effective magnetic field that acts on the nuclear magnetic moment, but as seen from Eq. (1) this would be a gross oversimplification.

The electric quadrupole interaction for $^{57}Fe$ is given by

$$\mathcal{H}_Q = P\{I_z^2 - \tfrac{1}{3}I(I + 1) + \tfrac{1}{3}\eta(I_x^2 - I_y^2)\}, \qquad \eta = (V_{xx} - V_{yy})/V_{zz}, \quad (2)$$

where $P = \tfrac{1}{4}e^2QV_{zz}$; $Q$ = quadrupole moment of $^{57}Fe$, $V_{zz}$ is the electric field gradient (EFG) in the $z$ direction, and $\eta$ is a measure of the distortion from axial symmetry of the electric charge cloud about the $^{57}Fe$ nucleus.

## IV. Paramagnetic Hyperfine Structure (PHS)

### A. Kramers and Non-Kramers Ions

In paramagnetic iron molecules, the unpaired electrons bound to the iron can give rise to a magnetic interaction with the nucleus that splits the nuclear states. The resulting magnetic structure in a Mössbauer spectrum together with the electric quadrupole interaction is determined by the electronic state and gives information about the symmetry of the iron site, as well as some indication of the strength of the interaction with the local environment. This information can be obtained even from randomly oriented samples, such as powders or frozen solutions.

For a non-Kramers paramagnetic ion (an even number of electrons) in a site of low symmetry, the electronic ground state is expected to be a singlet.

In this case, there is no electronic magnetic moment in the absence of an external field or a cooperative magnetic interaction. This is to say that the expectation values of $\langle s \rangle$ and $\langle I \rangle$ are zero and therefore the expectation value of Eq. (1) is zero and the nuclear magnetic moment has nothing to which it can couple. This is why there is no PHS observed in the zero field spectra of high-spin ferrous ions ($S = 2$), although it is certainly paramagnetic. However, a magnetic moment can be created in each electronic state by the application of an external field, which in turn will produce a hyperfine field that acts on the nucleus.

For Kramers ions (odd number of electrons), there is a nonzero magnetic moment (the expectation value of $\langle S \rangle$ is not zero), and magnetic hyperfine structure can be observed even in zero external magnetic field. The details of the magnetic and electric hyperfine interactions will vary with the electronic configuration, and this will be discussed briefly below.

### B. Effects of Electronic Spin Relaxation

If the electronic spin changes direction, then so does the hyperfine field seen by the $^{57}$Fe nuclear moment. When the spin direction changes rapidly compared to the Larmor precession of the $^{57}$Fe nucleus, then the field sensed by the nucleus is time-averaged to zero.

Electronic spin relaxation takes place through two mechanisms. One is due to thermal excitations of the normal vibrational modes in the molecule. These oscillations cause the electronic spins to flip from one quantum state to another. This electronic spin–lattice relaxation can be slowed by going to low temperatures to reduce the vibrations of the lattice, which can flip the electronic spin. The other mechanism is spin–spin interactions between a pair of neighboring paramagnetic centers. The paramagnetic iron centers can be separated by large distances in these macromolecules so that the spin–spin relaxation effects can be made small. In the limit where the electronic relaxation time $\tau$ is very long compared to the Larmor precession time $\tau_L$ ($\sim 10^{-7}$ sec) of the $^{57}$Fe nucleus, the nuclear magnetic moment can couple strongly to the electronic moment, giving rise to sharp magnetically split Mössbauer patterns. However, as the electronic relaxation time $\tau$ is shortened, the Mössbauer resonances become less sharp and a very broad background appears (the area of the absorption is independent of the relaxation time). This is because the magnetic field sensed by the nucleus changes its orientation during the precession. At faster relaxation rates ($\tau \ll \tau_L$), the magnetic hyperfine field is time-averaged to zero so that only the electric quadrupole interaction is left to split the Mössbauer spectrum. One of the first examples of this behavior was observed in the sideramine ferrichrome A (Fig. 2), and the essentials of the time-dependent behavior

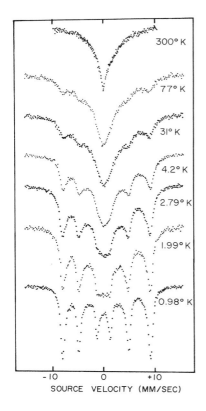

**Fig. 2.** Mössbauer spectra of ferrichrome A in zero applied field at different temperatures as reported by Wickman *et al.* The hyperfine pattern collapses as the relaxation rate increases with increasing temperature. (Reproduced by permission.)

were put forth by Wickman *et al.* (1966). The details of this calculation can also be found in a nice review paper by Wickman (1966). This is an approximate treatment, however, in that the off-diagonal terms of the hyperfine interaction were neglected. A complete treatment has been put forth by Clauser and Blume (1971). We are concerned with the long relaxation time limit, since the iron transport compounds are often magnetically dilute and the spins are slowly relaxing at low temperatures. Under these conditions, the nuclear magnetic moment is strongly coupled to the electronic moment and affords the opportunity to determine the electronic state.

If the data taken with zero magnetic field applied to isolated iron ions are only partly resolved at low temperatures as is the case with ferrichrome A and mycobactin *P*, the size of the molecule may not be large enough to separate the paramagnetic centers and prevent spin–spin relaxation. In such cases, it is sometimes possible to dilute these molecules in a diamagnetic solvent to achieve the necessary separation between the iron sites of neighboring molecules. The heme proteins are large enough to achieve the necessary separation naturally. An external magnetic field can help to give better resolution in the spectrum (Afanasev and Kagan, 1968; Wickman and Wertheim, 1966) if it is strong enough to overcome the spin–spin interactions with the paramagnetic neighbors. If these relaxation time conditions are met, then it is quite possible to determine the electronic energy states from the Mössbauer spectrum. It should be noted that PHS can be observed in those paramagnets with extremely rapid spin relaxation if a magnetic field is applied at low enough temperatures to create population differences between the magnetically split electronic states. This technique has been applied successfully in several cases.

### C. Spin Hamiltonians

The Mössbauer spectrum depends directly on the electronic state through the hyperfine interactions, and thus the electronic wavefunction can be determined from these experiments. It is usual to parameterize the wavefunctions in terms of a few coupling constants that arise in the spin Hamiltonian description of the electronic states. One is dealing with a small number of electronic states $N$, and an *effective spin* $S'$ is defined such that $N = 2S' + 1$. This spin Hamiltonian has the virtue that each interaction of the electrons can be described in terms of this effective spin $S'$.

The spin Hamiltonian with effective electronic spin $S'$ and nuclear spin $I$ is written as

$$\mathcal{H} = \mathbf{S}' \cdot \mathsf{D} \cdot \mathbf{S}' + \beta_e \mathbf{H} \cdot \mathsf{g} \cdot \mathbf{S}' + \mathbf{I} \cdot \mathsf{A} \cdot \mathbf{S}' + \mathbf{I} \cdot \mathsf{P} \cdot \mathbf{I} - g_N \beta_N \mathbf{H} \cdot \mathbf{I}.$$

The Zeeman interaction between the applied magnetic field $\mathbf{H}$ and the electronic magnetic moment $\boldsymbol{\mu} = \beta_e(\mathbf{L} + 2\mathbf{S})$ can be written as $\mathcal{H}_z = -\boldsymbol{\mu} \cdot \mathbf{H} = \beta_e \mathbf{H} \cdot (\mathbf{L} + 2\mathbf{S}) = \beta_e \mathbf{H} \cdot \mathsf{g} \cdot \mathbf{S}'$ where $\mathsf{g}$ is a tensor that makes up for any discrepancies caused by neglecting the orbital moment $\mathbf{L}$ in using an effective spin.

The magnetic hyperfine interaction (Eq. 1) can be represented by $\mathcal{H}_M = \mathbf{I} \cdot \mathsf{A} \cdot \mathbf{S}'$ where $\mathbf{I}$ is the nuclear spin, $\mathbf{S}'$ is the effective spin, and $\mathsf{A}$ is a tensor that represents Eq. (1) and may be anisotropic ($A_x \neq A_y \neq A_z$), depending on the symmetry of the local environment and the electronic wavefunction. In a similar manner, the quadrupole interaction, $\mathcal{H}_Q = \mathbf{I} \cdot \mathsf{P} \cdot \mathbf{I}$ and the crystal field interaction $\mathcal{H}_e = \mathbf{S}' \cdot \mathsf{D} \cdot \mathbf{S}'$ reflect the asymmetry of the electronic environment. The tensors $\mathsf{A}$, $\mathsf{g}$, $\mathsf{P}$, and $\mathsf{D}$ can be calculated from a knowledge of the electronic wavefunctions or, equivalently, the crystalline electric fields and the spin–orbit coupling that acts on the electrons. Examples of effective spin values characterizing commonly occurring electronic configurations are given below.

## V. Magnetic Structure in Different Electronic Configurations of Iron

### A. $S' = 0$ (*A Diamagnet with No Unpaired Electrons* $t_{2g}^6$)

Iron is commonly found in the diamagnetic low-spin ferrous state (a $t_{2g}^6$ configuration). In some cases (Ure and Flinn, 1971), such as $K_4Fe(CN)_6 \cdot 3H_2O$, there is a near cubic environment and only a single unsplit line is observed. In other examples, such as oxygenated hemoglobin ($HbO_2$) (Lang and Marshall, 1966), there is a large distortion of the electronic charge cloud giving rise to an electric field gradient, which causes a splitting of the $^{57}Fe$ nuclear excited state with a two-peak spectrum. By observing the spectrum in an applied magnetic field, the sign of the field gradient can be determined even for a powder in which the field gradient is randomly oriented with respect to the applied field. This is quite easy to do for a diamagnet since the complication of the electronic magnetic moment does not occur (Collins and Travis, 1967).

### B. $S' = \frac{1}{2}$ (*One Unpaired Electron,* $t_{2g}^5$ *or* $t_{2g}^6 e_g^1$)

The classic covalent iron salt is $K_3Fe(CN)_6$, which has a $t_{2g}^5$ electronic configuration (a hole in the filled $t_{2g}$ subshell), which is strongly influenced by the crystal fields and the spin–orbit interaction. An exhaustive treatment by Oosterhuis and Lang (1969a) for the concentrated salt and for $Fe^{3+}$ diluted in isomorphic diamagnetic $K_3Co(CN)_6$ illustrates the information that can be obtained from Mössbauer data. Although the iron has a near cubic environment there is a high degree of asymmetry in the A, P, and g tensors. This is due to the extreme sensitivity of the electronic state to small asymmetric crystal fields. Even more asymmetric examples are found in the heme proteins HbCN, $HbN_3$ (Lang and Marshall, 1966; Oosterhuis, 1971a), and cytochrome c (Lang *et al.*, 1968).

One rather unusual situation arises in reduced sodium nitroprusside and in HbNO (Oosterhuis and Lang, 1969b), in which there is one unpaired electron in a $3z^2 - r^2$ orbital from a ($t_{2g}^6 e_g^1$) configuration. This leads to very asymmetric hyperfine interactions that are characteristic of that electronic state.

The theory of paramagnetic hyperfine structure in $S' = \frac{1}{2}$ iron complexes has been worked out in terms of the crystal fields and spin–orbit coupling in materials where several examples of PHS are found with different degrees of asymmetry in the wavefunctions (Oosterhuis and Lang, 1969a, b; Oosterhuis, 1971a; Lang and Oosterhuis, 1969).

## C. $S' = 1$ (*Two Unpaired Electrons*, $t_{2g}^4$ *and* $e_g^2$)

The theory for the magnetic interactions in a low-spin Fe(IV) complex was worked out by Oosterhuis and Lang (1973) after several examples of this highly unusual chemical state were shown to exist. The list of these examples is dominated by biological iron materials such as complexes of cytochrome c peroxidase (Yonetani *et al.*, 1966), horseradish peroxidase (Moss *et al.*, 1969), and Japanese radish peroxidase (Dolphin *et al.*, 1971). A synthetic iron(IV) complex was studied by Paez *et al.* (1972) in a large magnetic field so that the paramagnetic hyperfine interactions in the fast relaxation regime could be used to determine the electronic state. A series of low-spin $t_{2g}^n$ complexes in the same crystalline environment was studied recently by Oosterhuis *et al.* (1974) using Mössbauer spectroscopy.

Another unusual chemical state with two unpaired electrons for iron is the Fe(VI) ion with an $e_g^2$ configuration. This paramagnetic ion was diluted, and a small magnetic field was shown (Oosterhuis and Barros, 1972) to induce a large spin polarization so that PHS could be observed in this non-Kramers ion in the slow relaxation regime.

## D. $S' = \frac{3}{2}$ (*Three Unpaired Electrons*, $t_{2g}^3 e_g^2$)

Some novel pentacoordinate $Fe^{3+}$ complexes have been studied by Wickman and Trozzolo (1967) using Mössbauer effect spectroscopy. These complexes have ligand fields strong enough and with the right symmetry to cause one of the five 3d electrons to flip its spin so that only three of the electrons are unpaired, giving rise to a spin quartet. Paramagnetic hyperfine structure is observed and interpreted in terms of a spin Hamiltonian with effective spin $S' = \frac{3}{2}$. No biological iron complexes with $S' = \frac{3}{2}$ have been found yet, although the met–hemoglobin complex is also pentacoordinate.

## E. $S' = 2$ (*Four Unpaired Electrons*, $t_{2g}^4 e_g^2$)

The high-spin ferrous ion is one of the very common oxidation states of iron and has four unpaired electrons and an orbital moment as well, which complicates the paramagnetism. Paramagnetic hyperfine structure can be easily observed in Mössbauer spectra of these complexes if a magnetic field is applied. A good example of this is shown by Johnson (1967) in ferrous fluorosilicate, where the spin and orbital effects are considered. Lang and Marshall (1966) have studied several biological iron complexes in the high-spin ferrous state, the best example being deoxyhemoglobin. An analysis of the paramagnetic hyperfine interactions for ferrous complexes is included in a paper by Oosterhuis (1971b) on crystal symmetry effects, by Lang and Marshall (1966), and by Johnson (1967).

## F. $S' = \frac{5}{2}$ (*Five Unpaired Electrons*, $t_{2g}^3 e_g^2$)

The high-spin ferric ion is the other common oxidation state of iron. It has no orbital moment, it is spherically symmetric, and it usually interacts strongly with the nuclear magnetic moment. There are many examples of high-spin ferric ions in biological materials, such as some of the heme proteins (Lang and Marshall, 1966), some model compounds such as hemin (Johnson, 1966), and the iron transport materials to be considered later in this chapter. Johnson (1966) has observed paramagnetic hyperfine structure in the spectra of hemin in the fast relaxation regime, but most other examples of PHS in ferric complexes occur in the slow relaxation limit. A detailed description of the analysis of PHS in high-spin ferric complexes as applied to the iron transport compounds follows below.

## VI. Spin Hamiltonian with $S' = \frac{5}{2}$ for High-Spin Ferric Ions

The detailed structure of the Mössbauer spectra for the iron transport compounds can be described in terms of a spin Hamiltonian $\mathscr{H}$ with effective spin $S' = \frac{5}{2}$ for the high-spin ferric ion. These parameters can give information about the site symmetry of the iron and the energy level structure of the electronic states. Since there is no orbital angular momentum in the $^6S$ state, the effective spin is the same as the real spin of the iron ion. There is spherical symmetry in the $^6S$ state to first order, but spin–orbit coupling to excited (nonspherical) orbital states gives rise to asymmetries about the iron site that are reflected in the spin Hamiltonian. The general form of the spin Hamiltonian which we will use here with **S** as the electronic spin operator and **I** as the nuclear spin operator is

$$\mathscr{H} = D[S_z^2 - \tfrac{35}{12} + \lambda(S_x^2 - S_y^2) + (\mu/6)(S_x^4 + S_y^4 + S_z^4 - \tfrac{707}{16})]$$
$$+ 2\beta\mathbf{H}\cdot\mathbf{S} + A\mathbf{I}\cdot\mathbf{S} + P[I_z^2 - \tfrac{5}{4} + \lambda(I_x^2 - I_y^2)] - g_N\beta_N\mathbf{H}\cdot\mathbf{I} \qquad (3)$$

The terms in $D$, $\lambda$, and $\mu$ give the zero field splitting of the six components of the electronic $^6S$. In terms of the conventional spin Hamiltonian parameters, we have defined $\lambda = E/D$, $\mu = a/D$. The quadratic interaction described by $D$ and $\lambda$ usually dominates except when the applied field makes the Zeeman term larger (Fig. 3). In several of the iron transport compounds, the zero field splitting is comparable to the Zeeman energy for a few kilogauss, so that there is considerable mixing of the states and perturbation theory is not applicable. Usually, the quartic term has been neglected ($\mu = 0$), but there have recently appeared some cases that are strongly dependent on $\mu$.

The splitting in zero field reflects the symmetry of the local molecular environment, and it arises out of the spin–orbit interaction that couples the ground ($^6S$) state with higher orbital states, which in turn are more sensitive to distortions in the local environment.

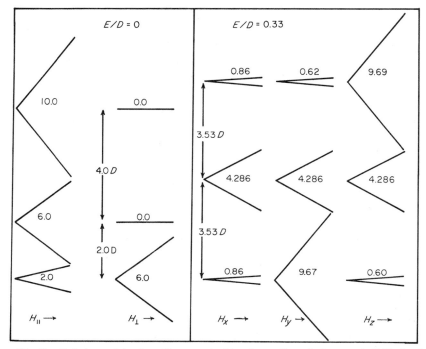

**Fig. 3.** Splittings of the three Kramers doublets in small applied magnetic fields for the cases of axial symmetry ($\lambda = E/D = 0$) and complete rhombic symmetry ($\lambda = E/D = \frac{1}{3}$). The effective g' tensor components are shown for each case. (From Oosterhuis, 1974.)

The term with $\mu$ as a coefficient represents the cubic crystal field energy. This contribution can usually be neglected except for those ions in nearly cubic symmetry.

The Zeeman interaction $2\beta\mathbf{H}\cdot\mathbf{S}$ is isotropic and describes how the energy levels behave in an external field. Small corrections due to admixtures of excited states can destroy the isotropy. Similarly, the magnetic hyperfine interaction, Eq. (1), can be written as $A\mathbf{I}\cdot\mathbf{S}$ if the A tensor is isotropic, although anisotropy can arise from the orbital and dipole contributions due to admixtures with higher electronic states.

The electric field gradient (EFG) is zero in the $^6$S state with spherical symmetry, but of course there are contributions from the charge asymmetries about the iron site and from the anisotropic admixtures mentioned above. Since the EFG has the same origin as $D$ and $\lambda$, the asymmetry parameter $\eta$ can be constrained by $\eta = 3\lambda = 3E/D$. The last term in Eq. (3) is the Zeeman interaction of the nuclear magnetic moment with the externally applied field. All these interactions combine to yield a unique spectral profile in the

Mössbauer data in powder samples or frozen solutions, and they can give valuable information because of the simultaneous interaction of the EFG and magnetic hyperfine coupling. An example of these interactions are shown in the Mössbauer spectra (Fig. 4) for mycobactin $P$, a growth factor (sideramine) for *Mycobacterium phlei*, which is slightly soluble in ethanol. The spin Hamiltonian parameters $D$, $\lambda$, $\mu$, $A$, and $P$ listed in Table I were found to give reasonably good simulation of the data (Spartalian *et al.* 1973). This simulation is the solid curve in Fig. 4.

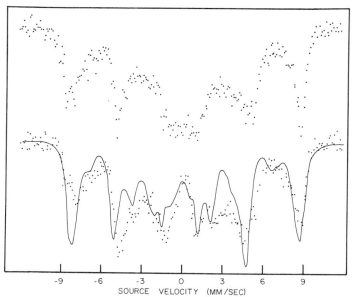

SOURCE VELOCITY (MM/SEC)

**Fig. 4.** Mössbauer spectra of mycobactin $P$ in ethanol at 4.2°K in zero field (top) and 1.3 kOe transverse (bottom). The applied field sharpens the paramagnetic hyperfine structure.

**TABLE I**

Spin Hamiltonian Parameters for Some $^6$S Iron Complexes
as Determined by the Mössbauer Effect [a]

| Complex | $D$ (cm$^{-1}$) | $\lambda$ | $\mu$ | $A$ (kOe/spin) | $P$ (mm/sec) |
|---|---|---|---|---|---|
| Ferrichrome A | 0.50 | 0.28 | — | — | — |
| Deferoxamine | 0.50 | 0.46 | — | 217 | −0.30 |
| Enterobactin | 0.48 | 0.46 | −0.27 | 222 | −0.15 |
| Mycobactin $P$ | 0.22 | 0.45 | −0.26 | 221 | −0.20 |
| Transferrin | 0.15 | 0.31 | — | 221, 215, 213 | +0.37 |

[a] For references see text.

## VII. Basis States and Calculation of Mössbauer Spectra in Small Fields

We assume that the electronic spin relaxation is very slow so that we do not have to consider time dependent effects in the spectrum. The electronic basis states are the six states $|m_s\rangle = |-\frac{5}{2}\rangle, |-\frac{3}{2}\rangle, \ldots, |+\frac{5}{2}\rangle$. These, together with the nuclear basis states $|m_I\rangle$, form product states $|m_s\rangle |m_I\rangle$, and the spin Hamiltonian (Eq. 3) operates on these product states. The various terms mix the basis states and the $(n \times n)$ matrix $[n = (2S + 1)(2I + 1)]$ is diagonalized to find the energies and eigenstates.

This procedure is followed for the nuclear excited state $(I = \frac{3}{2})$ and the ground state $(I = \frac{1}{2})$, and then the transition probabilities and line positions are calculated with a Lorentzian line shape of width $\Gamma$ folded in to give the Mössbauer spectrum. This has been done on the computer, assuming the applied magnetic field and the $\gamma$ beam in some fixed direction relative to the crystal axes. The same calculation is then repeated for a number of field and $\gamma$ beam directions equally distributed over the unit sphere, and the spectra are added with equal weights to simulate the spectrum from the randomly oriented molecules of a powder or frozen solution.

The profile of the resulting spectrum depends on the spin Hamiltonian parameters, which are adjusted to maximize the agreement between the calculated spectrum and the experiment. Detailed descriptions of this type of calculation exist in the literature. (Wickman and Wertheim, 1966; Lang and Marshall, 1966; Lang et al., 1971; Viccaro et al., 1972).

In the slow relaxation situation, there is one Mössbauer spectrum associated with each electronic state, and the spectra due to the six electronic states are superimposed on each other with weights given by their Boltzmann factors. These states have separations typically of $0.75–3.5$ cm$^{-1}$. Thus, one can get significant changes in the Mössbauer spectrum by depopulating some of the electronic states at low temperatures. An example of this is in enterobactin (Spartalian, et al., 1975) shown in Fig. 5. If the applied field is small enough so that there is no significant mixing of the electronic states by the applied field, then each member of a Kramers doublet will yield the same Mössbauer spectrum so that there are essentially only three different contributions, one from each Kramers doublet.

In the heme proteins, there is a large axial crystal field described by a $D$ that is large enough to dominate all the other terms in the spin Hamiltonian. These give spectra such as those of Lang et al., (1969) for cytochrome c peroxidase.

In each of the iron transport proteins, there is a considerable rhombic crystal field and a relatively weak axial field, so that the net result is to have $E$ comparable to $D$. In fact, $E = D/3$ or $\lambda = \frac{1}{3}$ is the condition for which

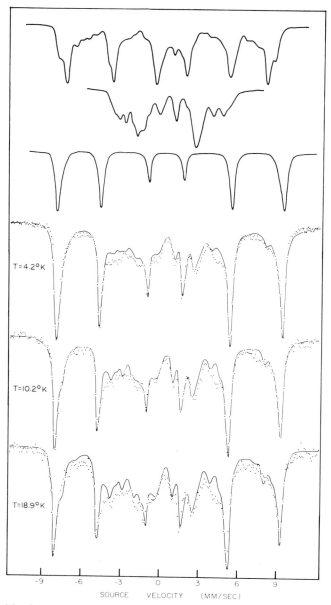

**Fig. 5.**   Mössbauer spectra of enterobactin in methanol at various temperatures in 1.3 kOe transverse. The contributions from each of the three Kramers doublets not weighted by their Boltzmann factors are shown at the top of the figure. (From Oosterhuis, 1974.)

complete rhombic symmetry exists. A value of $\lambda > \frac{1}{3}$ merely implies that the principal axis of the zero field splitting is no longer the $z$ axis (Blumberg, 1967).

Of the Kramers doublets (Fig. 3) in rhombic symmetry ($\lambda \simeq \frac{1}{3}$), the middle doublet has an isotropic $g$ value ($g = 4.3$), which means that the magnetic moment in those states will follow the field in any direction, and so the contribution of that doublet to the Mössbauer spectrum indicates the hyperfine field to be antiparallel to the applied field. A strong resonance at $g = 4.3$ in ESR is found in most of the iron complexes considered here. The Mössbauer spectra calculated for a frozen solution of enterobactin for each Kramers doublet assuming $\lambda = E/D = 0.46$ and $\mu = -0.27$ are shown in Fig. 5, where there is a small field of 1.3 kOe in a direction perpendicular to the $\gamma$ beam. The reader can observe the differences in the contribution from each Kramers doublet in the composite spectrum.

In Fig. 6 are presented some computer generated spectra with different degrees of rhombic symmetry $\lambda = E/D < \frac{1}{3}$ and $\mu = 0$, so that the variation of $\lambda$ can be seen as indicated by the arrows in the figure.

Although it has been the usual practice, it must not be assumed that the quartic term in the spin Hamiltonian can be neglected, especially in these iron transport complexes in which the magnitude of $D$ is small and more nearly comparable to $a$. If one constrains $\mu = 0.75(1 - 3\lambda)$ then $g = 4.3$ even for $\lambda \ll \frac{1}{3}$ (Vanderven et al., 1974). Thus, one can obtain a "rhombic" ESR signal even in near axial symmetry.

Many of the iron transport compounds have hydroxamic acid ligands that bind the ferric ion, and so synthetic iron(III) hydroxamic acid compounds have been studied with the expectation of gaining insight into the natural compounds. Experiments on these synthetic trihydroxamic acids have shown the rhombic symmetry effects in the Mössbauer data (Oosterhuis and Spartalian, 1974). Other experiments using Mössbauer spectroscopy on rhombic iron sites have been published by Lang et al., (1971) on EDTA, and by Zabinski et al., (1972) on protocatechuic acid 4,5-oxygenase.

### A. Method for Analyzing Mössbauer Spectra with Paramagnetic Hyperfine Structure

When one has a sample that yields a magnetically split Mössbauer spectrum with many peaks in it, one might wonder how to proceed with the analysis and how to determine the spin Hamiltonian parameters.

The Mössbauer spectra for a given molecular species at high temperature and concentrations will probably not exhibit magnetic hyperfine interaction. Thus, under these conditions, the quadrupole splitting and the isomer shift are easily observed.

By changing the conditions to low temperatures and low concentrations

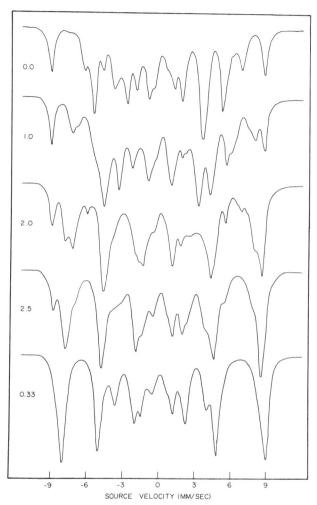

SOURCE   VELOCITY  (MM/SEC)

**Fig. 6.**  Computer-simulated spectra showing the effects of the rhombic parameter $\lambda$. The other assumed parameters are $D = 0.5 \text{ cm}^{-1}$, $\mu = 0$, $A = 225 \text{ kOe/spin}$, $P = -0.35 \text{ mm/sec}$, $T = 4.2°\text{K}$, $H = 1.3 \text{ kOe transverse}$. (From Oosterhuis, 1974.)

of paramagnetic ions, it is usually easy to observe the magnetic hyperfine structure in the Mössbauer spectrum, as well as the quadrupole interaction.

For the high-spin ferric chelates, spectra in large applied fields ($\sim 50 \text{ kOe}$) are often useful for determining the hyperfine-coupling parameter $A$, because in this case, the electronic Zeeman energies overwhelm the zero-field splitting energies that define the three Kramers doublets ($\beta_e H_{app} \gg D$) so that the spin Hamiltonian can be written as $\mathcal{H} = g_N \beta_N (\mathbf{H}_{int} - \mathbf{H}_{app}) \cdot \mathbf{I}$, where the internal field is $\mathbf{H}_{int} = A\mathbf{S}/g_N \beta_N$.

The negative sign occurs because the Fermi contact interaction opposes the applied field. Once the hyperfine-coupling parameter is determined, the low-field ($\sim 1$ kOe) data can be used to obtain parameters $D$, $\lambda$, and $\mu$. Moreover, since $D$ is essentially a scaling factor for the zero-field splittings, data at temperatures slightly higher and lower than 4.2°K (1.5–15°K) are helpful for a more accurate determination of $D$ as the Kramers doublets undergo changes in their Boltzmann populations.

This step in the analysis of the Mössbauer data for the spin Hamiltonian parameters is done with the data taken in a small applied field of several hundred oersteds. This is done in order to break the (Back–Goudsmit) coupling between the electronic spin and the ligand nuclear moment, which can have drastic effects on the Mössbauer spectrum. We see this in the case of transferrin (Spartalian and Oosterhuis, 1973) (Fig. 7), where the spin Hamiltonian parameters ($D$, $A$, $P$, $\mu$, $\lambda$) used to fit the data in an applied field can also be

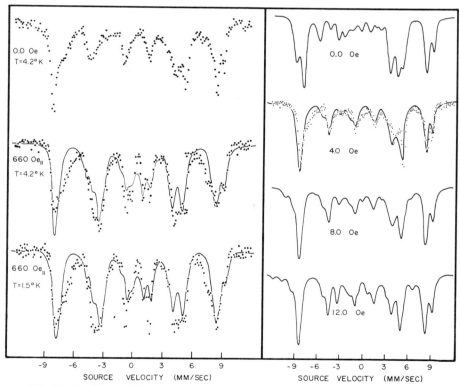

**Fig. 7.** (*Left*) Mössbauer spectra of human serum transferrin in water with applied fields and temperatures as shown. The solid line is a computer simulation. (*Right*) The zero field experimental spectra are best simulated using the same spin Hamiltonian parameters but with a randomly oriented field of 4 Oe designed to simulate neighboring nuclei effects.

used to calculate the spectrum expected in zero applied field. The neglect of the Back–Goudsmit coupling leads to a calculated spectrum, in sharp disagreement with the experimental results in zero field, even though the spectrum from the same sample in an applied field gives good agreement between calculation and experiment.

### B. Neighbor Nucleus Effects in Zero Field

The ligand nuclear moments $I_j$ interact with the electronic spin, which is covalently shared to some extent between $^{57}Fe$ and the ligands. The proper term to add to the Hamiltonian (Eq. 3) is $\sum_j I_j \cdot B_j \cdot S$ where $B_j$ is a coupling tensor for the $j$th ligand nucleus.

This could be done on a computer and has been fairly successful (Lang, 1968). However, if there are several ligand nuclear moments, then the computation rapidly gets out of hand because the dimension of the matrix to be diagonalized goes as $(2I_{Fe} + 1)(2S + 1) \prod_j (2I_j + 1)$. Thus, it is useful to approximate the sum of the ligand interactions acting on the electron spin $S$ by a random magnetic field $H_R$ with strength of several oersteds. Then one simply adds the term $g\beta H_R \cdot S$ to Eq. (3) and calculates the spectrum. This is done for varying field strengths $H_R$, and indeed, this does improve the agreement with experiment (Oosterhuis and Viccaro, 1972) (Fig. 7). Sometimes it is difficult to achieve good agreement between the calculated spectra and the zero field data, even though excellent agreement exists in the cases of applied magnetic fields.

## VIII. Experimental Studies of Iron Transport Proteins

Some of the practical and necessary details for the preparation of iron absorbers is discussed by G. Lang in Chapter 4. We mention here some of the typical operations used in connection with the iron transport complexes. In each of these siderochromes, the Mössbauer absorbers were easily prepared in solution, mostly due to the very large binding constant for the ferric complex. The general procedure is to find a solvent that dissolves both the iron-free sideramine and the inorganic compound. The inorganic iron salt will usually be enriched to about 90% $^{57}Fe$ so that the signal-to-noise ratio for the Mössbauer spectrum will be optimized. An excess of iron-free sideramine is added to ensure that each iron atom is complexed, since any non-specifically bound iron will contribute an undesirable signal in the spectrum. The solvent used will vary from one complex to another because of the wide range of solubilities. It is hoped that a glass-forming organic solvent can be found for the ferric complex which then can be frozen so as to maintain the separation between the paramagnetic iron ions. If powder samples are desired, then the solution is readily evaporated leaving a powder residue in the container.

### 1. Ferrichrome A

Ferrichrome A, which is comprised of three hydroxamic acid residues, is an inactive growth factor (sideramine) for several microorganisms. This iron compound is found in low-iron fermentations of bacteria, yeasts, and fungi. The Mössbauer effect experiments done by Wickman *et al.* (1966) not only were among the first Mössbauer spectra in biological iron compounds that exhibited paramagnetic hyperfine structure, but these authors also formulated a theory to describe effects of the time-dependent hyperfine fields in the Mössbauer spectrum. The data (Fig. 2) show an increase in the electronic relaxation time as the temperature is lowered and as the dilution of the magnetic species is increased. The spin Hamiltonian parameters found for this compound are given in Table I and have the typical "rhombic" implications for the iron site environment, although the crystal structure is known to be nearly trigonal (Zalkin *et al.*, 1966).

### 2. Deferoxamine

Deferoxamine is a growth factor (sideramine) in which there are three hydroxamic acid residues to bind ferric iron. This iron chelating agent is produced by *Microbacterium lacticum* and various species of streptomyces to compete for the available iron supply. The ratio $\lambda = E/D = 0.46$, while $\lambda = \frac{1}{3}$ represents the extreme rhombic distortion. $\lambda > \frac{1}{3}$ means that the axis of principal distortion is the $y$ direction instead of the $z$ direction (Blumberg, 1967). Deferoxamine (Desferal) is sometimes used in treatment of iron poisoning. Mössbauer effect studies have been published by Bock and Lang (1972) and show the typical features of these iron transport compounds.

### 3. Enterobactin

Enterobactin is derived from certain enteric bacteria such as *Escherichia coli* and is classified as a siderochrome. The iron is bound by three bidentate catechol ligands, in contrast to the hydroxamate ligands employed in ferrichrome A and deferoxamine. Mössbauer spectra for $^{57}$Fe enterobactin show unusually sharp absorption lines as compared to the spectra of the other natural compounds. The data in Fig. 5 are well represented by a calculation (solid curves) with the spin Hamiltonian parameters $D = 0.48 \text{ cm}^{-1}$, $\lambda = 0.46$, $\mu = -0.27$, $P = 0.15$, $A = 1.50$ mm/sec. Again we have the data extremely well fit by a spin Hamiltonian with rhombic symmetry, whereas NMR data indicate trigonal symmetry (Llinás *et al.*, 1973).

### 4. Mycobactin P

The iron in the mycobactin is ligated to two hydroxamic acid residues and two more atoms (nitrogen and oxygen) capable of binding the metal. It thus differs from the complexes mentioned above, which have three bidentate ligands at the iron site. Nevertheless, the ferric iron is bound at least as tightly. The mycobactins are growth factors for a class of mycobacteria that have been studied by Snow (1970) and Ratledge (1971). The Mössbauer spectra for mycobactin *P* are shown in Fig. 4 and the spin Hamiltonian parameters are in Table I. The magnetic hyperfine coupling constant *A* is larger, indicating less delocalization of the 3d electrons than in the other materials. The crystal structure of mycobactin *P* has been determined recently by Hough and Rogers (1974) and showed a distorted environment for the iron ion in agreement with the Mössbauer results.

### 5. Transferrins

Transferrin is an iron chelating agent found in the blood serum of animals and its function is the transport of iron from the iron storage areas for incorporation into immature red blood cells, or perhaps into other iron proteins as well. This molecule has two iron sites, which from the Mössbauer data (Spartalian and Oosterhuis, 1973; Tsang *et al.*, 1973), appear to be identical, but recent ESR work by Aasa (1972) has shown the two sites to be slightly different. The spectra (Fig. 7) show the magnetic hyperfine interaction. There are no hydroxamic acid ligands in transferrin, yet the iron is bound as tightly, as in the previous cases. There are, however, three bidentate ligands that grip the iron as in the hydroxamic acid complexes.

### 6. Conalbumin

Conalbumin is similar to transferrin, it has two iron sites, and it is found in egg white. Its function is possibly to chelate iron and prevent its uptake from the bacterial growth factors that are sure to be present. Lactoferrin is a similar molecule found in milk and probably has the same function in preventing the growth of bacteria by depriving them of iron. Mössbauer data for conalbumin taken by Aisen *et al.*, (1973) show evidence for two different iron sites.

The iron transport compounds described above have a common ability to bind ferric iron very strongly, yet they have little affinity for ferrous iron. They also have rhombic symmetry in their electronic states and ligand field splittings of less than 1 cm$^{-1}$. It is not known whether these compounds bind the iron outside the cell and then gain entrance within the cell wall to give up the iron or whether these compounds might be attached to the membrane. If the spectra of these compounds are characterized, it may be possible to use this information in the study of how iron gains entrance through a cell wall.

## IX. Iron Storage Proteins—Introduction

In addition to producing molecules for the accumulation of iron, many organisms produce molecules for the storage of iron, and in some systems these storage proteins prevent excess uptake of iron. In humans, there is ferritin, which is found in the liver and spleen, as is hemosiderin, which is like ferritin without some of the protein shell, and gastroferrin (Webb et al., 1973), found in the gastric juice and may have a role in the ingestion of iron. Of the total iron contained in the body (about 4 gm) 65% is in the form of hemoglobin, 30% in ferritin, 5% in myoglobin, 0.1% in plasma as transferrin, and 0.3% in the form of various enzymes. There is some speculation that iron is taken from ferritin by transferrin to the bone marrow for incorporation into the red blood cells.

Ferritin is also found in such diverse organisms as mammals, plants (Robards and Humpherson, 1967; Hyde et al., 1963), and fungi (David and Easterbrook, 1971). The ferritin molecule has a molecular weight of about 500,000, of which it contains up to 20% iron by weight. The iron is in the form of ferric oxyhydroxide (FeO—OH) with up to about 4000 atoms enclosed by a protein shell. When this shell is saturated with iron, the iron core is nearly spherical with a diameter of about 70 Å, while the protein shell has an outer diameter of about 120 Å (Blaise et al., 1965). Because of the limit on the volume of iron in the ferritin the possibility exists of obtaining a uniform core-particle size. There are some interesting physical properties of these particles (between the microscopic and macroscopic domains) in that they become antiferromagnetically ordered at about 200°K (Blaise et al., 1965; Boas and Window, 1966), although the ordering temperature probably depends on the particle size, as well as on the surface and shape effects. A review dealing exclusively with ferritin has been published recently by Crichton (1973).

## X. Superantiferromagnetism—Néel Theory

When one is dealing with small particles composed of a limited number of iron atoms close to each other, these atoms can be magnetically ordered. We will consider such particles with antiferromagnetic ordering in which the direction of the antiferromagnetic axis can change, but the individual iron spins maintain the same relation to this axis. The temperature at which the thermal energy becomes comparable to the anisotropy energy is the so-called blocking temperature $T_B$. Above this temperature, the relaxation rate begins to increase very rapidly. Mössbauer effect experiments on ferritin extracted from horse spleen have been reported by Blaise et al. (1965) and Boas and Window (1966). The spectra obtained in these experiments have been interpreted on the basis of Néel's theory of superparamagnetism of ultrafine particles (Néel, 1962). Ultrafine particles of sizes less than 100 Å in diameter of an antiferromagnetic material are expected to have poor lattice structure. They would not have a perfect geometrical shape, as they would rather exhibit a "spongy" structure with a number of crevices and badly shaped surfaces. Consequently, at temperatures below the Néel point of the bulk material, the opposing spins in each sublattice will not be expected to cancel exactly. There will be a net magnetization, and a potential barrier will be formed of the type

$$U = \tfrac{1}{2}CV \sin^2 \alpha, \tag{4}$$

where $\alpha$ is the angle between the easy axis and the direction of magnetization, $C$ is a characteristic constant of the material, and $V$ is the volume of the particle in question. Obviously, the energy is minimized for $\alpha = 0$ or $\alpha = \pi$, so that the magnetization will remain stable in the direction of the easy axis if there are no perturbations acting on the system. Thermal agitation may provide such perturbations so that, if $V$ is small enough, $kT$ may be large enough to reverse the magnetization direction by supplying enough energy to take it over the potential barrier $U$. The rate at which the magnetization changes direction in this case is given by

$$f = f_0 \exp(-CV/2kT), \tag{5}$$

where $f_0$ is a constant (Néel, 1962) of order $CV/h = 10^{12} \text{ sec}^{-1}$.

The above considerations are sufficient to enable an understanding of the Mössbauer spectra of ferritin. Each $Fe^{3+}$ nucleus in the ferritin core experiences a hyperfine field of about 490 kOe due to the Fermi contact interaction. The nuclear Larmor precession frequency in this field is $\omega_N = 2.5 \times 10^6 \text{ sec}^{-1}$, but the hyperfine field reverses its direction with a frequency $f$ given by Eq. (5). Two extreme limits are distinguished:

1. If $kT \gg CV$, then $f \gg \omega_N$, which means that the hyperfine field changes direction too rapidly for the nucleus to follow. The result is a spectrum characteristic of fast electronic relaxation rates, namely a quadrupole split two-line pattern, even though the electronic spins are magnetically ordered.

2. If $kT \ll CV$, then $f \ll \omega_N$, which means that the hyperfine field is essentially constant during the precession time of the nucleus, and as a

result, the Mössbauer spectrum will be a Zeeman split six-line pattern. Since the electronic relaxation rate $f$ can be controlled by the experimenter by controlling the temperature for a fixed particle volume, spectra characteristic of values of $f$ intermediate to the above limits can also be obtained.

## XI. Experimental Studies of Iron Storage Proteins

### 1. Horse Ferritin

The horse ferritin data are shown in Fig. 8. The relaxation rate increases with temperature. There is a drastic change in the spectra around 35°K at which the relaxation rate of the electronic spins is comparable to the nuclear Larmor precession rate. At higher temperatures, the electronic relaxation

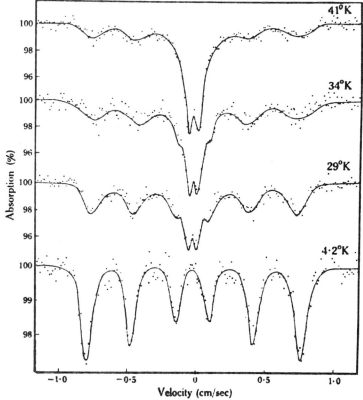

**Fig. 8.** Mössbauer spectra of horse ferritin in zero field at various temperatures. (Boas and Window, 1966; reproduced by permission.)

rate exceeds this precession rate of the $^{57}$Fe nucleus so that the magnetic hyperfine interaction is washed out. If the volume of the particle could be made smaller by a certain factor, then one should observe the transition temperature to be lowered by about the same factor.

### 2. Phycomyces

We are afforded this opportunity in the case of the fungus *Phycomyces* (Fig. 9), which also uses ferritin to store iron but in smaller-size particles. We take the ferritin in the spores and examine the temperature dependence of the spectra (Spartalian *et al.*, 1974), and indeed, the transition, or blocking, temperature is about 17°K, about half of that for horse ferritin, which leads to the suggestion that the phycomyces ferritin is about half the size of the horse ferritin molecule. This is in agreement with determination of particle size by a sedimentation technique (David and Easterbrook, 1971). It was quite easy to prepare samples in this case where no biochemical separation was involved, only the separation of the spores from the stems with subsequent cutting of the stems with scissors. The spectra of Fig. 9 were obtained in half a day with a 50 mCi source.

The technique of growing organisms in an enriched medium and then using the Mössbauer radiation that is specific for $^{57}$Fe to follow the biochemical state of the iron as it participates in the life of the organism is certainly possible.

An example of the dependence of the chemical state of iron upon the location in an organism is found also in the phycomyces (Spartalian *et al.*, 1974). Spectra of different parts of the fungus are shown in Fig. 10; they correspond to the spores, top third of the sporangiophores (stems), middle third, bottom third, the mycelia (roots), and the growth medium. The iron in the chemical state of the growth medium (ferric chloride) is present in all spectra as the inner doublet in each spectrum, but the proportion is seen to diminish as one looks closer to the top of the phycomyces. It might be interpreted as the digestion of iron as it rises in the fungus. Again, no chemical separation was done on these samples.

Since iron is used in such diverse ways in a given organism, one very heavily weighted chemical state; e.g., ferritin in the phycomyces will obscure the Mössbauer patterns due to the other chemical states. Thus, to observe these lightly populated states such as cytochrome c in phycomyces, biochemical separation is necessary.

### 3. Hemosiderin

Iron is also found in other types of storage proteins such as hemosiderin and gastroferrin, which have also been investigated with Mössbauer spectroscopy. These materials give similar Mössbauer patterns to those observed

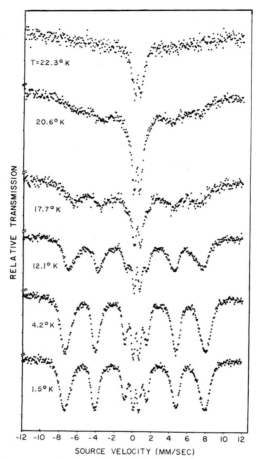

**Fig. 9.**   Zero field spectra of phycomyces spores at various temperatures.

in ferritin and in fact have nearly the same hyperfine field, quadrupole splitting, and isomer shift. As with the phycomyces ferritin, the temperature dependence of the Mössbauer spectrum provides a good indication of the particle size.

Mössbauer spectra for hemosiderin have been reported by Fischbach *et al.* (1971), who made comparisons between the spectra for ferritin and hemosiderin. The Mössbauer, x-ray, and electron microscopy results suggest the chemical state of the individual iron in hemosiderin to be nearly identical to that in ferritin. Some speculation is made that ferritin is the precursor of hemosiderin.

Johnson (1971) has shown Mössbauer spectra of healthy lung and material

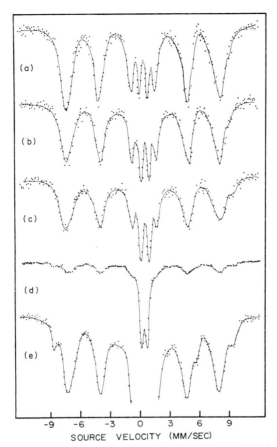

**Fig. 10.**   Zero field spectra at 4.2°K of (a) top, (b) middle, and (c) bottom sections of sporangiophores. The mycelia spectrum is shown in (d) and also in (e) expanded vertically to show the well-resolved ferritin peaks.

from the lung of a hemosiderosis victim showing an unusually large amount of iron in the diseased lung. The spectra show a magnetic hyperfine splitting similar to that observed in horse ferritin (Blaise *et al.*, 1965; Boas and Window, 1966).

## 4. Gastroferrin

Another iron-binding protein is gastroferrin, recently investigated by Webb *et al.* (1973), for which the Mössbauer data are reported to have a quadrupole doublet at 77°K. Since low-temperature data are not reported for gastroferrin, one can only speculate that the polynuclear iron complex will have a similar spectrum to ferritin. Further studies at 4°K would be illuminating.

There have been several synthetic polynuclear iron complexes formed recently, such as iron(III) oxide–hydrate gel by Giessen (1967), iron-containing polymers formed from bicarbonate hydrolyzed ferric nitrate solutions (Brady *et al.*, 1968), and hydrolyzed iron citrate solutions (Spiro *et al.*, 1967) and Fe(III)–alanine (Holt *et al.*, 1974), which exhibit similar isomer shifts and quadrupole splittings to those found for ferritin.

## XII. Summary

The potential of the Mössbauer effect as a tool for the investigation of biological macromolecules can best be realized when its capabilities and limitations are understood. As with any advanced technique, in order to extract the wealth of information often present in the Mössbauer spectra of biomolecules, the basic methods of experimentation and data analysis must be conducted at an appropriate level of sophistication. Beyond this, there may be problems with the preparation of the samples, as the small yield of biological compounds extracted from organisms often makes sample enrichment with $^{57}$Fe a necessity. In most cases, this problem can be solved by exchanging the natural iron with $^{57}$Fe using biochemical methods or by growing the organism in a $^{57}$Fe-rich environment.

Although it is true that the biological applications of the Mössbauer effect are essentially limited to iron-containing compounds, this limitation becomes a bonus in that the Mössbauer effect can selectively probe and is sensitive to only the iron, which, after all, is the main idea behind the existence of these biological compounds. Determination of the spin state of the iron atom is probably the most important contribution of the Mössbauer effect to the biologist. Other methods which focus on the iron center such as ESR and paramagnetic susceptibility measurements cannot often distinguish the different iron spin states as unambiguously as the Mössbauer effect, although they may provide corroborating evidence to the interpretation of the Mössbauer spectra. In this respect, the Mössbauer effect is particularly useful to the study of the iron in the diamagnetic state, which is invisible to ESR and susceptibility measurements. Molecules such as the cytochromes, which undergo cyclic changes of states as part of their biological function, can be followed through their entire cycle with the Mössbauer effect, as the iron is always detectable regardless of its spin state.

When more than one iron atoms are present in a single biomolecule, the Mössbauer effect may prove to be useful in determining (a) whether the iron atoms form distinct species and (b) whether there are any interactions between the iron atoms within the molecule. Furthermore, the distribution of the iron and its state within an organism or within parts of an organism may also be investigated with the Mössbauer effect. A necessary condition for this is that sufficient amounts of $^{57}$Fe be present to give rise to an appreciable signal, a situation which is the case with the iron storage compounds. Thus, the Mössbauer effect can be used to determine the content of an organism's iron reservoirs.

The application of the Mössbauer effect to biological systems is almost as old as the effect itself. This long history has permitted the classification of biological compounds in terms of their Mössbauer spectra profiles, which, in most cases, are quantified by parameters according to methods reported in the literature. The success of the unambiguous interpretation of the Mössbauer spectra of the simpler biomolecules has provided the background and confidence to the Mössbauer spectroscopist to attack more complex cases such as cytochrome P450, cytochrome oxidase, and the nitrogenases.

### References

Aasa, R. (1972). *Biochem. Biophys. Res. Commun.* **49**, 806–812.

Afanasev, A. M., and Kagan, Yu. M. (1968). *Sov. Phys.-JETP Lett.* **8**, 382–383.

Aisen, P., Lang, G., and Woodworth, R. (1973). *J. Biol. Chem.* **248**, 649–653.

Blaise, A., Chappert, J., and Giradet, J. (1965). *C. R. Acad. Sci. Paris* **261**, 2310–2313.

Blumberg, W. (1967). *In* "Magnetic Resonance in Biological Systems" (A. Ehrenberg, B. M. Malmström, and T. Vängard, eds.), pp. 119–133, Pergamon, Oxford.

Boas, J. F., and Window, B. (1966). *Aust. J. Phys.* **19**, 573–576.

Bock, J. L., and Lang, G. (1972). *Biochim. Biophys. Acta* **264**, 245–251.

Brady, G. W., Kurkjian, C. R., Lyden, E. F. X., Robin, M. B.., Saltman, P., Spiro, T., and Terzis, A. (1968). *Biochemistry* **7**, 2185–2192.

Clauser, M. J., and Blume, M. (1971). *Phys. Rev. B* **3**, 583–589.

Collins, R. L., and Travis, J. C. (1967). *In* "Mössbauer Effect Methodology" (I. J. Gruverman, ed.), Vol. 3, pp. 123–161. Plenum Press, New York.

Crichton, R. R. (1973). *Struct. Bonding* **17**, 67–134.

David, C. N., and Easterbrook, K. (1971). *J. Cell Biol.* **48**, 15–28.

Dolphin, D., Forman, A., Borg, D. C., Fajer, J., and Felton, R. H. (1971). *Proc. Nat. Acad. Sci. U.S.* **68**, 614–618.

Feeney, R. E., and Komatsu, St. K. (1966). *Struct. Bonding* **1**, 149–206.

Fischbach, F. A., Gregory, D. W., Harrison, P. M., Hoy, T. G., and Williams, J. M. (1971). *J. Ultrastruct. Res.* **37**, 495–503.

Giessen, A. A. (1967). *J. Phys. Chem. Solids* **28**, 343–346.

Holt, E. M., Holt, S. L., Tucker, W. F., Asplund, R. O., and Watson, K. J. (1974). *J. Amer. Chem. Soc.* **96**, 2621–2623.

Hough, E., and Rogers, D. (1974). *Biochim. Biophys. Res. Commun.* **57**, 73–77.

Hyde, B. B., Hodge, A. J., Kahn, A., and Birnstiel, M. L. (1963). *J. Ultrastruct. Res.* **9**, 248–258.

Johnson, C. E. (1966). *Phys. Lett.* **21**, 491–492.

Johnson, C. E. (1967). *Proc. Phys. Soc.* **92**, 748–757.

Johnson, C. E. (1971). *Phys. Today* **24**, 35–40.

Lang, G. (1968). *Phys. Lett.* **26A**, 223–224.

Lang, G., and Marshall, W. (1966). *Proc. Phys. Soc.* **87**, 3–34.

Lang, G., and Oosterhuis, W. T. (1969). *J. Chem. Phys.* **51**, 3608–3614.

Lang, G., Herbert, D., and Yonetani, T. (1968). *J. Chem. Phys.* **49**, 944–950.

Lang, G., Asakura, T., and Yonetani, T. (1969). *J. Phys. C* **2**, 2246–2261.

Lang, G., Aasa, R., Garbett, K., and Williams, R. J. P. (1971). *J. Chem. Phys.* **55**, 4539–4548.

Llinás, M., Wilson, D. M., and Neilands, J. B. (1973). *Biochemistry* **12**, 3836–3843.

Moss, T. H., Ehrenberg, A., and Bearden, A. J. (1969). *Biochemistry* **8**, 4159–4162.

Néel, L. (1962). *J. Phys. Soc. Japan Suppl. BI* **17**, 676–685.

Neilands, J. B. (1966). *Struct. Bonding* **1**, 59–108.

Oosterhuis, W. T. (1917a). *In* "Mössbauer Effect Methodology" (I. J. Gruverman, ed.), Vol. 7, pp. 97–121. Plenum Press, New York.

Oosterhuis, W. T. (1971b). *Phys. Rev. B* **3**, 546–552.

Oosterhuis, W. T., and Barros, F. de S. (1972). *J. Chem. Phys.* **57**, 4304–4307.

Oosterhuis, W. T., and Lang, G. (1969a). *Phys. Rev.* **178**, 439–456.

Oosterhuis, W. T., and Lang, G. (1969b). *J. Chem. Phys.* **50**, 4381–4387.

Oosterhuis, W. T., and Lang, G. (1973). *J. Chem. Phys.* **58**, 4757–4765.

Oosterhuis, W. T. (1974). *Struct. Bonding* **20**, 59–99.

Oosterhuis,.W. T., and Spartalian, K. (1974). *Bull. Amer. Phys. Soc.* **19**, 373.

Oosterhuis, W. T., and Viccaro, P. J. (1972). *Biochim. Biophys. Acta* **264**, 11–16.

Oosterhuis, W. T., Weaver, D. L., and Paez, E. A. (1974). *J. Chem. Phys.* **60**, 1018–1025.

Paez, E. A., Weaver, D. L., and Oosterhuis, W. T. (1972). *J. Chem. Phys.* **57**, 3709–3715.

Ratledge, C. (1971). *Biochem. Biophys. Res. Commun.* **45**, 856–862.

Robards, A. W., and Humpherson, P. G. (1967). *Planta* **76**, 169–178.

Snow, G. A. (1970). *Bacteriol. Rev.* **34**, 99–138.

Spartalian, K., and Oosterhuis, W. T. (1973). *J. Chem. Phys.* **59**, 617–622.

Spartalian, K., Oosterhuis, W. T., and Window, B. (1973). *In* "Mössbauer Effect Methodology" (I. J. Gruverman, ed.), Vol. 8, pp. 137–150. Plenum Press, New York.

Spartalian, K., Smarra, N., and Oosterhuis, W. T. (1974). *AIP Conf. Proc. No. 18* 1326–1329.

Spartalian, K., Oosterhuis, W. T., and Neilands, J. B. (1975). *J. Chem. Phys.* **62**, 3538–3543.

Spiro, G. T., Pape, L., and Saltman, P. (1967). *J. Amer. Chem. Soc.* **89**, 5555–5559.

Tsang, C. P., Boyle, A. J. F., and Morgan, E. H. (1973). *Biochim. Biophys. Acta* **328**, 84–94.

Ure, M. C. D., and Flinn, P. A. (1971). *In* "Mössbauer Effect Methodology" (I. J. Gruverman, ed.), Vol. 7, pp. 245–262. Plenum Press, New York.

Vanderven, N. S., Spartalian, K., Oosterhuis, W. T., and Ashkin, J. (1974). *Bull. Amer. Phys. Soc.* **19**, 373.

Viccaro, P. J., Barros, F. de S., and Oosterhuis, W. T. (1972). *Phys. Rev. B* **5**, 4257–4264.

Webb, J., Multani, J. S., Saltman, P., and Gray, H. B. (1973). *Biochemistry* **12**, 265–267.

Wickman, H. H. (1966). *In* "Mössbauer Effect Methodology" (I. J. Gruverman, ed.), Vol. 2, pp. 39–66. Plenum Press, New York.

Wickman, H. H., and Trozzolo, A. M. (1967). *Symp. Faraday Soc.* **1**, 21–30.

Wickman, H. H., and Wertheim, G. K. (1966). *Phys. Rev.* **148**, 211–217.

Wickman, H. H., Klein, M. P., and Shirley, D. A. (1966). *Phys. Rev.* **152**, 345–357.

Yonetani, T., Scheyer, H., and Ehrenberg, A. (1966). *J. Biol. Chem.* **241**, 3240–3243.

Zabinski, R., Munck, E., Champion, R. M., and Wood, J. M. (1972). *Biochemistry* **11**, 3212–3219.

Zalkin, A., Forrester, J. D., and Templeton, D. H. (1966). *J. Amer. Chem. Soc.* **88**, 1810–1814.

# 6

# Enzyme Systems

**Peter G. Debrunner**

Physics Department
University of Illinois
Urbana, Illinois

## I. Introduction

The chemistry of life is controlled by enzymes, highly specialized proteins that catalyze the biochemical reactions taking place in living cells. To unravel the structure and function of these biocatalysts has long been a challenge for chemists and physicists alike.

The number of different enzymes found in nature is very large; even a simple organism such as a bacterium is estimated to produce 600–800, potentially 2000 different types (Watson, 1970). On the other hand, it is not unusual that bacteria can be induced to synthesize up to $10^4$ copies of the same enzyme molecule. From cultures of such bacteria, the biochemists can, under favorable conditions, isolate an enzyme and prepare it in quantity and high purity. This in turn is a prerequisite for detailed physicochemical studies.

Although proteins are species specific, the existence of common metabolic pathways in all species implies that the enzymes catalyzing them are functionally equivalent. If our limited knowledge can be generalized, it would appear that, as a rule, the active sites of functionally related enzymes are strikingly similar (Stroud, 1974; Dayhoff *et al.*, 1972).

The three-dimensional structures of a number of enzymes and enzyme–substrate complexes have been determined by x-ray diffraction (Dickerson and Geis, 1969; Dayhoff, 1972). Although the conformation of proteins in solution may differ from that in a crystal, the x-ray studies certainly confirm an old postulate of biochemists, namely that enzymes contain stereospecific binding sites for their substrates. This observation provides at least a qualitative explanation for the enormous substrate specificity exhibited by many enzymes.

The most intriguing and least understood question, however, concerns the catalytic mechanism of an enzyme; in other words, the specific sequence of events that lead to the breaking and/or making of a covalent bond. Knowledge of the three-dimensional structure of the catalytic center and of the electronic configuration of the active groups may allow one to guess the reaction mechanism; a complete understanding, though, requires knowledge of all intermediate states and of the dynamic properties of the system.

Many of the known enzymes contain transition-metal atoms (Vallee and Coleman, 1964), either directly bound to the protein or incorporated in a prosthetic group such as the heme complex (iron protoporphyrin IX) (Falk, 1964) or the cobalt complex vitamin $B_{12}$ (Smith, 1965). The iron–sulfur proteins (Lovenberg, 1973) occupy an intermediate position, since their iron atoms bind to an inorganic bridging sulfur, S*, as well as to the sulfur of the amino acid cysteine, which is part of the protein.

Transition metals and their complexes play an important role as industrial catalysts (Halpern, 1968), and their activity in these processes is certainly based on the same properties that make them useful as biocatalysts. The partially filled d orbitals of the transition metals are energetically close to the s and p orbitals, and as a consequence, electrons readily change orbitals and adjust to a variety of different $\sigma$- and $\pi$-bonding schemes. Iron complexes, for instance, can have four, five, six, seven, or eight ligands, and the two normal charge states, ferrous and ferric, commonly exist in a low-spin and high-spin configuration, $Fe^{2+}(3d)^6$, $S = 0$ or $S = 2$, and $Fe^{3+}(3d)^5$, $S = \frac{1}{2}$ or $S = \frac{5}{2}$, respectively. Mössbauer spectroscopy allows differentiation between these four cases, and whenever there is a measurable hyperfine splitting, it provides further information about the electronic ground state of the iron. Knowledge of the symmetry and strength of the hyperfine interactions allows us to draw conclusions about the local environment of the iron, and even though it is not possible to deduce, without further information, the nature and arrangement of the iron ligands, one can frequently arrive at a plausible model that is compatible with all the data.

Mössbauer spectroscopy is ideally suited for the study of iron enzymes, since it measures directly the properties of the active center. Electron paramagnetic resonance (EPR) may be equally informative, but it is limited in practice to systems of half integer spin. Electron nuclear double resonance (ENDOR) is superior to Mössbauer and EPR spectroscopy because it allows accurate resolution of the hyperfine interactions of a paramagnetic center with various nuclei, the nature of which, moreover, can positively be identified. The method consists in monitoring an EPR signal while inducing nuclear resonance transitions with an additional strong radiofrequency field. Under favorable conditions, these nuclear transitions produce a change in the amplitude of the EPR signal. The effect depends critically on the spin relaxation mechanism, however, and the technique is therefore much more restricted in its applications than EPR.

So far, Mössbauer studies of enzymes have been limited to the isotope $^{57}$Fe. The other Mössbauer elements of biological significance (Frieden, 1972; O'Dell and Campbell, 1970) are potassium, the major cellular cation; zinc, which is required in over fifty different enzymes; tin, an essential element of unknown function; and iodine, which is present in the thyroid hormones. The resonances of $^{40}$K and $^{67}$Zn are difficult to observe. The isotopes $^{119}$Sn and $^{127}$I and $^{129}$I, for which the Mössbauer effect is readily observable, have not been studied in an enzyme, although the $^{129}$I resonance has been used to study thyroxine (Groves et al., 1973). Cobalt can be observed indirectly by Mössbauer emission spectroscopy (Nath et al., 1968) using $^{57}$Co, the parent isotope of $^{57}$Fe. Cobalt is the central atom of the important coenzyme vitamin $B_{12}$ and is found in several other enzymes (Vallee and Coleman, 1964).

In the following, we discuss the $^{57}$Fe Mössbauer spectra of three bacterial enzymes that illustrate the three levels of complexation mentioned earlier. In the enzyme of the first type, discussed in Section II, the iron is bound directly to the protein. Section III deals with the second type, an iron–sulfur protein, and Section IV finally treats a heme protein as an example of the third type.

Experimental considerations are discussed by Lang in Chapter 4. Since all Mössbauer measurements are carried out under nonphysiological conditions, usually in frozen solution, it is important to verify that freezing neither damages the samples nor introduces artifacts. Douzou (1975) describes procedures for stabilizing proteins at low temperatures.

## II. Protocatechuic Acid 4,5-Oxygenase

### A. Native Enzyme

Protocatechuic acid 4,5-oxygenase (PCAO) (Dagley and Patel, 1957) is isolated from the bacterium *Pseudomonas testosteroni* and catalyzes the incorporation of molecular oxygen into the substrate, protocatechuic acid, according to the formula

PCAO has a molecular weight of 140,000 Daltons and contains four iron atoms per molecule, which are bound directly to the protein. The reaction mechanism is not understood yet, and about the role of the iron in particular little else is known other than that it is required for enzymatic activity. Below we discuss the Mössbauer studies of Zabinski *et al.* (1972); the unexpected results of this work should stimulate further research.

The enzyme used in these experiments was enriched in $^{57}Fe$ by first removing the iron from the protein and then reconstituting the resultant apoenzyme with $^{57}Fe^{2+}$ ions. Interestingly enough, the Mössbauer spectrum of the enzyme prepared in this way (Fig. 1) consists of two broad peaks only with a separation of approximately 0.6 mm/sec and a center of gravity at 100°K of $\delta_{Fe} = 0.51$ mm/sec[†]. Since high-spin ferrous ions were used in the reconstitution, and since typical amino acid side groups that might coordinate to the iron are weak-field ligands, one should expect to observe the large quadrupole splitting, ($\Delta E_Q \gtrsim 2$ mm/sec) and isomer shift ($\delta_{Fe} \gtrsim 0.6$ mm/sec) of high-spin ferrous iron. The spectrum of Fig. 1 rules out such an assignment, and it furthermore implies that the state of the iron in the four binding sites is quite similar. Magnetic susceptibility measurements showed that the native enzyme is diamagnetic, a result that might indicate a low-spin ferrous, $S = 0$, state for all four iron atoms. The authors offer another explanation, however, suggesting that all iron atoms are in the high-spin ferric state, $S = \frac{5}{2}$, but pairs of iron atoms are supposed to couple antiferromagnetically to a resultant spin of zero. Such pairs are known to exist in the nonheme oxygen carrier hemerythrin (Dawson *et al.*, 1972) and in many of the four-coordinated iron–sulfur proteins (Dunham *et al.*, 1971). The strongest argument in favor of this hypothesis is the easy exchange of

---

[†] All isomer shifts, $\delta_{Fe}$, are quoted with respect to metallic iron.

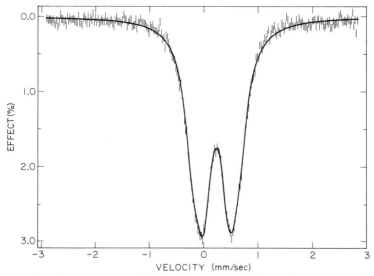

**Fig. 1.** Mössbauer spectrum of PCA 4,5-oxygenase in its resting state at 110°K. The solid line is the result of a least-squares fit assuming four lines of equal intensity and $\Gamma = 0.35$ mm/sec FWHM. A source of $^{57}Co$ in copper was used. (From Zabinski et al., 1972.)

the iron in PCAO which is typical of ionic, high-spin rather than of covalent, low-spin complexes. Further research is required, however, to settle the questions of solubility and dissociation constants for ferrous and ferric ions (Neilands, 1966) and of the implied oxidation $Fe^{2+} \rightarrow Fe^{3+}$, in PCAO.

### B. Inhibitor Complex

In the presence of protocatechuic acid or certain inhibitors, the native diamagnetic enzyme changes into a paramagnetic species with an EPR resonance at $g = 4.3$, which indicates that the iron is in the high-spin ferric state (Aasa, 1970). Figure 2 shows the Mössbauer spectrum of an inhibitor complex measured at 4.2°K in a small transverse magnetic field. An intricate pattern of magnetic splittings with an overall width of 17 mm/sec is observed, but in spite of its complexity, the spectrum is well reproduced by the computer simulation shown as a solid line in Fig. 2. The model used in the simulation is based on earlier work by Wickman et al. (1965) and has been applied to similar spectra by Lang et al. (1971).

The magnetic properties of high-spin ferric iron can be interpreted in terms of the spin Hamiltonian $\mathscr{H}_S$

$$\mathscr{H}_S = D(S_z^2 - \tfrac{1}{3}S(S+1)) + E(S_x^2 - S_y^2) + g_0\beta\mathbf{H}\cdot\mathbf{S} \qquad (S = \tfrac{5}{2}). \qquad (1)$$

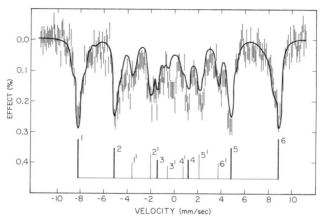

**Fig. 2.** Mössbauer spectrum of the protocatechuic aldehyde complex of PCA 4,5-oxygenase at 4.2°K in a perpendicular field of 150 G. The solid line is a computer simulation using the parameters $\Gamma = 0.35$ mm/sec, $\delta = 0.5$ mm/sec, $A_0 = -30.2$ MHz, $|\Delta E_Q| = 0.55$ mm/sec, $\eta = 1$, $|D| = 0.7$ cm$^{-1}$, $E/D = \frac{1}{3}$. The stick diagram with lines labeled 1 to 6 refers to the ground state Kramers doublet, while the lines 1' to 6' refer to the intermediate doublet with isotropic g tensor, $g = 4.29$. (From Zabinski *et al.*, 1972.)

The zero-field splitting represented by the first two terms gives rise to three Kramers doublets. These may be further split by the Zeeman interaction, $g_0\beta\mathbf{H} \cdot \mathbf{S}$, where $g_0 \cong 2$ is the electron spin $g$ factor, $\beta$ is the Bohr magneton and $\mathbf{H}$ is the applied field. An EPR signal at $g = 4.3$ is observed for $E/D \cong \frac{1}{3}$, $D \gg \beta H$, and this condition completely specifies the eigenstates of Eq. (1). The Mössbauer spectra can be derived from the nuclear Hamiltonian $\mathscr{H}_N$,

$$\mathscr{H}_N = A_0\mathbf{S} \cdot \mathbf{I} + (\tfrac{1}{4}eQV_{zz})(I_z^2 + (I_x^2 - I_y^2)\tfrac{1}{3}\eta - \tfrac{5}{4}) - g_N\beta_N\mathbf{H} \cdot \mathbf{I}, \quad (2)$$

where $S$ equals $\frac{5}{2}$ and all the symbols have their usual meaning. As long as the hyperfine interaction—$A_0\mathbf{S} \cdot \mathbf{I}$ of Eq. (2)—is much smaller than the electronic Zeeman term—$g_0\beta\mathbf{H} \cdot \mathbf{S}$ of Eq. (1)—the electron spin $S$ is decoupled from the nuclear spin $I$. To a good approximation the expectation values $\langle\mathbf{S}\rangle_i$ of $\mathbf{S}$ calculated for the six eigenstates of Eq. (1) can then be substituted for the operator $\mathbf{S}$ in Eq. (2). With this substitution, the nuclear Hamiltonian (Eq. 2) depends on the nuclear spin $I$ and the parameters $\langle\mathbf{S}\rangle_i$, $A_0$, $V_{zz}$, and $\eta$ only, and the Mössbauer spectrum can be calculated following standard procedures (Münck *et al.*, 1973). This simplification is a major reason that Mössbauer studies of paramagnetic species are usually performed in a magnetic field. Since the expectation values $\langle\mathbf{S}\rangle_i$ are different for the three Kramers doublets, the magnetic hyperfine splittings observed in the Mössbauer spectrum will be different, too. This effect is clearly visible in Fig. 2, where the lines marked 1–6 are due to the lower and upper Kramers doublets, which give almost identical spectra as long as $E/D \approx \frac{1}{3}$, whereas the lines marked $1'-6'$ are due to the intermediate doublet.

Using the method of analysis outlined above, Zabinski *et al.* (1972) draw the following conclusions from the Mössbauer spectra of the inhibitor complex. Within the limited resolution of the experiment no differences are discernible among the four iron atoms. They are all in a high-spin ferric state with $E/D \approx \frac{1}{3}$, $|D| = 0.7 \pm 0.3$ cm$^{-1}$, and a large hyperfine constant, $A_0 = -30.2$ MHz, which indicates a highly ionic character. The fact that the quadrupole splitting and isomer shift values are very close to those of the native enzyme supports the assumption that the latter contains pairs of spin-coupled ferric ions. Substrate or inhibitor binding thus leaves the iron in the same charge and spin state, but it breaks the exchange interaction that leads to antiferromagnetic spin coupling in the native enzyme.

## III. Nitrogenase

### A. Previous Studies

Several microorganisms are capable of reducing atmospheric nitrogen to ammonia according to the reaction

$$N_2 + 6H^+ + 6e^- \rightarrow 2NH_3.$$

As far as we know, the enzyme system catalyzing this reaction, a nitrogenase, is basically the same in all nitrogen-fixing bacteria. Great progress has been made over the last few years in the preparation of this enzyme (Burris, 1971), and as will be shown below, Mössbauer spectroscopy has made some important contributions to its characterization.

Nitrogenase consists of two parts, a molybdenum–iron–protein, molybdoferredoxin, of molecular weight $\sim 250,000$ Daltons and a 4Fe–4S*–protein, azoferredoxin. Together, the two proteins are biologically active if they are supplied with ATP (adenosine 5′-triphosphate), $Mg^{2+}$, and a reducing agent. Molybdoferredoxin contains two molybdenum atoms, approximately twenty-four iron atoms, and an equal number of acid-labile sulfur atoms, S*. It consists of two identical subunits, each of which may again be made up of smaller parts.

The first detailed Mössbauer measurements (Kelly and Lang, 1970; Smith and Lang, 1974) were done with the nitrogenase from *Klebsiella pneumoniae*. In this work, the molybdoferredoxin as well as the azoferredoxin and some of their complexes were studied. In the following, we will discuss complementary results of Münck *et al.* (1975) obtained with the molybdoferredoxin of the nitrogenase from *Azotobacter vinelandii*.

## B. Molybdoferredoxin

The samples studied by Münck *et al.* (1975) were prepared under strictly anaerobic conditions from bacteria grown on an $^{57}$Fe-enriched medium. As might be expected in a protein with twenty-four iron atoms, the Mössbauer spectra of molybdoferredoxin are quite complex, Figs. 3–6. By judicious use

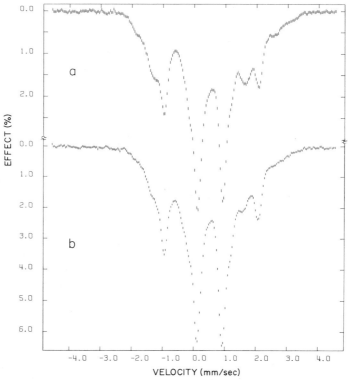

**Fig. 3.** Mössbauer spectra of the molybdenum–iron component, molybdoferredoxin, of the nitrogenase from *Azotobacter vinelandii* at 1.5°K. A magnetic field $H$ of 360 G was applied (a) parallel and (b) perpendicular to the beam. (From Münck *et al.*, 1975.)

of applied magnetic fields and of different temperatures, it is possible to divide the spectra into only four components, which can be assigned to distinct iron complexes.

It should be pointed out that the 1:1 atomic ratio of iron and inorganic sulfur, S*, in molybdoferredoxin is reminescent of the 2Fe–2S* (Dunham *et al.*, 1971; Münck *et al.*, 1972) and 4Fe–4S* clusters (Gersonde *et al.*, 1974) known to exist in the ferredoxins. All these clusters show antiferromagnetic

spin coupling, and frequently the Mössbauer spectra arising from the individual iron atoms are indistinguishable. It is to be expected, therefore, that
similar clusters exist in molybdoferredoxin. Nevertheless, the Mössbauer
spectrum taken at 30°K (Fig. 4) is surprisingly simple. Using a least-squares
fit, it can be decomposed quantitatively into four quadrupole doublets that
are labeled, in order of decreasing intensity, D (9–11 atoms), $M_{EPR}$ (8–10
atoms) $Fe^{2+}$ (3–4 atoms) and S (1 atom). The parameters derived from the
fit are listed in Table I.

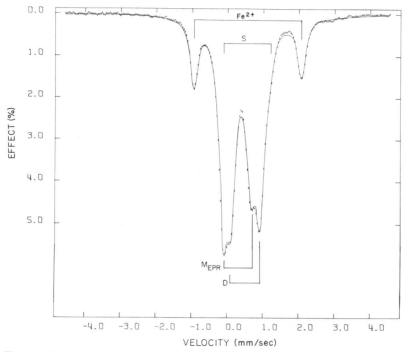

**Fig. 4.**  Mössbauer spectrum of the sample of Fig. 3 at 30°K. The solid line represents a
least-squares fit to the data using four doublets, labeled $Fe^{2+}$, S, $M_{EPR}$, and D. The two lines
of each doublet were constrained to have equal areas. (From Münck *et al.*, 1975.)

Doublet D [identical with species M5 of Smith and Lang (1974)] has a
quadrupole splitting that is practically independent of temperature; measurements in an applied field up to 13.3 kG indicate that the complex is
diamagnetic. The Mössbauer parameters of D are compatible with low-spin
ferrous iron, but Münck *et al.* (1975) do not rule out the possibility of spin-
coupled clusters.

The next component, $M_{EPR}$, is associated with an EPR-active center and
will be discussed in detail in the next section.

**TABLE I**

Mössbauer Parameters of the Four Quadrupole Doublets Observed in the Native Molybdoferredoxin from *Azotobacter vinelandii* at 30°K (Fig. 4).[a]

| Spectral component | $\Delta E_Q$ (mm/sec) | $\delta_{Fe}$ (mm/sec) | Fractional area |
|---|---|---|---|
| D | 0.81 ± 0.03 | 0.64 ± 0.03 | 0.425 ± 0.03 |
| $M_{EPR}$ | 0.76 ± 0.03 | 0.40 ± 0.03 | 0.385 ± 0.03 |
| $Fe^{2+}$ | 3.02 ± 0.02 | 0.69 ± 0.02 | 0.14 ± 0.01 |
| S | 1.4 ± 0.1 | 0.6 ± 0.1 | 0.05 |

[a] From Münck *et al.* (1975).

The species labeled $Fe^{2+}$ [M4 of Smith and Lang (1974)] strongly resembles reduced rubredoxin in its isomer shift and quadrupole splitting, but in contrast to the latter, it behaves like a diamagnetic complex. The $Fe^{2+}$ species, therefore, must arise from spin-coupled, sulfur-coordinated high-spin ferrous iron.

The small component S may be due to a low-spin ferrous impurity, but seems to be present in the molybdenum–iron–protein from *Klebsiella pneumoniae* as well.

## C. The EPR-Active Center

The most interesting species is the one labeled $M_{EPR}$. At 1.5°K a magnetically split spectrum is observed (Fig. 3) in place of the quadrupole doublet seen at 30°K (Fig. 4). The characteristic pattern of the magnetic hyperfine interaction is clearly discernible in Fig. 5, which was obtained from Fig. 3 by subtraction of the doublets D, $Fe^{2+}$, and S.

It has been known for some time that at low temperature molybdoferredoxin shows an EPR resonance with g tensor (3.65, 4.32, 2.01) (Orme-Johnson *et al.*, 1972). Hyperfine broadening of the signal at $g = 2.01$ observed in $^{57}Fe$-enriched samples proves that the iron is involved in the paramagnetic center.

In the presence of azoferredoxin, Mg-ATP, and a reductant, the EPR signal of molybdoferredoxin decreases by an order of magnitude, and is restored to its original intensity when the reductant is exhausted. The reduction in the intensity of the EPR-active center has been monitored also by Mössbauer spectroscopy, and the result is shown in Fig. 6. The magnetically split component ($M_{EPR}$ of Figs. 3 and 5) has practically disappeared, and in its place a broad quadrupole doublet with a splitting of 0.95–1.3 mm/sec and a slightly more positive isomer shift is found. In a strong magnetic

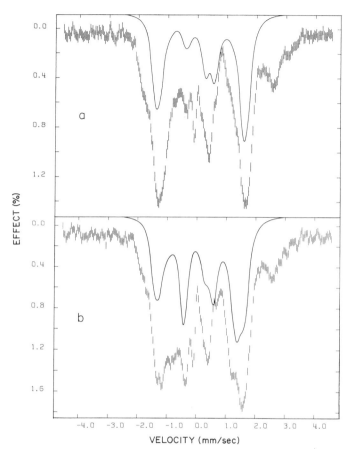

**Fig. 5.** Mössbauer spectra of the paramagnetic component of molybdoferredoxin at 1.5°K in a field of 360 G, applied (a) parallel and (b) perpendicular to the $\gamma$ rays. Figure 5 was obtained from Fig. 3 by subtracting the doublets $Fe^{2+}$, D, and S as determined by computer fitting the data of Fig. 4. (From Münck *et al.*, 1975.)

field, the spectrum broadens considerably, indicating that the complex is still paramagnetic. The original spectrum reappears when the reducing agent is exhausted. There can be no doubt, therefore, that the EPR signal of the native protein arises from the component $M_{EPR}$ seen in the Mössbauer spectrum.

The presence at low temperature of an EPR signal and of magnetic hyperfine interaction suggests a complex of half integer spin; at higher temperatures the magnetic splitting of this complex averages out to zero due to rapid spin relaxation, and the EPR resonance vanishes. During nitrogen

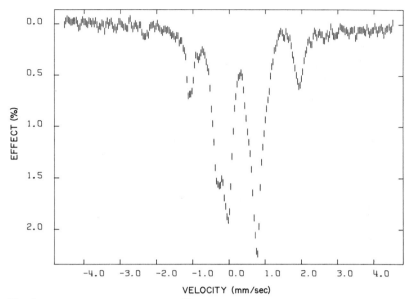

**Fig. 6.** Mössbauer spectrum of the ($^{57}$Fe-enriched) molybdoferredoxin in the presence of unenriched azoferredoxin and the fixing mixture (Mg-ATP and $Na_2S_2O_4$). The measurement was done at 4.2°K in a field of 360 G perpendicular to the $\gamma$ ray beam. (From Münck *et al.*, 1975.)

fixation, most of the component $M_{EPR}$ exists in a reduced state of integer spin; it is a paramagnetic, non-Kramers complex and therefore shows neither an EPR signal nor, in the absence of an applied field, any magnetic hyperfine interaction in the Mössbauer spectrum.

According to Palmer *et al.* (1972), the unusual *g* values of native molybdoferredoxin may arise from a spin quartet, $S = \frac{3}{2}$, in a spin Hamiltonian of the form of Eq. (1). The first two terms split the quartet into two Kramers doublets separated in energy by $\Delta = 2D[1 + 3(E/D)^2]^{1/2} \gg g_0\beta H$. With a value of $E/D = 0.055$, the model reproduces the experimentally observed *g* values for the lower doublet and predicts *g* values for the upper doublet that preclude experimental observation. Taking into account the depopulation of the lower doublet with temperature, the intensity of the EPR signal was estimated to correspond to a spin concentration of 0.91 times the molybdenum concentration, and $\Delta$ was found to be $\Delta = 7.5$ cm$^{-1}$ (Münck *et al.*, 1975). These results suggest that there are two EPR centers per molecule, each center associated with four iron atoms coupled to a spin $S = \frac{3}{2}$. If the spin coupling scheme were known, the magnetic hyperfine spectrum shown in Fig. 5 could be simulated. The solid line in Fig. 5 represents a first step in that direction. An isotropic hyperfine interaction, $A_0\mathbf{S} \cdot \mathbf{I}$, was assumed as it is typically found in high-spin ferric compounds. With a value $A_0 = 12.1$ MHz and a quadrupole interaction $\Delta E_Q = 0.78$ mm/sec, $\eta = 0.6$, the simulations shown in Fig. 5 were obtained. They reproduce certain features of the spectra, but the need for some other component is obvious.

To summarize, Mössbauer spectroscopy has allowed us to identify four distinct species of iron in molybdoferredoxin. Three of them very likely represent spin-coupled iron–sulfur clusters of the ferredoxin type. One of these is identical with the EPR center observed earlier; so far, it is the only component for which a redox function has been demonstrated in the enzymatic cycle. It is very likely, though, that the other iron clusters are involved in electron transfer, too, and if suitable conditions can be found to stabilize them in the reduced state, more can be learned about their spin configuration from EPR and Mössbauer spectroscopy.

## IV. Cytochrome P450

### A. Relation to Other Heme Proteins

Many of the known heme proteins are specialized for one of three functions, which are electron transfer (e.g., cytochrome c), reversible oxygen binding (e.g., hemoglobin, myoglobin), or catalysis (e.g., catalase, peroxidase). Cytochrome P450 is a particularly interesting example because its reaction sequence, shown in Fig. 7, combines all three functions.

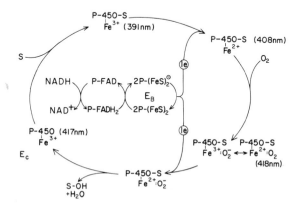

**Fig. 7.** Proposed reaction mechanism for the hydroxylation of camphor in *Pseudomonas putida*. NADH (NAD⁺) is a biological reducing agent in its reduced (oxidized) state, P-FAD is a flavoprotein, P-(FeS)₂ is the 2Fe–2S* protein putidaredoxin, S and S-OH represent the substrate D-camphor and product 5-*exo*-hydroxycamphor, respectively.

The structure of the heme group (iron protoporphyrin IX) has long been known (Falk, 1964); it is a large, conjugated ring with a central iron atom coordinated to four pyrrole nitrogens. The protein, wrapping around the heme group, usually provides a fifth, axial ligand to the iron such as a histidine nitrogen in myoglobin and hemoglobin. It is generally believed (Hoard, 1971) that five-coordinated iron is in the high-spin state and that it is located between the heme plane and the fifth ligand. Six-coordination, on the other hand, would imply essentially in-plane location and would lead to a low-spin state. Either an amino acid side group or a small molecule like $O_2$ could bind as a sixth ligand. In the case of cytochrome P450, it is not known yet what residues are coordinated to the heme iron. It is clear, though, that the heme group is at the active center of the enzyme and reflects the changes taking place in the course of the reaction cycle. In fact, all the steps indicated in Fig. 7 are accompanied by changes in the optical, EPR, and Mössbauer spectra.

Cytochromes of the P450 type are found in many different organisms from bacteria to mammals (Hayaishi, 1974). Their name derives from a peculiar optical absorption band near 450 nm that they exhibit in the reduced, carbon monoxide-complexed, state. All of them are monoxygenases, which means that they incorporate one atom of oxygen derived from atmospheric dioxygen, while they reduce the second oxygen atom to water. Their specific functions are varied, ranging from steroid metabolism in the adrenal cortex to drug detoxification in liver microsomes (Omura *et al.*, 1966). In mammals these enzymes are membrane-bound, and harsh methods are needed to solubilize them. The subject of the following discussion is a bacterial cytochrome P450, however, which resembles the mammalian enzymes very closely but has the great advantage of being soluble and more easily accessible. This model system is the camphor hydroxylase, cytochrome $P450_{CAM}$, which is isolated from the bacterium *Pseudomonas putida* (Hedegaard and Gunsalus, 1965; Gunsalus *et al.*, 1974).

As shown in Fig. 7 the complete system consists of three proteins, a flavoprotein P-FAD, an iron–sulfur protein $P-(FeS)_2$ or $E_B$, and the heme protein cytochrome $P450_{CAM}$.[†] The first two components transfer electrons from the biological reducing agent NADH to the heme protein. Cytochrome P450 is the actual catalyst; it binds a molecule of the substrate, D-camphor, then accepts one electron, next binds molecular oxygen, and after a further reduction step, catalyzes the hydroxylation of the substrate and releases the product, the 5-exo alcohol of camphor. All the reaction intermediates shown in Fig. 7 except for the last one have been studied by Mössbauer spectroscopy, and it will be demonstrated below that the results provide a very detailed

[†] For the rest of this section the subscript CAM will be dropped.

picture of the changes taking place at the heme iron (Sharrock *et al.*, 1973; Sharrock, 1973). The samples used in these experiments were obtained from bacteria grown on an $^{57}$Fe-enriched medium.

### B. *Cytochrome P450 in Its Oxidized States*

In the resting state of cytochrome P450, the heme iron is in the low-spin ferric state, with spin $S = \frac{1}{2}$ and a g tensor $(g_x, g_y, g_z) = (1.91, 2.26, 2.45)$ (Tsai *et al.*, 1970). Its low-temperature Mössbauer spectra measured in a weak applied field are shown in Fig. 8. They were analyzed on the basis of a model originally suggested by Griffith (1957) to explain the g tensors of low-spin ferric heme proteins. Oosterhuis and Lang (1969) expanded it to account for Mössbauer spectra. In this model the iron is supposed to have

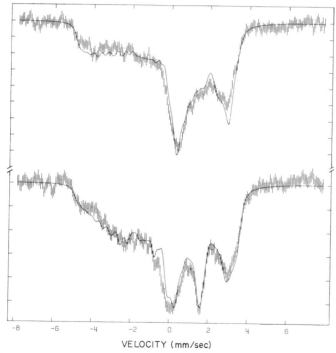

VELOCITY (mm/sec)

**Fig. 8.**  Mössbauer spectra of substrate-free oxidized cytochrome P450 at 4.2°K in a small applied field parallel (top) and perpendicular (bottom) to the $\gamma$-ray beam. The solid lines are simulations obtained with the following parameters: $(g_x, g_y, g_z) = (1.91, 2.26, 2.45)$; $(A_x, A_y, A_z)/g_N\beta_N = (-450, 102, 191)\,\mathrm{kG}$; $\frac{1}{4}eQV_{zz} = 0.99\,\mathrm{mm/sec}$; $\eta = -1.8$; $\delta_{\mathrm{Fe}} = 0.38\,\mathrm{mm/sec}$; linewidth (full width at half maximum) $\Gamma = 0.3\,\mathrm{mm/sec}$. Note that $V_{xx}$ exceeds $V_{zz}$ in magnitude, $\frac{1}{4}eQV_{xx} = -1.39\,\mathrm{mm/sec}$. (After Sharrock, 1973.)

the $(t_{2g})^5 \, ^2T_{2g}$ ground state configuration. A strong crystal field potential with components of tetragonal and rhombic symmetry combined with the spin–orbit interaction gives rise to three widely separated Kramers doublets, the wavefunctions of which can be calculated from the electronic g tensor. These wavefunctions were used to calculate the spectra shown in Fig. 8. To obtain a satisfactory simulation, the delocalization of the various $t_{2g}$ orbitals had to be taken into account. Relative to the principal axis of the g tensor, the orbital of lowest energy $d_{xy}$ has a covalency factor of $0.96 \pm 0.02$, whereas the two higher orbitals $d_{xz}$ and $d_{yz}$ were assigned a covalency factor of $0.83 \pm 0.02$ (Sharrock, 1973).

Upon addition of the substrate camphor, cytochrome P450 is partially converted to the high-spin ferric state with $S = \frac{5}{2}$. The low-temperature Mössbauer spectra are quite complex, being a superposition of two magnetically split spectra, but near 200°K, the magnetic hyperfine interaction averages out to zero due to rapid spin relaxation, and two pairs of broad lines emerge (Fig. 9a). Since low-spin ferric iron is known to have a larger quadrupole splitting than high-spin $Fe^{3+}$, the outer pair of lines must be assigned to $S = \frac{1}{2}$, the inner pair to $S = \frac{5}{2}$. The low-temperature Mössbauer spectra of the high-spin ferric component can be calculated as outlined in Section II,B. The quadrupole splitting $\Delta E_Q$ is obtained from spectra like Fig. 9a, and the g tensor, $g = (8, 4, 1.8)$ is known from EPR measurements of Tsai et al. (1970). The only additional parameters required are the hyperfine constant $A_0$ and the asymmetry parameter $\eta$ of Eq. (2). From extensive simulations, Sharrock (1973) found the values $A_0 = 22.8 \pm 0.2$ MHz, $\eta = 0.7 \pm 0.1$. All these parameters of cytochrome P450 are quite unusual for a high-spin ferric heme protein; they indicate a very strong deviation from axial symmetry and considerable delocalization of the 3d electrons.

### C. Spin Transitions—Cytochrome P450 and Chloroperoxidase

According to the proposed reaction mechanism of cytochrome P450 (Fig. 7), the iron–sulfur protein $E_B$ interacts with the cytochrome at two points of the enzymatic cycle. In fact, the two proteins form a well-defined complex, as demonstrated by chemical, Mössbauer, and other spectroscopic methods.

To study the effect of complex formation on the active centers of either $E_B$ or cytochrome P450, the two proteins are selectively enriched in $^{56}Fe$ and $^{57}Fe$, respectively. Figure 9 shows the changes in the Mössbauer spectrum of [$^{57}Fe$]cytochrome P450 in the presence of [$^{56}Fe$]$E_B$. A comparison of the spectra a, b, and c of Fig. 9 indicates that the low-spin fraction is larger in the $E_B$ complex, and that it decreases with increasing temperature.

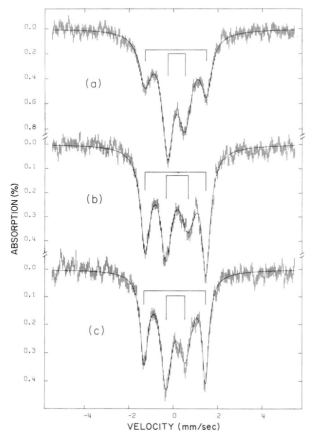

**Fig. 9.** Mössbauer spectrum of oxidized cytochrome P450 in the presence of a saturating concentration of camphor (a) measured at 193°K. Spectra (b) and (c) are from the same sample as (a) after addition of oxidized effector protein $^{56}$Fe-putidaredoxin, measured at (b) 197°K and (c) 233°K, respectively. The solid curves represent least-squares fits with two pairs of Lorentzians each, both lines of a pair having equal areas; the vertical lines indicate the positions and relative intensities of the doublets. (From Sharrock, 1973.)

The mechanism and significance of this partial spin transition is not clear yet; it is interesting to note, though, that the camphor complex with a standard potential of $E_0' = -170$ mV is more easily reduced than the substrate-free enzyme with $E_0' = -270$ mV. The standard potential of the electron transfer and effector protein $E_B$ is $E_0' = -240$ mV (Gunsalus et al., 1974). Incidentally, the Mössbauer spectrum of oxidized iron–sulfur protein $E_B$ (Münck et al., 1972) is not noticeably affected by complex formation with cytochrome P450 (Sharrock, 1973).

Temperature-dependent spin transitions are found in other ferric heme proteins also, as the example of Fig. 10 illustrates. Figure 10 shows the Mössbauer spectra of chloroperoxidase (Champion *et al.*, 1973) measured at different temperatures, and if the spectra are assumed to consist of two quadrupole doublets, the fractional area of the outer doublet is found to

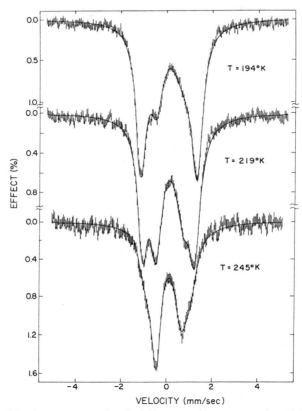

**Fig. 10.** Mössbauer spectra of native chloroperoxidase measured at 194°K, 219°K, and 245°K. The solid curves were obtained by least-squares fitting two pairs of Lorentzians to the data. (From Champion *et al.*, 1973.)

vary with temperature as plotted in the lower half of Fig. 11. Chloroperoxidase is a heme enzyme from the mold *Caldariomyces fumago* (Morris and Hager, 1966); it is functionally unrelated to cytochrome P450, but nevertheless has striking similarities with the latter in its optical, EPR, and Mössbauer spectra.

It should be pointed out that not only the spin fraction of chloroperoxidase depends on temperature, but also the quadrupole splitting $\Delta E_Q$ of the low-

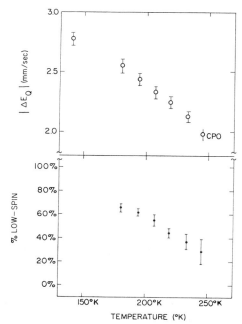

**Fig. 11.**   The quadrupole splitting $|\Delta E_Q|$ (upper part) and the low-spin fraction (lower part) of native chloroperoxidase as a function of temperature. (After Champion *et al.*, 1973.)

spin component (upper half of Fig. 11). Since the standard model (Oosterhuis and Lang, 1969) predicts a much smaller temperature dependence of $\Delta E_Q$, some other mechanism has to be invoked to explain the data, such as a reversible change in conformation or rearrangement of water molecules.

## D. Reduced States—Comparison with Other Heme Proteins

The Mössbauer spectrum of reduced, camphor-complexed cytochrome P450 consists of a single quadrupole doublet with practically temperature-independent splitting. The positive isomer shift, $\delta_{Fe} = 0.82 \pm 0.02$ mm/sec at $4.2°K$, and the size of the splitting, $\Delta E_Q = 2.42 \pm 0.02$ mm/sec at $4.2°K$, characterize the heme as a high-spin ferrous complex, the analog of deoxyhemoglobin. The quadrupole splitting of the latter, however, has a strong temperature dependence (Eicher and Trautwein, 1969), which arises from thermal population of low-lying excited states of the ferrous ion.

Champion *et al.* (1975a) measured low-temperature Mössbauer spectra of reduced cytochrome P450 in strong magnetic fields and observed a well-resolved structure (Figs. 12, 13). Again, deoxyhemoglobin shows an

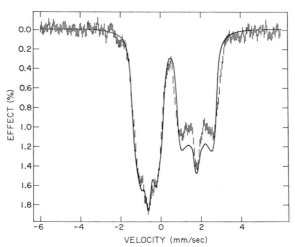

**Fig. 12.** Mössbauer spectrum of the reduced camphor complex of cytochrome P450 at 4.2°K in a field of 8.6 kG parallel to the transmitted beam. The solid curve is a simulation based on the following parameters, $D = 20°K$, $E/D = 0.15$, $(g_x, g_y, g_z) = (2.24, 2.32, 2.00)$; $(A_x, A_y, A_z)/g_N\beta_N = (-180, -125, -150)$ kG; $\frac{1}{4}eQV_{\zeta\zeta} = 1.1$ mm/sec; $\eta = 0.8$; $\delta_{Fe} = 0.82$ mm/sec; linewidth (full width at half maximum) $\Gamma = 0.25$ mm/sec. The principal axes $(\xi, \eta, \zeta)$ of the electric field gradient tensor are rotated with respect to the $(x, y, z)$ frame by the Euler angles $\varphi = 60°$, $\theta = 70°$, $\psi = 0°$. (From Champion *et al.*, 1975a)

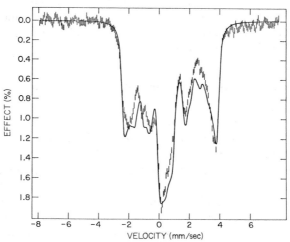

**Fig. 13.** Mössbauer spectrum of the reduced camphor complex of cytochrome P450 at 4.2°K in a parallel field of 25 kG. The simulation shown as a solid line is based on the parameters listed in Fig. 12. (From Champion *et al.*, 1975a)

entirely different behavior, its Mössbauer spectra, measured under the same conditions, are broad and featureless (Lang and Marshall, 1966). Using a spin Hamiltonian with $S = 2$ and assuming fast relaxation, Champion et al., (1975a) were able to parametrize the spectra, and the simulations shown in Figs. 12 and 13 prove that their approach was successful.

The spin Hamiltonian is obtained from Eq. (1) by generalizing the Zeeman term to read $\beta \mathbf{S} \cdot \mathbf{g} \cdot \mathbf{H}$, where g is the g tensor,

$$\mathscr{H}_S = D(S_z^2 - \tfrac{1}{3}S(S + 1)) + E(S_x^2 - S_y^2) + \beta \mathbf{S} \cdot \mathbf{g} \cdot \mathbf{H} \qquad (S = 2). \qquad (3)$$

Since the electron spin relaxation is assumed to be fast with respect to nuclear precession times, the magnetic hyperfine interaction is given by $\langle \mathbf{S} \rangle_T \cdot \mathbf{A} \cdot \mathbf{I}$, where $\langle \mathbf{S} \rangle_T$ is the spin expectation value thermally averaged over the five eigenstates of Eq. (3). The nuclear Hamiltonian then is

$$\mathscr{H}_N = \langle \mathbf{S} \rangle_T \cdot \mathbf{A} \cdot \mathbf{I} + (\tfrac{1}{4}eQV_{\zeta\zeta})(I_\zeta^2 + (I_\xi^2 - I_\eta^2)\tfrac{1}{3}\eta - \tfrac{5}{4}) - g_N\beta_N\mathbf{H} \cdot \mathbf{I}. \qquad (4)$$

The form of the quadrupole interaction in Eq. (4) explicitly shows that its principal axes need not coincide with those of the zero-field splitting.

The number of adjustable parameters in Eqs. (3) and (4) is quite large, and to find a solution that reproduces the data (Champion et al., 1975a) is a formidable problem. It is clear, though, that high-field low-temperature Mössbauer spectroscopy is an ideal probe for high-spin ferrous iron, because it provides more information than any other method. Cytochrome P450 is the first ferrous heme protein to be analyzed on the basis of the spin Hamiltonian formalism; Eq. (3) has also been used to interpret susceptibility measurements on deoxyhemoglobin (Nakano et al., 1972), yielding a value of $D \simeq 7.5°$K. In order to arrive at a consistent set of parameters for cytochrome P450, Champion et al., (1975a) analyzed spectra taken under many different conditions of temperature and applied field. The best set obtained in this way was used for the simulations displayed in Figs. 12 and 13; the parameter values are listed in the capture of Fig. 12. Some of the results are quite unexpected and deserve further comment.

The zero-field splitting, $D = 20°$K, is large in comparison with the value $D = 7.5°$K reported for hemoglobin (Nakano et al., 1972). The meaning of the parameter $D$ appears to be different in the case of hemoglobin, however, since an analysis of the temperature-dependent quadrupole splitting $\Delta E_Q$ of the latter (Huynh et al., 1974) indicates the presence of low lying levels of singlet and triplet character. Cytochrome P450 has a much weaker temperature dependence of $\Delta E_Q$, and any excited state must be well separated in energy from the $S = 2$ ground state described by Eq. (3).

The symmetry at the iron site is very low; model calculations indicate that a triclinic crystal field potential is required to explain the data (Champion et al., 1975a). The asymmetry parameter is close to unity, $\eta = 0.8$, and the largest component of the electric-field gradient $V_{\zeta\zeta}$ is almost perpendicular to the largest component of the zero-field splitting $D$. The orientation of the electric-field gradient relative to the heme plane is not known yet for P450, but it should be mentioned that Gonser *et al.* (1974) found the largest component $V_{\zeta\zeta}$ for myoglobin to lie essentially in the heme plane.

As noted earlier, cytochrome P450 in the oxidized state shares many spectral features with the functionally unrelated heme enzyme chloroperoxidase. Even stronger similarities are found in the optical and Mössbauer spectra of the reduced proteins. This point is illustrated in Fig. 14, which

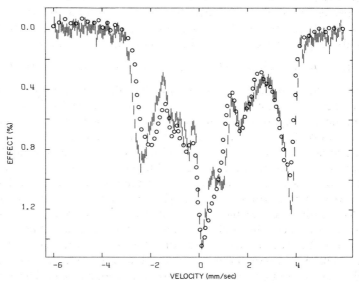

**Fig. 14.** Mössbauer spectrum of reduced chloroperoxidase at 4.2°K in a field of 25 kG parallel to the transmitted beam. For comparison, the spectrum of cytochrome P450 (Fig. 13) measured under identical conditions, has been superimposed (circles). (From Champion *et al.*, 1975b.)

shows a superposition of the almost identical high-field low-temperature spectra of P450 and chloroperoxidase. The fact that other heme proteins have totally different spectra under the same conditions suggests that the heme irons in ferrous P450 and chloroperoxidase have the same ligand geometry (Champion *et al.*, 1975b).

In the enzymatic cycle of cytochrome P450, reduction to the ferrous state is followed by oxygen binding. The Mössbauer spectrum of the oxygen adduct of P450 is strikingly similar to the one of oxyhemoglobin (Lang and Marshall, 1966) and a number of oxygenated heme model complexes (Collman *et al.*, 1974; Wagner and Kassner, 1974). A measurement made in strong magnetic field (Fig. 15) clearly shows that the complex is diamagnetic.

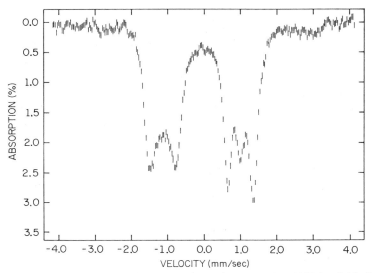

**Fig. 15.** Mössbauer spectrum of oxygenated cytochrome P450 at 4.2°K in a field of 28 kG parallel to the transmitted $\gamma$ rays. The oxygenated sample contained some unreacted high-spin ferrous protein as well as some autoxidized ferric P450. The spectrum of the latter extends beyond the velocity scale used and is rather broad. The spectrum of the ferrous heme, however, is very distinct (see the spectrum of Fig. 13 taken at 25 kG). In the spectrum shown, an appropriate fraction ($\sim 30\%$) of high-spin ferrous P450 measured under identical conditions has been subtracted. (From Sharrock, 1973.)

The electronic structure of the heme–oxygen adduct is still controversial, but the Weiss model (Weiss, 1964) of a superoxide anion coupling antiferromagnetically to low-spin ferric iron appears to be closest to the truth.

   In contrast to the first reduction step in the enzymatic cycle of cytochrome P450 (Fig. 7), which can be performed by any reducing agent, transfer of the second reducing equivalent occurs only in the presence of an effector, normally the iron–sulfur protein $E_B$. This second reduction is followed by the formation of product and, although the reaction must proceed through a number of steps, it has not been possible to isolate any intermediate states yet. After release of product cytochrome P450 is again in its native, low-spin ferric state, thus completing the enzymatic cycle.

### V. Conclusion

   A few additional remarks are in place to put the discussions of Sections II–IV into proper perspective. As pointed out repeatedly, many of the stable reaction intermediates of the enzymes considered have their counterpart in some other, better known iron protein, and much can be learned in these cases from a comparison of all the pertinent data. Since Mössbauer spectroscopy is unable to measure enzyme kinetics, but is limited rather to the observation of static properties, the distinction between enzymes and proteins without catalytic function is of no consequence at this level. Enzymes, on the other hand, provide the intriguing opportunity to study several reaction intermediates and, by putting together the various pieces of information, to draw some conclusions about the details of the reaction mechanism.

In addition to the examples given in Sections II–IV, a number of other iron enzymes have been studied by Mössbauer spectroscopy. The reader interested in this field will find the complete and up-to-date bibliography edited by May (1968) most useful. Among the heme enzymes, catalase (Maeda *et al.*, 1973) and peroxidases (Maeda, 1968; Moss *et al.*, 1969) have been studied in greatest detail. Of particular interest are compounds I and II of peroxidases, which supposedly contain iron in the unusual charge state $Fe^{4+}$. The Mössbauer spectra of compounds I and II measured in zero magnetic field are quite similar, but differ from those of the native, ferric enzyme. According to Moss *et al.* (1969), their isomer shifts are compatible with a charge state $Fe^{4+}$, but only a measurement in strong magnetic field will make it possible to determine the spin states of these complexes.

Throughout this chapter, special emphasis was placed on those systems for which the Mössbauer method can yield unique information. Specifically, these are the EPR-silent integer-spin states, $Fe^{2+}$, $Fe^{4+}$ and a variety of spin-coupled clusters. For clusters of half-integer spin, the magnetic hyperfine splitting observed in the Mössbauer spectrum allows, under favorable conditions, determination of the spin residing on each iron atom. The measurement of other quantities such as isomer shift and quadrupole splitting is straightforward, but their interpretation is generally more difficult. Taken all together, the Mössbauer parameters do not reveal the ligands of the iron yet, but they greatly help in developing suitable models for the active site.

### References

Aasa, R. (1970). *J. Chem. Phys.* **52**, 3919.
Burris, R. H. (1971). *In* "The Chemistry and Biochemistry of Nitrogen Fixation" (J.R. Postgate, ed.), Chapter 4. Plenum Press, New York.
Champion, P. M., Lipscomb, J. D., Münck, E., Debrunner, P., and Gunsalus, I. C. (1975a). *Biochemistry* **14**, 4151.
Champion, P. M., Chiang, R., Münck, E., Debrunner, P., and Hager, L. P. (1975b) *Biochemistry* **14**, 4159.
Champion, P. M., Münck, E., Debrunner, P. G., Hollenberg, P. F., and Hager, L. P. (1973). *Biochemistry* **12**, 426.
Collman, J. P., Gagne, R. R., Reed, C. A., Robinson, W. T., and Rodley, G. A. (1974). *Proc. Nat. Acad. Sci. U.S.* **71**, 1326.
Dagley, S., and Patel, M. D. (1957). *Biochem. J.* **66**, 227.

Dawson, J. W., Gray, H. B., Hoenig, H. E., Rossman, G. R., Schredder, J. M., and Wang, R. H. (1972). *Biochemistry* **11**, 461.

Dayhoff, M. O. (1972). "Atlas of Protein Sequence and Structure," Vol. 5. Nat. Biomed. Res. Foundation, Washington, D.C.

Dayhoff, M. O., Barker, W. C., and Hardman, J. K. (1972). *In* "Atlas of Protein Sequence and Structure," Vol. 5, p. 53. Nat. Biomed. Res. Foundation, Washington, D.C.

Dickerson, R. E., and Geis, I. (1969) "The Structure and Action of Proteins." Benjamin, New York.

Douzou, P. (1975). *Methods Biochem. Anal.* **22**, 401.

Dunham, R., Bearden, A., Salmeen, I., Fee, J., Petering, D., Sands, R. H., and Orme-Johnson, W. (1971). *Biochim. Biophys. Acta* **253**, 134.

Eicher, H., and Trautwein, A. (1969). *J. Chem. Phys.* **50**, 2540.

Falk, J. E. (1964). "Porphyrins and Metalloporphyrins." Elsevier, Amsterdam.

Frieden, E. (1972). *Sci. Amer.* **227**, #1,52.

Gersonde, K., Schlaak, H. E., Breitenbach, M., Parak, F., Eicher, H., Zgorzalla, M., Kalvius, M. G., and Mayer, A. (1974). *Eur. J. Biochem.* **43**, 307.

Gonser, U., Maeda, Y., Trautwein, A., Parak, R. and Formanek, H. (1974). *Z. Naturforsch.* **29B**, 241.

Gunsalus, I. C., Meeks, J. R., Lipscomb, J. D., Debrunner, P., and Münck, E. (1974). *In* "Molecular Mechanisms of Oxygen Activation" (O. Hayaishi, ed.), pp. 561–613. Academic Press, New York.

Griffith, J. S. (1957). *Nature (London)* **180**, 30.

Groves, J. L., Potasek, M. J., and Depasquali, G. (1973). *Phys. Lett.* **42A**, 493.

Halpern, J. (1968). *In* "Homogeneous Catalysis, Industrial Applications and Implications" (*Amer. Chem. Soc. Advan. Chem.*), Vol. 70, pp. 1–24. Amer. Chem. Soc., Washington, D.C.

Hayaishi, O. (1974). "Molecular Mechanisms of Oxygen Activation." Academic Press, New York.

Hedegaard, J., and Gunsalus, I. C. (1965). *J. Biol. Chem.* **240**, 4038.

Hoard, J. L. (1971). *Science* **174**, 1295.

Huynh, B. H., Papaefthymiou, G. C., Yen, C. S., Groves, J. L., and Wu, C. S. (1974). *J. Chem. Phys.* **61**, 3750.

Kelly, M., and Lang, G. (1970). *Biochim. Biophys. Acta* **223**, 86.

Lang, G., and Marshall, W. (1966). *Proc. Phys. Soc.* **87**, 3.

Lang, G., Aasa, R., Garbett, K., and Williams, R. J. P. (1971). *J. Chem. Phys.* **55**, 4539.

Lovenberg, W. (ed.) (1973). "Iron-Sulfur Proteins." Academic Press, New York.

Maeda, Y. (1968). *J. Phys. Soc. Japan* **24**, 151.

Maeda, Y., Trautwein, A., Gonser, U., Yoshida, K., Kikuchi-Torii, K., Homma, T., and Ogura, Y. (1973). *Biochim. Biophys. Acta* **303**, 230.

May, L. (1968). Index of Publications in Mössbauer Spectroscopy of Biological Materials. Dep. of Chem., The Catholic Univ. of America, Washington, D.C.

Morris, D. R., and Hager, L. P. (1966). *J. Biol. Chem.* **421**, 1763.

Moss, T. H., Ehrenberg, A., and Bearden, A. J. (1969). *Biochemistry* **8**, 4159.

Münck, E., Debrunner, P. G., Tsibris, J. C. M., and Gunsalus, I. C. (1972). *Biochemistry* **11**, 855.

Münck, E. Groves, J. L., Tumolillo, T. A., and Debrunner, P. G. (1973). *Comput. Phys. Commun.* **5**, 225.

Münck, E., Rhodes, H., Orme-Johnson, W. H., Davis, L. C., Brill, W. J., and Shah, V. K. (1975). *Biochim. Biophys. Acta.* **400**, 32.

Nakano, N., Otsuka, J. and Tasaki, A. (1972). *Biochim. Biophys. Acta.* **278**, 355.

Nath, A., Harpold, M., Klein, M. P. and Kündig, W. (1968). *Chem. Phys. Lett.* **2**, 471.

Neilands, J. B. (1966). *Struct. Bonding* **1**, 59.

O'Dell, B. L., and Campbell, B. J. (1970). *In* "Comprehensive Biochemistry" (M. Florkin and E. H. Stotz, eds), Vol. 21, pp. 179–266. Elsevier, Amsterdam.

Omura, T., Sato, R., Cooper, D. Y., Rosenthal, O., and Estabrook, R. W., (1966). *Fed. Proc.* **24**, 1181.

Oosterhuis, W. T., and Lang, G. (1967). *Phys. Rev.* **178**, 439.

Orme-Johnson, W. H. *et al.* (1972). *Proc. Nat. Acad. Sci. U.S.* **69**, 3142.

Palmer, G., Multani, J. S., Cretney, W. C., Zumft, W. G., and Mortenson, L. E. (1972). *Arch. Biochem. Biophys.* **153**, 325.

Sharrock, M. P. (1973). Thesis, Univ. of Illinois, Urbana, Illinois, unpublished.

Sharrock, M., Münck, E., Debrunner, P. G., Marshall, V., Lipscomb, J. D., and Gunsalus, I. C. (1973). *Biochemistry* **12**, 258.

Smith, E. L. (1965). "Vitamin $B_{12}$," 3rd. ed. Wiley, New York.

Smith, B. E., and Lang, G. (1974). *Biochem. J.* **137**, 169.

Stroud, R. M. (1974). *Sci. Amer.* **231**, #1,74.

Tsai, R. *et al.* (1970). *Proc. Nat. Acad. Sci. U.S.* **66**, 1157.

Vallee, B. L., and Coleman, J. E. (1964). *In* "Comprehensive Biochemistry" (M. Florkin and E. H. Stotz, eds.), Vol. 12, pp. 165–235. Elsevier, Amsterdam.

Wagner, G. C., and Kassner, R. J. (1974). *J. Amer. Chem. Soc.* **96**, 5593.

Watson, J. D. (1970), "Molecular Biology of the Gene," 2nd ed., pp. 435f. Benjamin, New York.

Weiss, J. J. (1964). *Nature (London)* **202**, 83.

Wickman, H. H., Klein, M. P., and Shirley, D. A. (1965). *J. Chem. Phys.* **42**, 2113.

Zabinski, R., Münck, E., Champion, P. M., and Wood, J. M. (1972). *Biochemistry* **11**, 3212.

# SOLID STATE CHEMISTRY

# 7

# *Monitoring of Solid State Reactions*

**P. K. Gallagher**

Bell Laboratories
Murray Hill, New Jersey

## I. Introduction

Determining and following the course of chemical reactions in the solid state or heterogeneous reactions involving solids is frequently a difficult task. Establishing the nature of transient or intermediate products during a complex reaction is seldom easy. One of the more obvious and commonly used techniques for this purpose is x-ray diffraction analysis. Frequently, however, the initial products are amorphous or so finely divided that patterns are absent or so broad as to be not readily identifiable. In some cases, the intermediates may be completely unknown without established patterns to facilitate identification. In still other cases, such factors as oxidation state, site preferences, and magnetic properties are inadequately determined without supplementary information.

If the atomic species involved in the reaction are suitable subjects for Mössbauer spectroscopy, this technique can be very valuable, particularly with respect to evaluating the oxidation state. Fortunately, $^{57}$Fe is quite suitable for Mössbauer spectroscopy and eminently desirable as a chemical subject from both a technological and academic standpoint. If one were given the choice of a metallic atom for study, certainly iron would be very high on the list. As a consequence, the material in this chapter deals primarily with iron compounds, although some relevant work has also been done on compounds of tin, europium, and antimony.

The title for this chapter is indicative of the subject to be discussed but so broad as to offer considerable overlap with nearly all of the other portions of this book. No effort will be made to comprehensively review the literature, but rather the usefulness of the technique will be demonstrated by a somewhat detailed discussion of selected examples. Emphasis will be placed on direct reaction in the bulk of the solid and particularly on the thermal decompositions of solids. Solid state reactions directly involved in metallurgy, geochemical phenomena, catalysis, corrosion, colloids, glassy systems, and frozen solutions are deferred to other chapters herein or in future volumes. Simple $solid_1 \rightleftharpoons solid_2$ phase transitions are also excluded.

## II. Solid–Solid Reactions

As one might anticipate, the foremost examples of the application of the Mössbauer effect to the study of this class of solid state reactions are in the area of the formation of ferrites. It is a common approach in the technology of reacting solids to facilitate the process by forming at least one of the reactants in a highly reactive state, i.e., having a fine particle size; high concentration of defects and strain, etc.; by the thermal decomposition of a precursor compound, e.g., hydroxide, carbonate, oxalate. Those situations in which the compound containing the Mössbauer active atom has been obtained in this manner are arbitrarily placed in Section IV on thermal decompositions.

### A. Ferrite Formation Reactions

#### 1. $ZnFe_2O_4$

The formation of $ZnFe_2O_4$ has been studied by Duncan and Stewart (1967) and shown to follow the straightforward reaction

$$ZnO + Fe_2O_3 \rightarrow ZnFe_2O_4 \qquad (1)$$

without the formation of distinct intermediate compounds. Duncan *et al.* (1968) followed this with a detailed study of the kinetics of formation utilizing the Mössbauer effect (Figs. 1 and 2) and compared the results with those obtained by x-ray diffraction techniques.

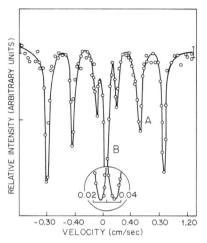

**Fig. 1.** Mössbauer spectrum at room temperature of (A) mixed ZnO and $Fe_2O_3$ and (B) $ZnFe_2O_4$. Inset is the peak due to $ZnFe_2O_4$ enlarged to show the quadrupole splitting. Isomer shift is relative to iron, and error bars represent $\pm\sigma$. (Duncan *et al.*, 1968.)

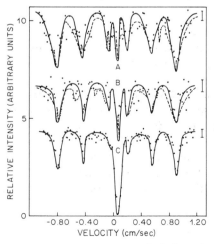

**Fig. 2.** Mössbauer spectra at room temperature and relative to iron of equimolar mixtures of ZnO and $Fe_2O_3$ heated at 760°C for the following times: (A) 3 min, (B) 6 min, and (C) 9 min. (Duncan *et al.*, 1968.)

Calculations of the fraction reacted were done by several methods. It was found that the mean intensity of either the outer pair or next to the outer pair could be used for the hematite phase, while the unresolved doublet was satisfactory for the spinel phase. Figure 3 compares the results based on

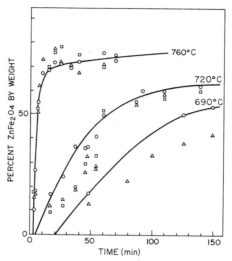

**Fig. 3.**    Percent of $ZnFe_2O_4$ formed from an equimolar mixture of ZnO and $Fe_2O_3$ as a function of time. Determined by □ = integrated Mössbauer intensity; △ = peak Mössbauer intensity; ○ = x-ray diffraction technique. (Duncan *et al.*, 1968.)

Mössbauer spectra and those derived from x-ray data. There is a further comparison within the Mössbauer approach based on integrated line intensities versus peak heights. There appears to be little to choose from in this case between the use of peak intensities or peak areas. Near completion, x-ray and Mössbauer methods yield similar results; however, at lower conversions, i.e., early stages of the reaction, the two techniques are somewhat divergent. Duncan *et al.* (1968) have explained this in terms of the difference inherent in the two approaches. The crystal structure, and hence x-ray pattern, is largely determined by the spacing within the oxygen lattice, while the Mössbauer effect is directly dependent on the precise environment of the iron ion and is therefore more subject to subtle variations than is the x-ray pattern.

### 2. $(Y, Gd)_3(Fe, Al)_5O_{12}$

Levinson *et al.* (1975) have demonstrated the usefulness of the Mössbauer effect for studying the degree of homogeneity achieved during the synthesis of a ferrite by determining the width of the temperature range corresponding to the magnetic transition. Using samples of $(Y, Gd)_3(Fe, Al)_5O_{12}$ ($T_c \approx$ 185°C) prepared from mixed oxides and carbonates at 1300°C and then fired for various times at 1450°C, they observed that the temperature range for the conversion from 10 to 90% varied from 9° to 3°C with increasing homogeneity resulting from prolonged firing. Jet milling prior to firing was highly effective changing it from 9° to 2.3°C for the same firing time.

### 3. $CaFe_2O_4$ and $Ca_2Fe_2O_5$

The formation of both $CaFe_2O_4$ and $Ca_2Fe_2O_5$ from physical mixtures of CaO and $Fe_2O_3$ has also been studied utilizing the Mössbauer effect by Yusfin et al. (1974a, b). Initial work at 500°C with a starting mixture of $CaO \cdot Fe_2O_3$ revealed the formation of 8.7% $Ca_2Fe_2O_5$ after 40 min but no $CaFe_2O_4$. After 38 hr, however, 3.8% $CaFe_2O_4$ had formed, and the $Ca_2Fe_2O_5$ had increased to 12.1%. Starting with a $2CaO \cdot Fe_2O_3$ mixture, $Ca_2Fe_2O_5$ formed in a comparable time, i.e., 10.9% after 40 min at 500°C. The later work determines the phases present after firings at various temperatures from 800° to 1100°C with a claimed accuracy of $\pm 2\%$. Kinetic models and constants for the solid state reaction are calculated. This is a tenuous process, even using more accurate techniques.

Other examples of the formation of ferrites may be found (see below) in the discussions of the final stage in the thermal decomposition of some mixed hydroxides, oxalates, and cyanides.

### B. Other Reactions

Reactions to form calcium antimonates have also been investigated using the Mössbauer effect. Lebeder et al. (1974) have studied the formation of $CaSbO_x$ starting from $Sb_2O_3$. Not only is there the formation of a new compound, but there is also some oxidation of $Sb^{3+}$.

From a very practical standpoint, the hydration or setting of Portland cement has been followed by use of the Mössbauer effect by Wittman (1974) and by Tamas and Vertes (1973). The latter authors have suggested the following equation for this reaction:

$$4CaO \cdot Al_2O_3 \cdot Fe_2O_3 + 10H_2O$$
$$\rightleftarrows 3CaO \cdot Fe_2O_3 \cdot 6H_2O + \tfrac{1}{3}(3CaO \cdot Al_2O_3 \cdot 6H_2O) + \tfrac{4}{3}Al(OH)_3 \qquad (2)$$

Chemical changes at the surface of pyrite ores induced by floatation reagents have been studied by Solozhenkin et al. (1974). The formation of the appropriate organometalic compound of iron at the surface was verified by Mössbauer spectroscopy.

There is a host of solid state reactions in the field of glass technology, such as the devitrification of iron-containing borate glasses by Jach (1973a,b), which are more properly considered in the chapter on glassy systems.

A particularly nice piece of work showing the application of the Möss-bauer effect to solid state reactions is that of Gutlich and Hasselbach (1969, 1974). Iron(II) sulfate hydrates were reacted with KCN in pressed pellets. A 1:6 ratio of iron to potassium was used. The reaction is fast at room temperature, so that the kinetics were studied at 5° and 20°C by quenching the samples to $-96°C$ for measurement of the spectra.

Depending on the extent of the reaction, five lines were resolved corresponding to a doublet associated with the iron(II) sulfate hydrate, a doublet corresponding to a $K_3[Fe(CN_5)H_2O]$ intermediate, and a single line from the final product $K_4[Fe(CN)_6]\cdot3H_2O$. Rate constants and activation energies were determined for both steps of the overall reaction starting with $FeSO_4\cdot7H_2O$. If $FeSO_4\cdot2.5H_2O$ is used initially, there is no reaction after 30 days at 5°C, as compared with complete reaction in 30 hr with the heptahydrate. Clearly, replacing water molecules by sulfate ions in the first coordination sphere of the iron ion greatly inhibits the reaction. This conclusion, however, assumes that no liquid phase is formed by the extraction of water by KCN.

Kaufman and Hazel (1974) have also followed the reaction of KCN with iron compounds by means of Mössbauer spectroscopy. Different products were obtained when using a fresh colloidal sol formed by the hydrolosis of $FeCl_3$ in boiling water rather than with a gel. There is extensive reaction with the sol, but little or none with the gel at room temperature. They concluded that the hydrolized species in the sol was a tetramer having an Fe—O—Fe bond angle of 40°, but that the bonding in the gel was primarily by hydroxyl bridges.

Calusaru et al. (1973) have used the Mössbauer effect as well as other techniques to study the exchange of radioactive iron (*) between $Na_2[Fe(CN)_5NO]\cdot2H_2O$ and $Fe*(OH)_3$. Mössbauer spectra, infrared spectra, and thermoanalytical curves indicated that no new compound was formed in compressed pellets of the physically mixed compounds heated to 120°C, but chemical separation and radiochemical analysis indicated appreciable exchange of active iron. The extent of the exchange was less using the anhydrous pentacyanonitrosyl complex but still very significant. Mössbauer spectra at low temperature revealed differences in the $Fe(OH)_3$ portion of the spectra that were interpreted in terms of delocalization of the iron sites with concomitant atomic tunneling to account for the observed exchange.

The previous examples have demonstrated the usefulness of the Mössbauer effect in studies of reactions where the appropriate isotope is one of the major reactants. There are also numerous cases where the active isotope is present as an impurity or minor component of the reaction and yet provides useful information about the reaction. There have been many such studies to elucidate the nature and reactions of solid surfaces or behavior of impurities during the first- or second-order phase transitions, however, these are outside the scope of this chapter. An example of a study where iron was deliberately added as an impurity is that of Nozik (1972), who investigated the photochemical oxidation of $Fe^{3+}$ incorporated onto rutile. Ultraviolet excitation produces holes and electrons in the rutile. The former migrate to the surface and react with the added $Fe^{3+}$ to generate $Fe^{4+}$. Since the holes and $Fe^{4+}$ ions rapidly recombine with the electrons, the Mössbauer spectrum of $Fe^{4+}$ is only observed during irradiation when there is a steady state concentration formed.

In this section, it has been shown that the Mössbauer effect can be of considerable value in the study of a variety of solid–solid reactions. There is a particular sensitivity toward homogeneity and crystalline perfections indicated in the ferrite and Portland cement studies. There is also the ability to distinguish the different chemical environments and bonding as seen in the various cyanide reactions and the pyrite floatation investigation. This latter aspect lends itself to quantitative chemical analysis on occasions where more conventional techniques prove difficult. Finally, the technique is particularly of value in following the oxidation state of even small amounts of impurities, as demonstrated in the photochemical oxidation of $Fe^{3+}$ in rutile.

## III. Solid–Gas Reactions

This section is concerned with the reaction with a gas phase to form a new product rather than the evolution of a gas which is considered in Section IV. Obviously, reversible decompositions involving a gas phase can be considered in either fashion. Gain or loss of oxygen corresponding to changing defect structures within a single composition is treated in this section, while other reversible decompositions are deferred to Section IV.

The interaction between atmospheric oxygen and various oxides is of immense technological importance. The accompanying effects upon the semiconducting, magnetic, catalytic, and other properties are enormous. Similarly the use of the Mössbauer effect, such as in the work by Channing and Graham (1972), to study the oxidation of alloys and metals is of great importance but is more appropriately covered in Chapter 3 on corrosion.

Because of the interest in iron in the unusual valence state of $4^+$, investigation with the alkaline earth ferrates offers one of the best documented examples of the use of the Mössbauer effect for the study of such equilibria. Ichida (1973a) has extensively investigated the phase fields in the system $BaFeO_{2.5}$–$BaFeO_3$. Earlier Mössbauer work (Gallagher et al., 1965) had shown the existence of a hexagonal $BaFeO_{3-x}$ phase with as much as 90% $Fe^{4+}$ and a completely trivalent phase $BaFeO_{2.5}$, in which there were equivalent numbers of octahedral and tetrahedral iron sites. This earlier work was based on samples prepared by conventional ceramic techniques utilizing

$BaCO_3$ and $Fe_2O_3$. Ichida, in contrast, started with $BaFeO_4$ in which the iron was in an even higher oxidation state and fired if under differing conditions to effect varying degrees of reduction.

Figure 4 indicates the type of room-temperature spectra that result in the stability region of the tetragonal phase. The dashed lines correspond to

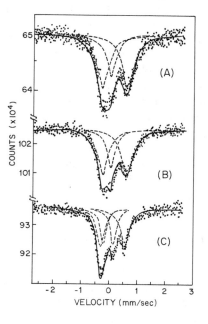

**Fig. 4.** Mössbauer spectra at room temperature using a copper source. Tetragonal compounds obtained by heating $BaFeO_4$ at 400°C for 50 hr in the following atmospheres: (A) 1 atm and air, (B) 1 atm of oxygen, (C) 1500 atm of oxygen. (Ichida, 1973a.)

Ichida's assignments, and the solid curve is the summation of the dashed lines. The outside pair of dashed lines corresponds to a quadrupole split doublet attributed to $Fe^{3+}$ (isomer shift 0.20 mm/sec and quadrupole splitting of 0.85 mm/sec). The interior line in (A) and (B) and the interior quadrupole doublet in (C) are attributed to $Fe^{4+}$. The trend in the quadrupole splitting of the $Fe^{4+}$ spectra is surprising. Because of the $d^4$ nature of $Fe^{4+}$, it is necessary to have high symmetry in order to have no quadrupole splitting. It is natural, therefore, to expect increased splitting with increasing oxygen vacancies, in contradiction to Fig. 4. Using areas under the respective curves, it was determined that the sample in 1 atm of oxygen had 22% $Fe^{4+}$, while that fired at 1500 atm had 42% $Fe^{4+}$.

Using both Mössbauer spectroscopy and x-ray diffraction techniques, the regions of stability associated with the various phase fields and their compositional ranges of oxygen were determined. Figures 5 and 6 summarize these results. The irreversibility of transitions from the tetragonal phase indicated in Fig. 6 strongly argues that this phase is metastable. Ichida,

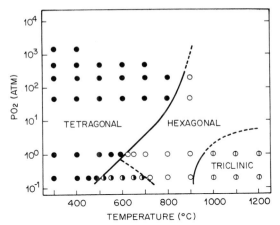

**Fig. 5.**   Formation diagram of the products obtained by heating $BaFeO_4$ at various oxygen pressures and temperatures; ● = tetragonal phase, ◑ = low-temperature $BaFeO_{2.5}$, ○ = hexagonal phase, ⊕ = triclinic phase. (Ichida, 1973a.)

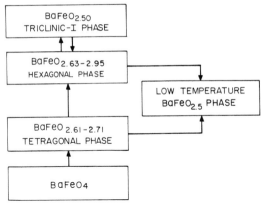

**Fig. 6.**   Phase transformation among several $BaFeO_x$ phases. Arrows indicate direction of phase transformation. (Ichida, 1973a.)

however, maintains that it is an equilibrium phase because of the reproducibility of the transition boundary indicated in Fig. 5 and that the reverse reactions do not occur because of a kinetic barrier to diffusion. Unfortunately, the Mössbauer effect can do nothing to resolve this question.

It is worthwhile to point out the stabilizing influence of the perovskite structure upon the existence of $Fe^{4+}$. In earlier work, Ichida (1973b) has shown that there were no intermediate compounds containing $Fe^{4+}$ formed by treating $K_2FeO_4$ where the cation ratios are unsuitable for formation of a perovskite phase. Instead, at temperatures between 170° and 250°C, an amorphous phase containing $Fe^{3+}$ is formed. Above 250°C this reacts to form $KFeO_2$ and $K_2O$. On the other hand, complete conversion to stoichiometric $Fe^{4+}$ has been achieved in $SrFeO_3$, where the radius of the alkaline earth ion is more favorable than that of $Ba^{2+}$ for the formation of a perovskite structure (Gallagher *et al.*, 1964). Ichida (1973c) also investigated the loss of oxygen from $SrFeO_4$. It also formed an amorphous trivalent phase below about 300°C. By 400°C, the perovskite phase was evident, and the $Fe^{4+}$ content went through a maximum with increasing temperature.

Considerable work has been done on the intercalation compounds of $FeCl_3$ in graphite. Novikov *et al.* (1970) used Mössbauer spectroscopy to investigate the amount of $FeCl_3$ vapor incorporated into graphite as a function of the many experimental variables, e.g. temperature, carrier gas, particle size. They also observed the reduction upon heating in hydrogen to give intercalated $FeCl_2$ compounds. Hooley *et al.* (1968, 1970) had also investigated the reduction upon heating in nitrogen or hydrogen. Where the Mössbauer spectra of the intercalated $FeCl_3$ showed evidence of bonding as $C_7FeCl_3$ or $C_{12}FeCl_3$ species, the intercalated $FeCl_2$ species resulting from the reduction did not. There were two sites for $Fe(II)$ formed. The more prevalent one gave Mössbauer parameters identical to those of anhydrous $FeCl_2$. The minor site had the same isomer shift but nearly twice as large a quadrupole splitting. Hooley *et al.* (1970) found that the concentration of this second species increased with increasing size of the graphite flake and speculated, therefore, that it was attributable to the presence of an adjacent trapped molecule of $Cl_2$. The magnetic properties of these two divalent sites have been studied in detail by Ohhashi and Tsujikawa (1974).

Other excellent samples of the use of the Mössbauer effect to study solid–gas equilibria are within the system $Fe \leftrightarrow FeO \leftrightarrow Fe_3O_4 \leftrightarrow \gamma Fe_2O_3 \leftrightarrow \alpha Fe_2O_3$. See, for example, studies of the disproportionation of $FeO$ (e.g., Schechter *et al.*, 1966; Greenwood and Howe, 1972) and the $\gamma \rightarrow \alpha$ rearrangement of $Fe_2O_3$ by Annersten and Hafner (1973). Greenwood and Howe describe the disproportionation by

$$(1 - 4z)Fe_{1-x}O \xrightarrow{300°-570°C} (1 - 4x)Fe_{1-z}O + (x - z)Fe_3O_4 \tag{3}$$

$$4Fe_{1-z}O \xrightarrow{570°-700°C} (1 - 4z)Fe + Fe_3O_4 \tag{4}$$

The structural changes in the $FeO$ phase were described in terms of the aggregation of vacancies. The presence of a vacancy or other charge defect in the nearest or next nearest neighbor to $Fe^{2+}$ leads to quadrupole splitting, which is otherwise absent.

Mössbauer spectroscopy and x-ray diffraction have also been used to follow the reduction of $Fe_3O_4$ to the metal with hydrogen and thus the subsequent reaction of this metal with $NH_3$ to form a mixture of $\varepsilon$ and $\gamma'$ $Fe_4N$ (Maksimov *et al.*, 1974).

A novel application of Mössbauer spectroscopy was made by Seregin *et al.* (1972). The oxidation of SnO or SnTe was observed during the grinding of these materials in air. Considerable Sn(IV) was evident in the Mössbauer spectra, and the quantity increased with increasing grinding time. Presumably, a layer of amorphous $SnO_2$ is formed, because there is no evidence for the formation of a new phase by x-ray techniques. If the samples are ground under liquid nitrogen, there is no formation of Sn(IV). The stability of SnS and SnSe is greater as evidenced by the lack of formation of any Sn(IV) upon grinding these materials at room temperature in air.

Before leaving the subject of solid–gas reactions some brief discussion of the extensive work on the deposition of iron-containing materials on the surface of various hosts such as silica gel, $Al_2O_3$, and zeolites should be made. The behavior of these deposits upon heating in various atmospheres has been the subject of considerable effort because of the implications for catalysis. The general topic is worthy of a chapter by itself, not only because of the mass of work, but also because much of it is conflicting.

Collins and Mulay (1968) applied $Fe(CO)_5$ or $FeCl_3$ to zeolite and obtained similar results with both compounds. They concluded, therefore, that the anion was probably not a factor. Malathi and Puri (1971) showed, however, that the choice of supporting materials was a factor, since they observed different results using silica, alumina, or zeolites. For samples heated to 800°C in air, the iron had entered substitutionally into the alumina lattice, was still identifiable as superparamagnetic $\alpha\text{-}Fe_2O_3$ on silica, and had grown to sufficient size on zeolite to give a pattern typical of bulk $\alpha\text{-}Fe_2O_3$.

Examples of some inconsistencies are that Wedd *et al.* (1969) concluded that $FeCl_3$ applied to zeolites became distributed between surface and interior sites. On the other hand, Collins and Mulay (1968) observed only the internal site in their study. They also found evidence for only superparamagnetic $\alpha\text{-}Fe_2O_3$ in samples heated to 500°C in air while Malathi and Puri (1971) found nearly equivalent amounts of superparamagnetic $\alpha\text{-}FeOOH$ and $\alpha\text{-}Fe_2O_3$ in samples heated at 400°–600°C in air on zeolite, silica, or alumina.

The behavior of such supported materials in vacuum or hydrogen is also interesting. Both Wedd *et al.* (1969) and Tachibana and Ohya (1969), respectively, observed the formation of Fe(II) upon heating supported materials on zeolite to 350°C and samples supported on silica to 475°C in vacuum. When fired at 450°C in hydrogen, the materials supported on silica had a similar spectrum with only slightly more Fe(II) formed. The samples on zeolites, however, yielded mostly iron metal upon heating to 350°C in hydrogen. Tachibana and Ohya made further studies that suggest that the partial reduction of $Fe_2O_3$ at such a surprisingly low temperature in only a moderate vacuum is due to the reaction with organic contaminants from the pumping system and is also related to the presence of surface hydroxyl ions.

In contrast, Rubashov *et al.* (1972) have recently observed that $Fe(NO_3)_3$ or $Fe(CO)_5$ adsorbed on silica will form fine particles when heated to moderate temperatures, but if $FeCl_3$ is adsorbed, the resulting particles of $Fe_2O_3$ are much larger at the same temperature. In addition, heating these samples to 500°C in vacuum will lead to the formation of significant amounts of $Fe_3O_4$ for the coarse particles but no reduction in the fine particles.

Gager *et al.* (1972, 1973) have studied the interaction of $H_2S$, $H_2O$, $CH_3OH$, $NH_3$, $O_2$, and $H_2$ with microcrystals of $\alpha$-$Fe_2O_3$ supported on silica gel. They show $O_2$ adsorption does not cause any change in the Mössbauer parameters and conclude therefore that the $O_2$ is physically absorbed rather than chemisorbed. In contrast, adsorption of $H_2O$, $CH_3OH$, $H_2S$, or $NH_3$ reduces the observed quadrupole splitting, implying chemisorption of these polar gases without any accompanying reduction in the valence of iron.

A particularly interesting situation was noted by Gager *et al.*, upon reduction of their $\alpha$-$Fe_2O_3$ microcrystallites by hydrogen. Figure 7 shows the Mössbauer spectra of such a sample. Two sites for Fe(II) are observed.

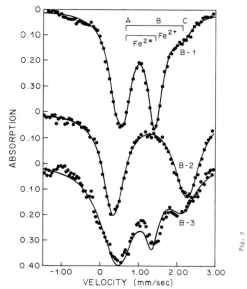

**Fig. 7.** Mössbauer spectra of $\alpha$-$Fe_2O_3$ microcrystallites after various treatments: (B–1) reduced in hydrogen and outgassed at 600°C; (B–2) reduced as above and exposed to $H_2O$ (23 Torr); (B–3) reduced as above and exposed to $H_2S$ (85 Torr). (Gager *et al.*, 1973.)

The asterisk denotes the site that is particularly surface active. As can be seen in Fig. 7 the adsorption of $H_2O$ eliminates the $Fe(II)^*$ spectrum. Adsorption of $H_2S$ leads to a reduction of the $Fe(II)^*$ spectrum and the formation of a sulfide.

Again, the aspects of the Mössbauer effect that are important, in this case for solid–gas reactions, are the ability to focus on small amounts of a selected material and help establish its bonding, environment, and valence. This was illustrated by the studies on intercalated compounds of graphite and by studies of the interactions of iron compounds with their supports and the atmosphere. Following the oxidation and formation of new phases has also been shown to be possible, as in the $BaFeO_x$ or $FeO_x$ systems and, more importantly, in the oxidation of some tin compounds where x-ray diffraction analysis was insensitive.

## IV. Thermal Decomposition of Solids

### A. Minerals

Most of the examples in this chapter are concerned with compounds in which the Mössbauer atom is a true component. Mineralogy is an area, however, in which the Mössbauer atom, generally iron, is usually present as an impurity. Nevertheless, much useful information can be gleaned about the nature of these clays and minerals (see, for example, Bancroft et al. (1967, 1968).

Duncan et al. (1968) studied the transformation of kaolinite clay to mullite. The reaction sequence determined from a combination of experimental techniques is

$$\underset{\text{kaolinite}}{Al_2O_3 \cdot 2SiO_2 \cdot 2H_2O} \xrightarrow[550°-650°C]{} \underset{\text{metakaolinite}}{Al_2O_3 \cdot 2SiO_2} + 2H_2O \tag{5}$$

$$\underset{\text{metakaolinite}}{2(Al_2O_3 \cdot 2SiO_2)} \xrightarrow[980°C]{} Si_3Al_4O_{12} + SiO_2 \tag{6}$$

$$\underset{\text{Al–Si–spinel}}{3(Si_3Al_4O_{12})} \xrightarrow[1100°C]{} \underset{\text{mullite}}{2(3Al_2O_3 \cdot 2SiO_2)} + 5SiO_2 \tag{7}$$

Figure 8 presents Mössbauer spectra utilizing the naturally occurring iron content in kaolinite. Curve A corresponds to kaolinite, curve B to (essentially) metakaolinite, curves C and D to primarily the spinel, and curve E to (basically) mullite. Conclusions drawn by Duncan et al. (1968) are summarized in Table I. The interpretations are far from unequivocal, as indicated in the general discussion accompanying the paper. Of particular concern is whether curves B and E represent possible quadrupole split doublets, and also the valence inferred from the indicated isomer shifts of

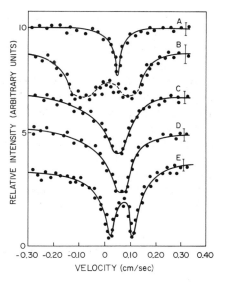

**Fig. 8.** Mössbauer spectra at room temperature (palladium source) of kaolinite heated for 30 min at (A) 20°C, (B) 650°C, (C) 980°C, (D) 1100°C, and (E) 1280°C. (Duncan *et al.*, 1968.)

**TABLE I**

Mössbauer Parameters for Kaolinite Absorbers at Room Temperature[a]

| Firing temperature (°C) | Isomer shift (cm/sec) | Assignment |
|---|---|---|
| Unfired | +0.03 | $Fe^{3+}$ high energy sites |
| 650 | −0.12, +0.12 | $Fe^{3+}$ many loosely bound sites |
| 980 | +0.03 | $Fe^{3+}$ fewer loosely bound sites |
| 1100 | +0.05 | $Fe^{3+}$ fewer tightly bound sites |
| 128 | +0.01, +0.09 | $Fe^{3+}$ two tightly bound sites |

[a] *Calibration*: Sodium nitroprusside has an isomer shift of −0.028 cm/sec. (From Duncan *et al.*, 1968.)

curve B. The relevant point here, however, is that the Mössbauer effect is sensitive to this sequence of reactions and can serve as another source of information and, it is hoped, of understanding for such processes. Even more complex and controversial spectra were obtained by doping kaolinite with $^{57}Co$ and performing similar experiments using these doped kaolinites as sources.

Montmorillonite and illite are two other claylike minerals that have been studied using the Mössbauer effect. Malathi *et al.* (1971) investigated the collapse of the clay structure at elevated temperatures, the effects of an acid pretreatment upon this collapse, and also the uptake of iron ions from aqueous solution and their subsequent migrations upon heating. Figure 9 shows their results for various samples of montmorillonite. There is very little change in the untreated material upon heating (Fig. 9a). The iron ions substituted for $Al^{3+}$. The simple acid treatment, however, has substantially

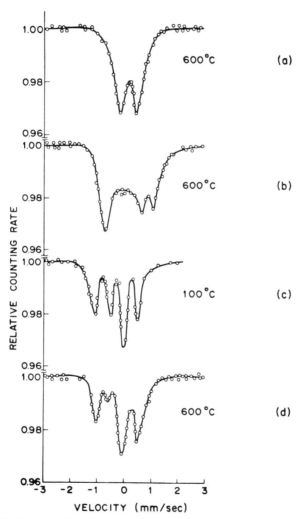

**Fig. 9.**  Mössbauer spectra at room temperature (palladium source) of variously treated montmorillonite heated in air for 6 hr at the indicated temperature. (a) untreated; (b) treated with HCl (10%) for 3 days and washed with dilute $HNO_3$; (c) and (d) as in (b) followed by soaking in aqueous $Fe(NO_3)_3$ (pH = 0.1) for 3 days and washed with dilute $HNO_3$. (Malathi et al., 1971.)

weakened the lattice and it decomposes by 600°C, forming a variety of other minerals (Fig. 9b). The spectra are interpreted as two quadrupole split doublets with their lower energy peaks overlapping. The Mössbauer parameters are consistent with the designated decomposition products.

The spectra of the material treated with a solution of $Fe(NO_3)_3$ is complex and interesting. Malathi et al. (1971) interpret the spectrum as indicating that the iron is adsorbed in three sites whose spectra are superimposed upon the naturally occurring doublet (see Fig. 9c). The three sites are in order of decreasing energy Fe(III) in the basal plane (isotope shift $= -0.12$ mm/sec), at the broken edges (isotope shift $= -0.50$ mm/sec), and in vacant silicon sites ($-1.00$ mm/sec). Figure 9d, then, shows a migration of the Fe(III) from the edges into the basal plane as the temperature is raised. This imparts some stability to the acid-treated material so that it does not break down until above 600°C. The final decomposition appears similar to that of the acid-treated sample without additional iron. Because of the structural difference between the two minerals, illite does not accept Fe(III) into silica vacancies and accepts very little in the basal plane. Other than that, the two materials behave similarly upon heating.

## B. Hydroxides and Hydrated Oxides

Much of the work on hydroxides falls more properly in the realm of corrosion (e.g., Pritchard and Mould, 1971; Pritchard et al., 1971) and is discussed in Chapter 3. Another aspect of hydroxides not covered here is their early stages of formation, i.e., by hydrolysis (Vertes et al., 1973) or as a gel (van der Giessen, 1967). Moreira et al. (1973) used Mössbauer spectroscopy to study the effects of hydrolysis or precipitating conditions (i.e., anions, rate of change of pH, time, temperature, etc.) in relation to whether $\alpha$- or $\beta$-FeOOH, an amorphous gel, or $\alpha$-$Fe_2O_3$ was formed.

There have been a number of studies that utilize the Mössbauer effect to investigate thermal decomposition of various crystalline modifications of FeOOH. Many of these have been primarily concerned with the formation and properties of superparamagnetic particles. A particularly thorough and informative study of the actual decomposition of $\beta$- and $\delta$-FeOOH was performed by Dezsi et al. (1967). They also correlated their results with conventional thermal analytical techniques such as thermogravimetry (TG) and differential thermal analysis (DTA). Figures 10 and 11 summarize the results of their Mössbauer measurements.

The thermoanalytical results for $\delta$-FeOOH are complementary to the Mössbauer results of Fig. 10. Below 350°K, there is no weight loss, but the material passes through its Néel temperature $T_N$ of 295°K. As the six-line

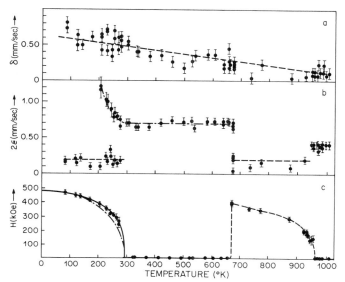

**Fig. 10.** Variations of the isomer shift $\delta$ (chromium source), the quadrupole splitting $2\varepsilon$, and the internal magnetic field $H$ as functions of temperature in the Mössbauer patterns of $\beta$-FeOOH. The dashed line is the Brillion Function for $S = \frac{5}{2}$. (Dezsi *et al.*, 1967.)

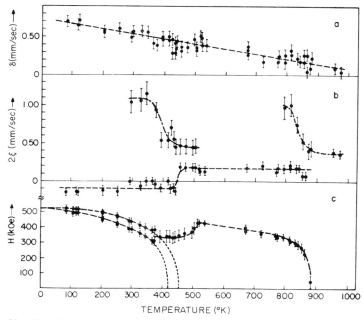

**Fig. 11.** Variations of the isomer shift (chromium source) $\delta$, the quadrupole splitting $2\varepsilon$, and the internal magnetic field $H$ as functions of temperature in the Mössbauer patterns of $\delta$-FeOOH. (Dezsi *et al.*, 1967.)

magnetic hyperfine pattern collapses, a quadrupole doublet begins to grow, and the two spectra coexist for a short range of temperature below $T_N$. This is evident from Fig. 10b, where two quadrupole splittings are shown from $\sim 200°-300°K$. The isomer shift does not exhibit any discontinuities and indicates that the iron remains trivalent. In the region of $300°-600°K$, there is a gradual loss of weight corresponding to the formation of $Fe_2O_3$. There is also an endothermic peak in the DTA around $500°K$ corresponding to this event. The $Fe_2O_3$ initially formed is amorphous or microcrystalline, and hence there is no magnetic hyperfine splitting but a rather large electrical quadrupole splitting instead.

Subsequently, there is a crystallization of this sample into well-formed $\alpha$-$Fe_2O_3$. This is evidenced by the discontinuities in Figs. 10b and c around $673°K$. The Mössbauer parameters assume the values associated with $\alpha$-$Fe_2O_3$. Upon further heating, the sample passes through its $T_N$. The temperature of the recrystallization is indicated at $673°K$ in Fig. 10, because the sample was held at that temperature for a prolonged time. Mössbauer spectra taken at 1 hr intervals showed a gradual transition over a period of 12 hr from the paramagnetic to antiferromagnetic state. The DTA and TG experiments performed at heating rates of $9°K/min$ indicated an abrupt exotherm and small weight loss at $743°K$ corresponding to the recrystallization. Such a displacement to higher temperatures would be expected from these dynamic techniques.

The case of $\delta$-FeOOH is more complex. The TG curve shows a gradual weight loss in the region of $370°-600°K$, and there is no real structure in the DTA trace. $\delta$-FeOOH is ferrimagnetic, with $Fe^{3+}$ ions on random octahedral or tetrahedral sites in proportions that are variable, depending upon sample history (Francombe and Rooksby, 1959). Consequently, two barely resolved magnetic hyperfine splittings are shown in Fig. 11c. The material begins to decompose, as indicated by the weight loss below the $T_N$ of $\delta$-FeOOH, and $\alpha$-$Fe_2O_3$ is formed. There may be some $\alpha$-FeOOH formed as an intermediate, and the second quadrupole splitting in the $400°-500°K$ region is attributed to this intermediate. There is a gradual conversion of the bulk of the $\delta$-FeOOH to $\alpha$-$Fe_2O_3$, as indicated by the magnetic hyperfine splitting shown in Fig. 11c. Again, the isomer shift indicates that the iron remains trivalent during the decomposition.

Jadhao et al. (1972) also investigated the dehydration of $\beta$-FeOOH using the Mössbauer effect. The linewidth was found to decrease significantly with increasing temperature for both the $\beta$-FeOOH and $\alpha$-$Fe_2O_3$ phases. This is presumably due to an increase in particle size and perfection. They also observed the decomposition of $Cr_{0.8}Fe_{0.2}(OH)_3$, for which there is not the complication of magnetic ordering of either the starting material or product. Here, the quadrupole splitting and linewidth decreased up to complete conversion to the oxide but did not change greatly beyond that. Vlasov *et al.* (1970, 1972) actually observed the recoil-free fraction to become virtually zero at the height of the decomposition of $\beta$- and $\delta$-FeOOH. The decomposition of FeOOH at high pressures has also been investigated (Pernet *et al.*, 1973). Srivastava and Singh (1974) have concluded from Mössbauer spectra and DTA that the decomposition of $Fe(OH)_3$ proceeds directly to $\alpha$-$Fe_2O_3$ and is nearly complete by 350°C.

Various hydroxide compounds of iron have served as intermediates in the preparation of ferrites. In these cases, the desired ratio of cations is coprecipitated as a hydroxide or mixed hydroxide–carbonate. On occasion, the Mössbauer effect has been used to follow the thermal conversion of such coprecipitates to the desired ferrite. Examples of this are the work of Haneda *et al.* (1974) on the formation of $BaFe_{12}O_{19}$, Robbins *et al.* (1972) on $LnFeO_3$ and $Ln_3Fe_5O_{12}$, and Schrey (1967) on $Mn_{0.5}Zn_{0.4}Fe_{2.1}O_4$. Schrey's work illustrates the increase in particle size and perfection on hydrothermal treatment of a coprecipitated mixed hydroxide. The well-crystallized spinel is formed around 300°C hydrothermally. X-ray and Mössbauer spectra also reveal the exsolution of iron oxide upon heating above 500°C in air and the subsequent reformation of the ferrite upon firing at temperatures above 1000°C.

Besides the extensive work on hydroxide and hydrated oxides of iron, there has been some interesting work on the thermal decomposition of $Sb_2O_5 \cdot nH_2O$, where $n = 3.5$–$3.7$. Stewart *et al.* (1970) checked the weight loss of this material, but it was inconclusive in determining the oxidation state of antimony because they were unable to establish the proportion of water to oxygen evolved. The Mössbauer effect was used, and they found no plateau in the weight loss that corresponded to pure $Sb_2O_5$. Examination of several commercial sources of $Sb_2O_5$, in fact, showed about 30–35% of the antimony was trivalent, corresponding to about $Sb_6O_{13}$ or $Sb_3O_6OH$. The latter was ruled out by infrared spectroscopy. Upon heating to $>935$°C $\beta$-$Sb_2O_4$ was formed and the ratio of Sb(III)/Sb(V) derived from Mössbauer spectra was 0.98. Above 1050°C, the material sublimed, and both $Sb_2O_3$ and $\beta$-$Sb_2O_4$ condensed.

## C. Cyanides

There has been considerable interest in the cyano complexes of iron because of the strong covalent bonding aspects involved. They also have some value as precursors in some syntheses (Gallagher, 1968). The thermal decomposition of these materials is generally complex, particularly in reducing or inert atmospheres, and consequently any information obtained by Mössbauer spectroscopy is highly welcome.

The hydrates of $Li_3Fe(CN)_6 \cdot xH_2O$ and $Na_3Fe(CN)_6 \cdot xH_2O$ were studied by Dominques and Danon (1972). They first measured the spectrum of $Li_3Fe(CN)_6 \cdot 4H_2O$ and found a doublet having a quadrupole splitting of 0.80 mm/sec and an isomer shift of $-0.30$ mm/sec relative to $^{57}CoPd$. Upon equilibrating this hygroscopic material with a moist atmosphere at room temperature, a second spectrum associated with a higher unknown hydrate rapidly replaced the original. The parameters of the new spectrum were a quadrupole splitting of only 0.25 mm/sec and the same isomer shift. This change was reversible, as evidenced by the observed spectra after various times in vacuum. The spectrum associated with the higher hydrate was gradually replaced by the spectrum of the tetrahydrates.

Solid higher hydrates of $Na_3Fe(CN)_6 \cdot H_2O$ were only formed by freezing aqueous solutions of the monohydrate. The quadrupole splitting was 0.958 mm/sec at 77°K for the monohydrate and completely collapsed for the frozen solution. This total disappearance of the quadrupole splitting for a low spin $d^5$ ion was explained as the result of a compensating arrangement of the $Na^+$ ions in the frozen phase.

There were no observed hydrates for the potassium salt even in frozen solutions. The general conclusions, therefore, were that the tendency for hydrate formation is $Li > Na > K$ and that increasing the degree of hydration within a particular compound of this series markedly reduces the electric field gradient at the iron site.

There have been a number of studies at higher temperatures beyond the simple dehydration reactions that made use of the Mössbauer effect. One of the more detailed and informative examples is that of the thermal decomposition of $EuFe(CN)_6 \cdot 5H_2O$ and $NH_4EuFe(CN)_6 \cdot 4H_2O$ by Gallagher and Schrey (1969) and Gallagher and Prescott (1970). In this instance, both iron and europium were studied by Mössbauer spectroscopy, and the results correlated with those of TG, evolved gas analysis (EGA), x-ray diffraction analysis, infrared spectroscopy, and surface area analysis.

The decomposition in air or oxygen is relatively straightforward and leads to the formation of $EuFeO_3$. Figure 12 shows some selected results starting from $NH_4EuFe(CN)_6 \cdot 4H_2O$. The uncalcined sample shows no quadrupole splitting and an isomer shift typical of $Fe(CN)_6^{4-}$. At 200°C, the sample has oxidized to an amorphous or micro-crystalline $Fe_2O_3$. As the temperature is raised, this reactive $Fe_2O_3$ reacts with the $Eu_2O_3$ formed in the initial stages of the reaction to yield $EuFeO_3$. At 600°C, the spectra of both $EuFeO_3$ and the unreacted $Fe_2O_3$ are discernable. By 800°C, the material has completely reacted to form $EuFeO_3$. The europium spectra for this sequence merely indicate that the ion remains trivalent during the course of the reaction.

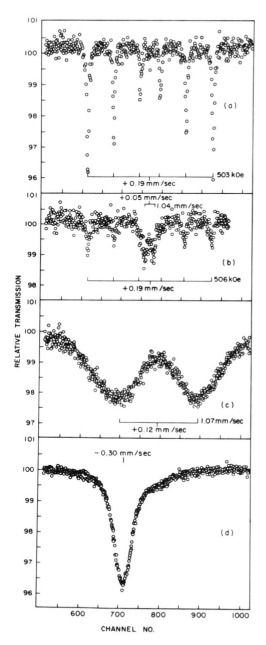

**Fig. 12.** Mössbauer spectra measured at room temperature relative to $^{57}CoPd$ of samples of $NH_4EuFe(CN)_6 \cdot 4H_2O$ calcined in air for $\frac{1}{2}$ hr at (a) 800°C, (b) 600°C, (c) 200°C, (d) 25°C. (Gallagher and Schrey, 1969.)

The analogous series starting with $EuFe(CN)_6 \cdot 5H_2O$ reveal the same general pattern, i.e., essentially the combustion of the hexacyano complex around $200°-300°C$ to form finely divided and intimately mixed $Fe_2O_3$ and $Eu_2O_3$, which react at $500°-700°C$ to form $EuFeO_3$. A physical mixture of these two oxides would not react nearly as quickly or completely at such relatively low temperatures. This, along with the precise europium-to-iron stoichiometric ratio indicates the value of these compounds for synthesis.

The decomposition of these compounds in inert atmosphere or vacuum is much more complex. Figure 13 shows selected spectra of samples of $EuFe(CN)_6 \cdot 5H_2O$ heated in vacuum. The unheated material exhibits the quadrupole splitting, which serves to differentiate $Fe(CN)_6^{3-}$ from $Fe(CN)_6^{4-}$ in this system. By $300°C$, then, $Fe(CN)_6^{3-}$ has been reduced to $Fe(CN)_6^{4-}$ with the evolution of cyanogen, $(CN)_2$, as indicated in the EGA experiments. This reduction is also verified by infrared spectroscopy.

At slightly higher temperatures, or perhaps simultaneously, the cyanide complex decomposes hydrolytically by the evolution of HCN. A divalent iron cyanide product is formed that is somewhat arbitrarily referred to as $Fe(CN)_2$.

The overall scheme of the decomposition of these compounds at intermediate temperatures based on the Mössbauer spectra and thermoanalytical measurements is summarized as

$$2EuFe(CN)_6 \cdot 5H_2O \rightarrow 2EuOOH + 2Fe(CN)_2 + (CN)_2 + 6H_2O + 6HCN \quad (8)$$

$$NH_4EuFe(CN)_6 \cdot 4H_2O \rightarrow EuOOH + Fe(CN)_2 + NH_3 + 2H_2O + 4HCN. \quad (9)$$

This divalent cyanide species persists to fairly elevated temperatures ($\sim 600°C$) in vacuum. It begins to break down above $600°C$ to form $Fe_3C$. This was identified from its Mössbauer pattern, particularly the magnetic hyperfine splitting of about 210 kOe. By $1000°C$ in vacuum, metallic iron is clearly evident. The Mössbauer spectra were of particular value during this sequence because the x-ray diffraction patterns were very poorly defined.

In vacuum or inert atmospheres, the europium ion also undergoes a reduction at the higher temperatures, as evident from Mössbauer spectra shown in Fig. 14. The peak around $-13$ mm/sec is due to $Eu^{2+}$. It can be detected around $300°C$ and grows continuously, becoming predominant at $700°C$ and above. This reduction is the result of the reaction with the various carbonaceous residues and decomposition products.

Borshagovskii *et al.* (1967) have similarly used the Mössbauer effect to study the thermal decomposition of some hexacyanoferrates with very similar conclusions regarding the chain of iron products. They point out the probable polymerization of the divalent cyanide as

$$3Fe(CN)_2 \rightarrow Fe_2[Fe(CN)_6]. \quad (10)$$

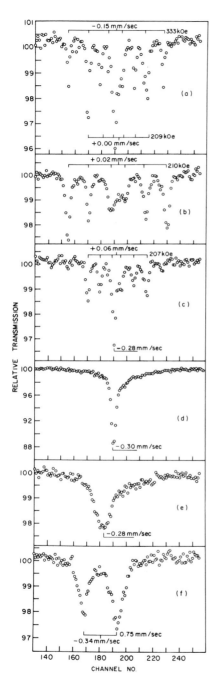

**Fig. 13.** Mössbauer spectra measured at room temperature relative to $^{57}$Co*Pd* for samples of EuFe(CN)$_6$·5H$_2$O calcined in vacuum for $\frac{1}{2}$ hr at (a) 1000°C, (b) 900°C, (c) 700°C, (d) 600°C, (e) 300°C, (f) 25°C. (Gallagher and Schrey, 1969.)

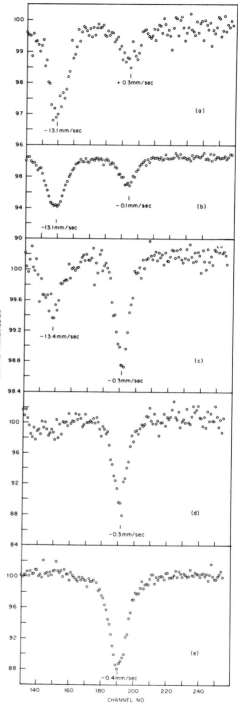

**Fig. 14.** Mössbauer spectra measured at room temperature relative to $^{151}Sm_2O_3$ of samples of $NH_4EuFe(CN)_6 \cdot 4H_2O$ calcined in vacuum for $\frac{1}{2}$ hr at (a) 900°C, (b) 700°C, (c) 600°C, (d) 300°C, (e) 25°C. (Gallagher and Schrey, 1969.)

222

There is, however, some disagreement in that Borshagovskii *et al.* (1967) propose the evolution of $(CN)_2$ from the decomposition of $Fe(CN)_6^{4-}$, which was not observed in the EGA work of Gallagher and Prescott (1970), and also suggest that the rare earth compounds formed initially are an equal molar mixture of the oxide and nitride, as opposed to the oxyhydroxide described in Eqs. (8) and (9).

The reduction of $Fe(CN)_6^{3-}$ to $Fe(CN)_6^{4-}$ upon heating was also observed by Ramshesh *et al.* (1973), starting with either $K_3Fe(CN)_6$ or $Na_2[Fe(CN)_5NO]\cdot 2H_2O$. They, however, propose a highly unexpected oxidation of iron as the first step in the decomposition of $Na_2NH_4[Fe(CN)_5NH_3]$. However, their proposed equation,

$$Na_2NH_4[Fe(CN)_5NH_3] \rightarrow Na_2[Fe(CN)_5NH_3] + NH_4^+, \tag{11}$$

is not charge balanced nor does it indicate to what species the electron is transferred. At higher temperatures, it is proposed that reduction of iron to $Na_4Fe(CN)_6$ occurs.

Use of the Mössbauer effect has been made by Cosgrove *et al.* (1973) to elucidate the ferric ferrocyanide $\rightleftharpoons$ ferrous ferricyanide equilibria. They showed that both Prussian blue and Turnbull's blue were ferric ferrocyanide composed of high-spin iron(III) and low-spin iron(II). Upon vacuum pyrolysis, both materials were found to convert to ferrous ferricyanide with the evolution of water. The interpretation of the Mössbauer spectra indicating this electron transfer was clearly simplified by the selective enrichment of various iron sites during synthesis. Figure 15 shows the spectrum of Prussian blue prepared with highly enriched $Fe^{3+}$. The spectra arising

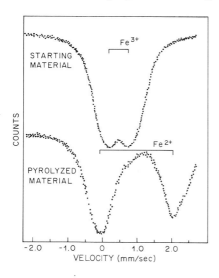

**Fig. 15.** Mössbauer spectra measured at room temperature relative to $^{57}CoCu$, illustrating $Fe^{3+}$ enrichment of Prussian blue and subsequent vacuum pyrolosis at 400°C. (Cosgrove *et al.*, 1973.)

from the two types of Fe(II) are essentially overwhelmed and lost in the background. Hence, the complexity of the spectrum is greatly reduced by essentially isolating that due to one valence and site. The mild vacuum pyrolysis at 400°C can be seen to lead to a clean almost complete conversion to $Fe^{2+}$. The electron transfer is attributed by Cosgrove *et al.* (1973) to the relaxation in internal pressure brought about by the release of considerable water. The reaction is reversible, as shown by equilibration of the pyrolized product with moist air or aqueous HCl. Both x-ray and Mössbauer spectra indicate such samples revert to ferric ferrocyanide.

A different interpretation of the decomposition of Prussian blue has recently been made by Allen and Bonnette (1974). The lack of agreement arises from their inability to detect high-spin $Fe^{2+}$ similar to that seen in Fig. 15 and, most importantly, from their chemical analysis. The carbon: nitrogen:iron ratio decrease from 2.9:2.9:1 for samples fired in vacuum at 219°C to 1.0:0.9:1.0 for samples fired at 370°C in vacuum. Cosgrove *et al.* (1973) do mention the evolution of "small amounts of $(CN)_2$" around 300°C in vacuum but certainly do not allow for any such gross change in composition.

Allen and Bonnette (1974) also use isotopic enrichment to facilitate their interpretation of spectra but in this case substitute with Mössbauer-inactive $^{56}Fe$. They conclude that there is a physical isotopic exchange of iron between the sites in carbon and nitrogen holes via either an atomic hop by iron or a $CN^-$ flip. Electron transfer is not considered, because they see no $Fe^{2+}$ spectra.

They extend their studies to higher temperatures in vacuum and also in air. In vacuum, the sequence is to $Fe_3C$ and finally iron. In air, the oxides are formed around 400°C. Both of these results are similar to the decompositions of $EuFe(CN)_6$ and $NH_4EuFe(CN)_6$ described earlier. Clearly, further work will be required to establish the lower temperature decomposition phenomena of Prussian blue in vacuum and their reversibility.

The subject of the thermal decomposition of cyanides leads nicely into the analogous effects of pressure and radiation upon these compounds. A detailed description of these effects is left to further chapters, but the parallel nature of the results should be briefly mentioned. For example, Fung and Drickamer (1969) had demonstrated that ferricyanides tend to convert rapidly to ferrocyanides at room temperature under modest pressures and that this conversion is readily reversible. The correlation between this and the thermal decomposition was described by Cosgrove *et al.* (1973).

Similarly, the effects of radiation frequently closely resemble those due to thermal decomposition. Saitovitch *et al.* (1972) show that electron irradiation of hydrated ferricyanides leads to the formation of ferrocyanide. The reduction does not occur with the anhydrous compounds, suggesting the free radicals arising from the radiolysis of the water molecules play an important role in the mechanism.

Kisynska *et al.* (1972) have investigated the effects of irradiation by 1.5 MeV protons upon $K_3Fe(CN)_6$ and $K_4Fe(CN)_6$. The reduction of $Fe(CN)_6^{-3}$ occurs as anticipated, but at higher beam currents or longer times, the reduction continues to $Fe_3C$ and finally iron, very analogous to the thermal decomposition in inert atmospheres. In fact, the distinction between the thermal and radiative effects may frequently be difficult to establish. The use of the Mössbauer effect to investigate radiation effects and defects in general has been extensively discussed by Wertheim *et al.* (1971).

Besides the qualitative analytical advantages offered by the Mössbauer effect in studying these decompositions, Lee *et al.* (1974) have shown that more reliable quantitative results can be obtained under certain circumstances. In studying the neutron-induced decomposition of $K_3Fe(CN)_6$, they determined from the Mössbauer spectra that approximately 3% of the sample had been reduced to $Fe(CN)_6^{4-}$ after 3 weeks of irradiation. Conventional chemical analysis of a sample similarly irradiated for only 100 hr implied that 18% had been reduced. It was subsequently determined that complication from the formation of $Fe(OH)_3$ and Prussian blue invalidated the chemical analysis but did not interfere with the Mössbauer analysis.

## D. Other Inorganic Salts

Much of the high quality $Fe_2O_3$ used in industry is derived from pickling liquors. Consequently, the thermal decomposition of $FeSO_4 \cdot xH_2O$ is of considerable interest. Several investigators (Gallagher *et al.*, 1970; Vertes and Zsoldos, 1970; Neto and Garg, 1973) have made use of the Mössbauer effect in their studies.

Figure 16 shows some of the Mössbauer spectra for samples of $FeSO_4 \cdot xH_2O$ at various temperatures in air. Figure 16a is of the unheated freeze-dried material. The composition based upon TG data is nearly $FeSO_4 \cdot H_2O$ (A). A small quantity of unknown hydrate having a smaller quadrupole splitting (B) is also present. The values of the Mössbauer parameters for the monohydrate agree reasonably well among the three groups of investigators. Because the weaker doublet disappears upon heating in a vacuum to about 244°C, it is attributed to a higher hydrate. The Mössbauer parameters, however, do not agree with those reported for $FeSO_4 \cdot 7H_2O$ or $FeSO_4 \cdot 4H_2O$ (Vertes and Zsoldos, 1970; Neto and Garg, 1973).

As the temperature is increased in air, a doublet (C) attributed to $FeOHSO_4$ appears in the spectrum at 205°C. This species is also postulated as a possible intermediate by Vertes and Zsoldos (1970); however, they

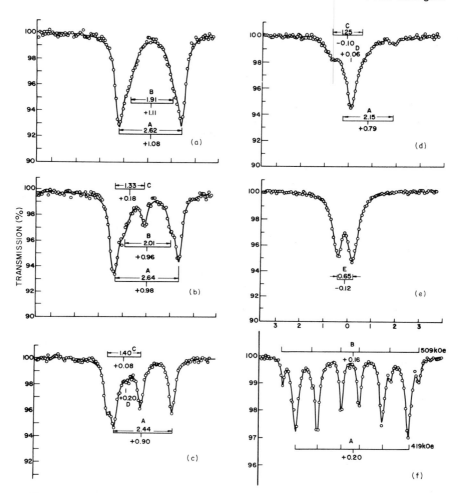

**Fig. 16.** Mössbauer spectra relative to $^{57}CoPd$ at 25°C of freeze-dried $FeSO_4 \cdot H_2O$ measured at (a) 25°C, (b) 205°C, (c) 317°C, (d) 499°C, (e) 562°C, (f) 25°C (sample cooled from 562°C). (Gallagher *et al.*, 1970a.)

assigned their doublet to $Fe_2O(SO_4)_2$. This species reaches a maximum in Fig. 16 at 317°C but is still present in the spectrum at 499°C.

An unsplit line (D) predominates at 499°C in Fig. 16d. A similar line is observed by Vertes and Zsoldos (1970), and they assign it to $Fe_2(SO_4)_3$ formed by the disproportion of $Fe_2O(SO_4)_2$ according to

$$3Fe_2O(SO_4)_2 \rightarrow 2Fe_2(SO_4)_3 + Fe_2O_3. \qquad (12)$$

This unsplit line (D) is attributed to some unknown intermediate by

Gallagher *et al.* (1970) because it was felt necessary to assign the quadrupole split doublet (E) observed at higher temperatures (562°C in Fig. 16e) to $Fe_2O(SO_4)_2$. This choice was made because the room-temperature spectrum (Fig. 16f) of the same material showed this sample to contain some $Fe_2O_3$ (B) but predominantly a different trivalent species (A), which is also magnetically ordered at room temperature. Since $Fe_2(SO_4)_3$ is not, it was concluded that this species (E and A) was most likely $Fe_2O(SO_4)_2$. Because $Fe_2(SO_4)_3$ shows significant quadrupole splitting (1.4 mm/sec at room temperature), it is also unlikely that the unsplit line (D) at lower temperature is a result of this compound. Neto and Garg (1973), on the other hand, claim yet another set of parameters for $Fe_2O(SO_4)$, i.e., a quadrupole splitting of 1.4 mm/sec at room temperature without magnetic hyperfine interaction. Clearly, the system is incompletely understood at this time, but studies utilizing the Mössbauer effect have been more informative than most and will undoubtedly play a major role in the final solution.

Tomov *et al.* (1971) have utilized the Mössbauer Effect to analyze the multiphase mixture of Fe, FeO, $Fe_3O_4$, $\gamma$-$Fe_2O_3$, and $\alpha$-$Fe_2O_3$ that finally arises from the calcination of $FeSO_4$ in air followed by treatment in hydrogen. They point out the suitability of this technique for evaluating iron oxide electrodes produced for alkaline batteries.

Vertes and Zsoldos (1970), Vertes *et al.* (1969), and Deszi *et al.* (1971) have also studied the decomposition of $FeCl_2 \cdot 4H_2O$. The sequence from the tetrahydrate, to the anhydrous salt was observed at 100°C without oxidation. At temperatures as low as 200°C, the formation of $Fe_2O_3$ is as great as 50% after only 2 hr. At 400°C, only $Fe_2O_3$ is present after 1 hr.

In the course of their studies on metal halide accelerators for the vulcanization of butyl rubber, Christov *et al.* (1971) followed the decomposition of $SnCl_2 \cdot 2H_2O$ by Mössbauer spectroscopy. They observed that the compound dehydrated during vulcanization at 170°C for 1 hr. Starting with the anhydrous salt in the mix initially to avoid decomposition proved technologically beneficial.

The Mössbauer effect has also been utilized to indicate the change in valence associated with the decomposition of some divalent compounds of europium (Gallagher, 1968). Figure 17 shows Mössbauer spectra of $EuCO_3$ in a vacuum ($\sim 10^{-5}$ Torr) at the indicated temperatures. There were significant amounts of Eu(III) present in the material originally. Oxidation is clearly evident in the spectrum at 400°C and is complete by 500°C. Similar spectra for $EuSO_4$ reveal the absence of any Eu(III) in the starting material, no significant oxidation at 400°C, but complete oxidation by 600°C.

The decomposition of Schlippe's salt, $Na_3SbS_4 \cdot 9H_2O$ was studied by Long and Bowen (1970). The isomer shift of the undecomposed compound is indicative of $Sb^{5+}$ with the anticipated quadrupole splitting associated

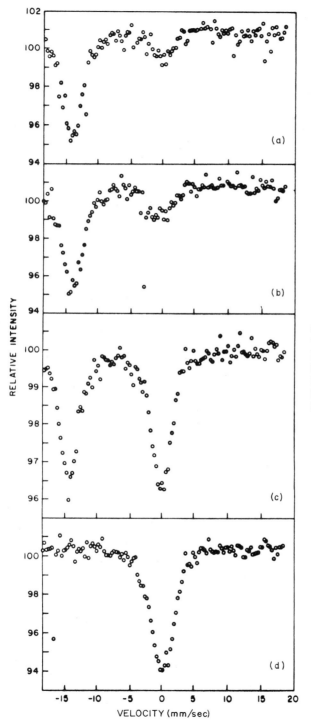

**Fig. 17.** Mössbauer spectra relative to $^{151}Sm_2O_3$ at 25°C of $EuCO_3 \cdot 0.5H_2O$ in vacuum ($\sim 10^{-5}$ Torr) measured at (a) 25°C, (b) 275°C, (c) 400°C, (d) 500°C. (Gallagher, 1968.)

with the surrounding tetrahedron of sulfur. They observe a broadening of the lines upon dehydration, which they attribute to distortions of the tetrahedral environment and concomitant variations of quadrupole splitting but with no change in oxidation state of the antimony. Upon exposure to air for long periods, however, a new peak forms that has a more positive isomer shift similar to those associated with Sb(V)—O bonds. These results clearly disprove the prior speculation that these decomposition products were sulfides of Sb(III).

### E. Organic Salts and Organometallics

Oxalates have been especially intriguing salts to chemists for many years. The complexing ability and sensitivity toward oxidation have made them a fascinating and fruitful subject. As a consequence, many of the early applications of the Mössbauer effect to the study of chemical decompositions were directed toward the decomposition of simple and complex oxalates of iron induced by heat and radiation.

Although one of the first such studies, the thermal decomposition of $Ba_3[Fe(C_2O_4)_3]_2$ and $Sr_3[Fe(C_2O_4)_3]_2$ is still highly informative and interesting due to the frequent changes in oxidation state that occur (Gallagher and Kurkjian, 1966). Figure 18 displays selected spectra for samples of $Sr_3[Fe(C_2O_4)_3]_2 \cdot 2H_2O$ that had been heated in air to the indicated temperatures. More detailed results are given in Table II. These results correlated well with previous thermoanalytical measurements (Gallagher, 1965).

The compound at room temperature (Fig. 18a) has an isomer shift indicative of trivalent iron. After having been heated to 300°C, the spectrum clearly indicates that the material has been reduced. Similar reduction was observed upon heating $Fe_2(C_2O_4)_3$ and $Ba_3[Fe(C_2O_4)_3]_2$ to 200°C. The parameters agree with the formation of $FeC_2O_4$ according to

$$Sr_3[Fe(C_2O_4)_3]_2 \rightarrow 3SrC_2O_4 + 2FeC_2O_4 + 2CO_2 \qquad (13)$$

There is a hint of some trivalent remnant or product concealed within the lower energy peak.

By 400°C in air (Fig. 18c), the oxalate has decomposed according to

$$2FeC_2O_4 + \tfrac{1}{2}O_2 \rightarrow Fe_2O_3 + CO + CO_2, \qquad (14)$$

which undoubtedly represents the summation of several process. In an inert atmosphere, Foerster *et al.* (1973) have shown that the decomposition proceeds to FeO and α-Fe. The FeO further disproportionates to $Fe_3O_4$ and α-Fe. In the presence of ample oxygen, these products are not evident. Because of the possible presence of $Fe_3O_4$, the initial $Fe_2O_3$ formed may

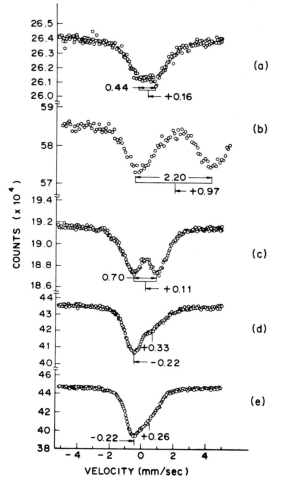

**Fig. 18.** Mössbauer spectra measured at room temperature relative to $^{57}CoCu$, of $Sr_3[Fe(C_2O_4)_3]_2 \cdot 2H_2O$ calcined in air to (a) 25°C, (b) 300°C, (c) 400°C, (d) 700°C, (e) 1000°C. (Gallagher and Kurkjian, 1966.)

be in the $\gamma$ state (Krauth *et al.*, 1967). Both Gallagher and Kurkjian (1966) and Krauth *et al.* (1967) agree that the initial $Fe_2O_3$ formed is so finely divided that it is superparamagnetic and gives rise to a quadrupole split doublet rather than the characteristic six-line patterns. Jach (1973b) also observes a trivalent doublet upon the initial decomposition but assigns it to some unknown intermediate. Table II indicates that this sample heated to 400°C in air does show a pattern characteristic of bulk $Fe_2O_3$ when cooled to 4°K.

This finely divided $Fe_2O_3$ intimately mixed with the $SrCO_3$ formed by the decomposition of $SrC_2O_4$ reacts by 700°C to form $SrFeO_{3-x}$ or $Sr_3Fe_2O_{7-x}$.

## TABLE II

Summary of the Mössbauer Parameters for $Sr_3[Fe(C_2O_4)_3]_2 \cdot 2H_2O$ at Various Points during Its Thermal Decomposition[a]

| Temperature of calcination (°C) | Temperature measurement (°K) | Primary | | | Secondary | | |
|---|---|---|---|---|---|---|---|
| | | Isomer shift[b] | Quad (mm/sec) | Mag (kOe) | Isomer shift[b] | Quad (mm/sec) | Mag (kOe) |
| 25 | 300 | +0.16 | 0.44 | 0 | — | — | — |
| 25 | 78 | +0.23 | — | 0 | — | — | — |
| 25 | 4 | +0.23 | — | 0 | — | — | — |
| 200 | 300 | +0.95 | 2.3 | 0 | +0.17 | 0.52 | 0 |
| 300 | 300 | +0.97 | 2.2 | 0 | — | — | — |
| 400 | 300 | +0.11 | 0.70 | 0 | — | — | — |
| 400 | 4 | +0.17 | — | 503 | — | — | — |
| 500 | 300 | +0.04 | 0.71 | 0 | — | — | — |
| 600 | 300 | −0.05 | 0.74 | 0 | — | — | — |
| 700 | 300 | −0.22 | 0 | 0 | +0.33[c] | — | 0 |
| 1000 | 300 | −0.26 | 0 | 0 | +0.26[c] | — | 0 |
| 1200 | 300 | −0.26 | 0 | 0 | +0.30[c] | — | 0 |
| 1000[e] | 300 | −0.27 | 0 | 0 | +0.20[c] | — | 0 |
| 1000[e] | 4 | −0.2[d] | — | ~228 | +0.1[d] | — | ~435 |

[a] Gallagher and Kurkjian (1966).

[b] Isomer shift (IS) is relative to a copper source and is in units of millimeters per second.

[c] These values may represent only a portion of a quadrupole split transition, half of which is concealed by the spectrum of another component.

[d] The values for these samples are based on just the outer lines of the magnetic hyperfine splitting and are only approximate.

[e] These samples have been slow cooled in contrast to the others, which were quenched.

These compounds of Fe(IV) were discussed at some length in Section III on solid–gas reactions. The predominance of Fe(IV) is evident in Fig. 18d. There is a significant amount of Fe(III) present at 700° (Fig. 18d), and this increases with increasing temperature (Fig. 18e). As mentioned earlier, the numerous changes in the valence of iron, i.e., III → II → III → IV → III, make this a particularly interesting example.

The Mössbauer effect has been used to follow the decomposition of solid solution of divalent oxalates to form $Ni_{0.6}Fe_{2.4}O_4$ spinel (Gallagher et al., 1969) or $Fe_{1-x}Mg_xO$ (Foerster et al., 1973). The two groups agree that substitution in solid solution by other divalent ions of comparable size does little to change the Mössbauer parameters of $FeC_2O_4 \cdot 2H_2O$ and that the lines broaden somewhat upon dehydration. Brady and Duncan (1964) similarly found that iron absorbed on the surface of $CoC_2O_4 \cdot 2H_2O$ did not appear chemically different from atoms in the bulk of $Fe_2C_2O_4 \cdot 2H_2O$. After decomposition of the solid solution the presence of $Mg^{2+}$ in the FeO was found to stabilize the compound against disproportionation (Foerster et al., 1973).

The introduction of iron into glassy matrices by reacting $FeC_2O_4 \cdot 2H_2O$ with $P_2O_5$ (Jach et al., 1972) and $H_3BO_3$ (Jach, 1973a,b) has also been investigated using the Mössbauer spectroscopy. The $Fe^{2+} \rightleftharpoons Fe^{3+}$ equilibrium in the resulting glasses was studied in some detail.

The thermal decomposition of $K_3Fe(C_2O_4)_3$ is similar to the decomposition of the strontium and barium complexes described earlier, except that the ionic radii and charges are not suitable for the formation of perovskite structure. In the absence of this stabilizing influence, Fe(IV) is not formed. Bancroft et al. (1970) have done an extensive investigation of this decomposition using the Mössbauer effect. Their conclusions are summarized

$$K_3Fe(C_2O_4)_3 \cdot 3H_2O \xrightarrow[110°C]{in\ air} K_3Fe(C_2O_4)_3 + 3H_2O \tag{15}$$

$$K_3Fe(C_2O_4)_3 \xrightarrow[260°C]{} \tfrac{1}{2}K_6Fe_2(C_2O_4)_5 + CO_2 \tag{16}$$

$$\tfrac{1}{2}K_6Fe_2(C_2O_4)_5 \xrightarrow[380°C]{} \tfrac{3}{2}K_2CO_3 + \tfrac{1}{3}Fe_3O_4 + xC + yCO_2 + zCO_2. \tag{17}$$

The actual ratio of C, CO, and $CO_2$ would be highly variable depending upon the access of oxygen during the decomposition and the uncertain extent of disproportionation of CO according to

$$2CO \rightleftharpoons C + CO_2 \tag{18}$$

At temperatures below about 700°C in the absence of oxygen, the disproportionation is thermodynamically favored. The stability of CO at these temperatures attests to the usual presence of kinetic barriers; however, the finely divided oxide surfaces can be expected to catalyze the reaction to an uncertain extent. As the reducing nature of the atmosphere is decreased, the intermediates containing reduced iron along with carbon formed during the disproportionation are oxidized to $Fe_2O_3$ and $CO_2$. If the decomposition is performed in vacuum, Bancroft *et al.* (1970) found that the final product was nearly pure $\alpha$-Fe.

As was the case with the cyanide complexes, the effects of irradiation of oxalates have been extensively studied using the Mössbauer effect. Dharmawardena and Bancroft (1968), Bancroft *et al.* (1969), and Temperley and Pumplin (1969) studied $K_3Fe(C_2O_4)_3$, and Saito *et al.* (1965) studied $Fe_2(C_2O_4)_3$ by this technique. All of the above studies indicated an initial reduction of iron with the release of $CO_2$. Buchanan (1970) also observed the reduction of $Fe^{3+}$ to $Fe^{2+}$ upon irradiation of the citrate with 1 MeV electron or ultraviolet light. Similarly, the electron capture involved in the decay of $^{57}Co$ has lead to a similar reduction of iron in oxalate sources—e.g., Sano and Ohnuma, (1974) in $[Co(NH_3)_6]_2(C_2O_4)_3 \cdot 4H_2O$ and $[Co(en)_3]_2(C_2O_4)_3 \cdot 9H_2O$, and Fenger *et al.* (1970) in cobalt-doped $K_3Fe(C_2O_4)_3 \cdot 3H_2O$ and $K_3Fe(C_2O_4)_3$. The presence of oxalate ions even in the second coordination sphere was demonstrated to have a pronounced reducing effect when compared with the analogous $[Co(NH_3)_6](NO_3)_3$ or $[Co(en)_3](NO_3)_3$ compounds (Sano and Ohnuma, 1974). Sano and Kanno (1969) observed a similar reduction, in this case for tin, during the decay of $^{119}Sn$ in $K_6Sn_2(C_2O_4)_7 \cdot 4H_2O$.

The use of the Mössbauer effect in the study of the thermal decomposition of $Eu_2(C_2O_4)_3 \cdot 10H_2O$ proved particularly valuable. In thermogravimetric studies by Glasner *et al.* (1964), a plateau in the weight loss curve corresponding to a stoichiometry of Eu–C–3O was observed. In view of the known reductions accompanying the decomposition of oxalate, it seemed reasonable to assume that $EuCO_3$ had formed corresponding to the reduction of $Eu^{3+}$ to $Eu^{2+}$. Thermodynamic considerations of oxalate decompositions as outlined by Dollimore *et al.* (1963) indicated such a reduction should not occur. Gallagher *et al.* (1970) confirmed the existence of such a plateau in the weight loss but utilized the Mössbauer effect to show that the intermediate must be a species of $Eu^{3+}$.

Uncertainty in the stoichiometry of $BaSnO_3$ is the primary drawback in its use as a standard Mössbauer source. Gallagher and Schrey (1972) developed a synthetic technique based upon the precipitation and subsequent thermal decomposition of $BaSn(C_2O_4)_2 \cdot 2.5H_2O$. This provided for the desired barium-to-tin ratio and allowed the preparation of $BaSnO_3$ at a relatively low temperature, where the loss of oxygen was insignificant. To demonstrate the properties of the resulting $BaSnO_3$ and to follow the course of the decomposition, they measured Mössbauer spectra at various temperatures during the decomposition in air. Figure 19 presents some of these spectra. The divalent nature of the tin is evident in Fig. 19a, and dehydration had little effect on the spectra (Fig. 19b). Between 300° and 400°C, the $Sn^{2+}$

**Fig. 19.** Mössbàuer spectra measured at room temperature relative to $BaSnO_3$ of $BaSn(C_2O_4)_2 \cdot 2.5H_2O$ calcined in air at (a) 25°C, (b) 300°C, (c) 400°C, (d) 1000°C. (Gallagher and Schrey, 1972.)

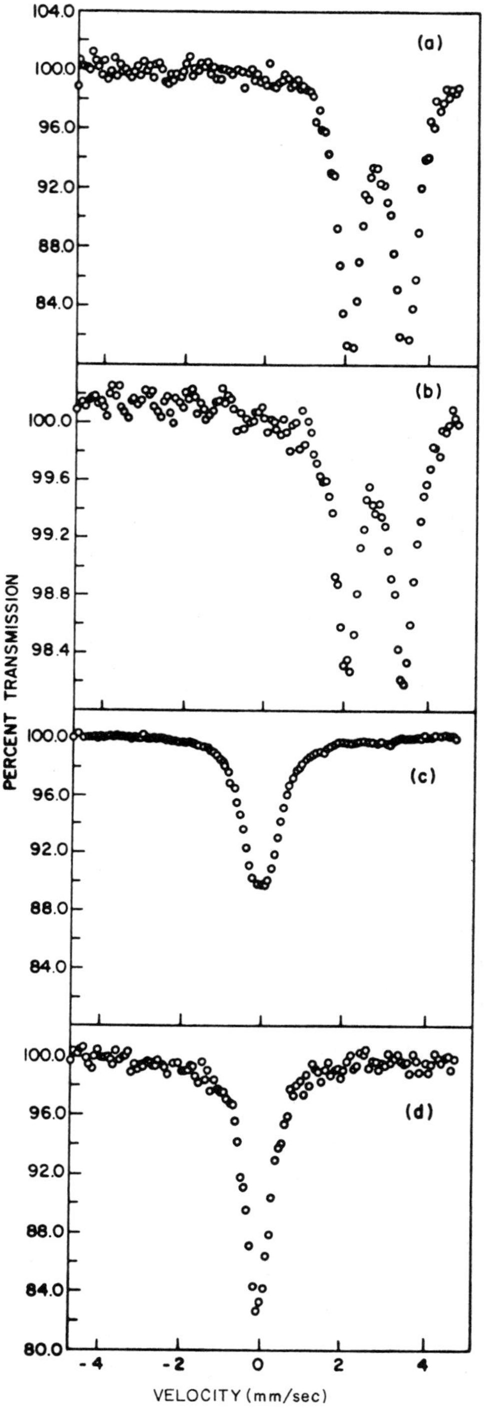

completely oxidized. At 400°C (Fig. 19c), the sample was an intimate mixture of $BaCO_3$ and $SnO_2$. By 1000°C (Fig. 19d), the mixture had reacted to form $BaSnO_3$. This latter reaction takes places at a much lower temperature than a physical mixture of $BaCO_3$ and $SnO_2$ would and proceeds without the formation of $Ba_2SnO_4$, which is a normal complication of the conventional synthesis (Von Wagner and Binder, 1958).

The decomposition of the tetravalent complex, $Ba_2Sn(C_2O_4)_4 \cdot 6H_2O$ was also studied. At 400°C, the products are $2BaCO_3 + SnO_2$. Around 700°C, $BaCO_3 + BaSnO_3$ are formed. Simple acid leaching will yield pure $BaSnO_3$ at this point. At higher temperatures, $Ba_2SnO_4$ forms.

Certainly the major effort in this section has been directed at oxalates, but some work on other organic complexes has been done. Mössbauer spectroscopy was used by Pompa and Van Diepen (1975) to study the formation of non-crystalline $Y_3Fe_5O_{12}$ from citrates. Films of amorphous highly imperfect $Fe_2O_3$ are deposited by the decomposition of $Fe(CO)_5$ on a hot surface. Gallagher et al. (1974) showed that these films contained significant amounts of $Fe^{2+}$ and that considerable $CO_2$ was evolved simultaneously with the crystallization of $\alpha$-$Fe_2O_3$. Tominga et al. (1966, 1970) used Mössbauer spectroscopy to follow the decomposition of some complexes of Fe(II) with pyridine and $\gamma$-picoline. Thermogravimetry indicated one to three steps in the loss of pyridine or $\gamma$-picoline from these complexes. Mössbauer spectra associated with these intermediates in the thermal decomposition of $Fe(Py)_4Cl_2$ are shown in Fig. 20. Although there is a large change in the quadrupole splitting initially, there is obviously no reduction

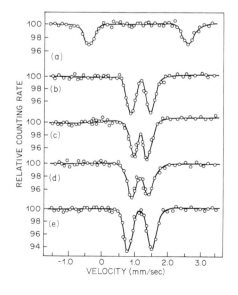

**Fig. 20.** Mössbauer spectra measured at 290°K relative to $^{57}CoCu$ of $Fe(py)_4Cl_2$ and intermediates in its thermal decomposition. (Tominaga et al.,1966.)

| Compound | $\delta$ | $2\varepsilon$ |
|---|---|---|
| (a) $Fe(py)_4Cl_2$ | $1.11 \pm 0.02$ | $3.01 \pm 0.06$ |
| (b) $Fe(py)_2Cl_2$ | $1.13 \pm 0.01$ | $0.52 \pm 0.01$ |
| (c) $Fe(py)Cl_2$ | $1.15 \pm 0.01$ | $0.38 \pm 0.02$ |
| (d) $Fe(py)_{2/3}Cl_2$ | $1.15 \pm 0.01$ | $0.51 \pm 0.01$ |
| (e) $FeCl_2$ | $1.16 \pm 0.01$ | $0.75 \pm 0.01$ |

in the valence of iron occurring through this progression of products. Similar conclusions were deduced from studies on $Fe(Py)_4(NCS)_2$ and $Fe(\gamma\text{-pic})_4Cl_2$, except that fewer intermediates were observed.

## V. Summary and Conclusions

The Mössbauer effect can be of considerable value whenever the reacting species contain or otherwise involve a suitable isotope. It is particularly useful in those instances when the material is microcrystalline or amorphous, where x-ray diffraction techniques are less useful. There is a high degree of sensitivity to the oxidation state, bonding, and local environment of the selected atom. Controlled isotopic substitution can sometimes be used to isolate particular sites or areas within the reacting system.

Examples have been selected that demonstrate some of the past uses of this technique. It now remains for the reader to exercise his imagination and background in order to recognize its value in helping to solve his problems.

### References

Allen, J. F., and Bonnette, A. K. Jr. (1974). *J. Inorg. Chem.* **36**, 1011.
Annersten, H., and Hafner, S. S. (1973). *Z. Krist.* **137**, 321.
Bancroft, G. M., Maddock, A. G., and Burns, R. G. (1967). *Geochim. Coschim. Acta* **31**, 2219.
Bancroft, G. M., Burns, R. G., and Stone, A. J. (1968). *Geochim. Coschim. Acta* **32**, 547.
Bancroft, G. M., Dharmawardena, K. G., and Maddock, A. G. (1969). *J. Chem. Soc. A* 2914.
Bancroft, G. M., Dharmawardena, K. G., and Maddock, A. G. (1970). *Inorg. Chem.* **9**, 223.
Borshagovskii, B. V., Gol'danskii, V. I., Seifer, G. B., and Stukan, R. A. (1967). *Russ. J. Inorg. Chem.* **12**, 1741.
Brady, P. R., and Duncan, J. F. (1964). *J. Chem. Soc.* 653.
Buchanan, D. N. E. (1970). *J. Inorg. Nucl. Chem.* **32**, 3531.
Calusaru, A., Morariu, M., Barb, D., and Rusi, A. (1973). *Radiochim. Acta* **19**, 203.
Channing, D. A., and Graham, M. J. (1972). *Corrosion Sci.* **12**, 271.
Christov, D., Bonchev, Z., Manouchev, T., Dimov, D., and Nenov, N. (1971). *Proc. Conf. Appl. Mössbauer Effect, Tihany, Hungary, 1969* p. 795.
Collins, D. W., and Mulay, L. N. (1968). *IEEE Trans.* **Mag-4**, 470.
Cosgrove, J. G., Collins, R. L., and Murty, D. S. (1973). *J. Amer. Chem. Soc.* **95**, 1983.
Dezsi, I., Keszthelyi, L., Kulgawczuk, D., Molinar, D., and Elissa, N. A. (1967). *Phys. Status Solidi* **22**, 617.
Dezsi, J., Ouseph, P. J., and Thomas, P. M. (1971). *Chem. Phys. Lett.* **9**, 390.
Dharmawardena, K. G., and Bancroft, G. M. (1968). *J. Chem. Soc. A* 2655.
Dollimore, D., Griffiths, D., and Nicholson, D. (1963). *J. Chem. Soc.* 2617.
Dominques, P. H., and Danon, J. (1972). *Chem. Phys. Lett.* **13**, 365.
Duncan, J. F., and Stewart, D. J. (1967). *Trans. Faraday Soc.* **63**, 1031.

Duncan, J. F., Mackenzie, K. J. D., and Stewart, D. J. (1968). *Symp. Faraday Soc. No. 1—Mössbauer Effect, 1967* p. 103.

Fenger, J., Siekierska, K. E., and Maddock, A. G. (1970). *J. Chem. Soc. A* 1456.

Foerster, H., Meyer, K., and Nagorny, K. (1973). *Z. Phys. Chem.* **87**, 64.

Francombe, N. H., and Rooksby, H. B. (1959). *Clay Min. Bull.* **4**, 45.

Fung, S. C., and Drickamer, H. G. (1969). *J. Chem. Phys.* **51**, 4253.

Gager, H. M., Hobson, M. C., and Lefelhocz, J. F. (1972). *Chem. Phys. Lett.* **15**, 124.

Gager, H. M., Lefelhocz, J. F., and Hobson, M. C., Jr. (1973). *Chem. Phys. Lett.* **23**, 386.

Gallagher, P. K. (1965). *Inorg. Chem.* **4**, 965.

Gallagher, P. K. (1968). Unpublished work.

Gallagher, P. K. (1969). *Mater. Res. Bull.* **3**, 225.

Gallagher, P. K., and Kurkjian, C. R. (1966). *Inorg. Chem.* **5**, 214.

Gallagher, P. K., and Schrey, F. (1969). In "Thermal Analysis" (R. W. Schwenker and P. D. Garn, eds.), Vol. II, pp. 929–952. Academic Press, New York.

Gallagher, P. K., and Prescott, B. (1970). *Inorg. Chem.* **9**, 2510.

Gallagher, P. K., and Schrey, F. (1972). In "Thermal Analysis" (H. G. Wiedemann, ed.), Vol. II pp. 623–634. Birkhäuser Verlag, Basel.

Gallagher, P. K., MacChesney, J. B., and Buchanan, D. N. E. (1964). *J. Chem. Phys.* **41**, 2429.

Gallagher, P. K., MacChesney, J. B., and Buchanan, D. N. E. (1965). *J. Chem. Phys.* **43**, 516.

Gallagher, P. K., O'Bryan, H. M., Jr., Schrey, F., and Monforte, F. R. (1969). *Amer. Ceram. Soc. Bull.* **48**, 1053.

Gallagher, P. K., Johnson, D. W., Jr., and Schrey, F. (1970a). *J. Amer. Ceram. Soc.* **53**, 666.

Gallagher, P. K., Schrey, F., and Prescott, B. (1970b). *Inorg. Chem.* **9**, 215.

Gallagher, P. K., Sinclair, W. R., Fastnacht, R. A., Luongo, J. P. (1974). *Thermochim. Acta* **8**, 141.

Glasner, A., Levy, E., Steinberg, M., and Bodenheimer, W. (1964). *Talanta* **11**, 405.

Greenwood, N. N., and Howe, A. T. (1972). *J. Chem. Soc. Dalton* 116.

Gutlich, P., and Hasselbach, K. M. (1969). *Angew. Chem.* **8**, 600.

Gutlich, P., and Hasselbach, K. M. (1974). *Ber Bun. Res.* **78**, 1017.

Haneda, K., Miyakawa, C., and Kajima, H. (1974). *J. Amer. Ceram. Soc.* **57**, 354.

Hooley, J. G., Bartlett, M. W., Liegme, B. V., and Sams, J. R. (1968). *Carbon* **6**, 681.

Hooley, J. G., Sams, J. R., and Liegme, B. V. (1970). *Carbon* **8**, 467.

Ichida, T. (1973a). *J. Solid State Chem.* **7**, 308.

Ichida, T. (1973b). *Bull. Chem. Soc. Japan* **46**, 79.

Ichida, T. (1973c). *Bull. Chem. Soc. Japan* **46**, 1591.

Jach, J. (1973a). *J. Nonmetals* **2**, 89.

Jach, J. (1973b). *J. Nonmetals* **1**, 155.

Jach, J., Borg, R. J., and Lai, D. Y. (1972). *J. Nonmetals* **1**, 79.

Jadhao, V. G., Singru, R. M., and Rao, C. N. R. (1972). *Phys. Status Solidi (a)* **12**, 605.

Kaufman, K. and Hazel, F. (1974). *Inorg. Nucl. Chem. Lett.* **10**, 595.

Kisynska, K., Kopcewicz, M., and Kotlicki, A. (1972). *Phys. Status Solidi (b)* **49**, 85.

Krauth, A., Tomandl, G., and Frischat, G. H. (1967). *Z. Angew. Phys.* **23**, 419.

Lebeder, R. A., Vertes, A., Ranagajec-Komor, M., Berend-Rom G., Babeskin, A. M., and Nesmeyanov, A. N. (1974). *Magy. Kem. Foly.* **80**, 23.

Lee, Y. J., Chen, J. W., Huang, C. H., Yeh, S. J. Cheng, H. S. (1974). *J. Chin. Chem. Soc. (Taipei)* **21**, 21.

Levinson, L. M., Jacobs, I. S., Greskovich, C., and Glover, G. H., (1975). *Proc. Ann. Conf. Magn. Mater., 20th* (in press).

Long, G. G., and Bowen, L. H. (1970). *Inorg. Nucl. Chem. Lett.* **6**, 837.

Maksimov, Y. V., Suzdalev, I. P., Kushnerev, M. Y., Arents, R. A. (1974). *Fiz. Metal. Metalloved.* **37**, 267.

Malathi, N., and Puri, S. P. (1971). *J. Phys. Soc. Japan* **31**, 1418.

Malathi, N., Puri, S. P., and Saraswat, I. P. (1971). *J. Phys. Soc. Japan* **31**, 117.

Moreira, J. E., Knudson, J. M., DeLima, G. G., and Dufresne, A. (1973). *Anal. Chim. Acta* **63**, 295.

Neto, K. S., and Garg, V. K. (1973). *Radiochem. Radioanal. Lett.* **15**, 357.

Novikov, Y. N. *et al.* (1970). *J. Struct. Chem.* **11**, 970.

Nozik, A. J. (1972). *J. Phys. C: Solid State Phys.* **5**, 3147.

Ohhashi, K., and Tsujikawa, I. (1974). *J. Phys. Japan* **37**, 63.

Pernet, M., Chenavas, J., Joubert, J. C., Meyer, C., and Gros, Y. (1973). *Solid State Commun.* **13**, 1147.

Popma, S. A., and Van Diepen, A. M. (1974). *Mater. Res. Bull.* **9**, 1119.

Pritchard, A. M., and Mould, B. T. (1971). *Corros. Sci.* **11**, 1.

Pritchard, A. M., Haddon, J. R., and Walton, G. N. (1971). *Corros. Sci.* **11**, 11.

Ramshesh, V., Venkateswarlu, K. S., and Shankar, J. (1973). *Ind. J. Chem.* **11**, 788.

Robbins, M., Wertheim, G. K., Storm, A. R., and Buchanan, D. N. E. (1972). *Mater. Res. Bull.* **7**, 233.

Ruboshov, A. M., Fabrichnyi, P. B., Strakhov, B. V., and Babeskin, A. M. (1972). *Russ. J. Phys. Chem.* **46**, 765.

Saitovitch, E. B., Raj, D., and Danon, J. (1972). *Chem. Phys. Lett.* **17**, 74.

Saito, N., Sano, H., Tominaga, T., and Ambe, F. (1965). *Bull. Chem. Soc. Japan* **38**, 681.

Sano, H., and Kanno, M. (1969). *Chem. Commun.* 601.

Sano, H., and Ohnuma, T. (1974). *Chem. Phys. Lett.* **26**, 348.

Schechter, H., Hillman, H., and Ron, M. (1966). *J. Appl. Phys.* **37**, 3043.

Schrey, F. (1967). *Bull. Amer. Ceram. Soc.* **46**, 788.

Seregin, P. P., Bondgrevskii, S. I., Shipatov, V. T., and Tarasov, V. A. (1972). *Inorg. Mater.* **8**, 497.

Solozhenkin, P. M., Voitkovskii, Y. B., Generalov, O. N., Sidorov, S. V. (1974). *Dokl. Akad. Nauk. Tadzh. SSR* **17**, 39.

Srivastava, B. N., and Singh, R. P. (1974). *Indian J. Pure Appl. Phys.* **12**, 311.

Stewart, D. J., Knop, O., Ayasse, C., and Woodhams, F. W. D. (1970). *Can. J. Chem.* **50**, 690.

Tachibana, T., and Ohya, T. (1969). *Bull. Chem. Soc. Japan* **42**, 2180.

Tamas, F., and Vertes, A. (1973). *Magg. Kem. Foly.* **79**, 266.

Temperley, A. A., and Pumplin, D. W. (1969). *J. Inorg. Nucl. Chem.* **31**, 711.

Tominaga, T., Morimoto, T., Takeda, M., and Saito, N. (1966). *Inorg. Nucl. Chem. Lett.* **2**, 193.

Tominaga, T., Takeda, M., Morimoto, T., and Saito, N. (1970). *Bull. Soc. Chem. Japan* **43**, 1093.

Tomov, T., Ruskov, T., and Geogiev, S. A. (1971). *Proc. Conf. Appl. Mössbauer Effect, Tihany, Hungary, 1969* p. 793.

Van der Giessen, A. A. (1967). *J. Phys. Chem. Solids* **28**, 343.

Vertes, A., Ranogajec-Komer, M., Gelencser, P. (1973). *Acta Chim.* **77**, 55.

Vertes, A., and Zsoldos, B. (1970). *Acta. Chim. Acad. Sci. Hungary* **65**, 262.

Vertes, A., Szekely, F., and Tarnoczy, T. (1969). *Magy. Kem. Foly.* **75**, 172.

Vlasov, A. Y., Loseva, G. V., Makrov, E. F., Murshko, N. V., Petukhov, E. P., and Povitskii, V. A. (1970). *Sov. Phys. Solid State* **12**, 1177.

Vlasov, A. Y., Loseva, G. W., Murshkov, N. V., and Petukhov, E. P. (1972). *Russ. J. Inorg. Chem.* **17**, 482.

Von Wagner, G., and Binder, H. (1958). *Z. Anorg. Allgem. Chem.* **297**, 334.

Wedd, K. W. J., Liengme, B. V., Scott, J. C., and Sams, J. R. (1969). *Solid State Commun.* **7**, 1091.

Wertheim, G. K., Hausmann, A., and Sander, W. (1971). "Electronic Structure of Point Defects as Determined by Mössbauer Spectroscopy and by Spin Resonance," Elsevier, Amsterdam.

Wittman, F. H. (1974). *Ind. Chim. Belg.* **39**, 693.

Yusfin, Y. S., Savitskaya, Y. B., and Generalov, O. N. (1974a). *Vyssh. Ucheb. Zaved. Cern. Met.* 17.

Yusfin, Y. S., Voitkovskii, Y. B., Savitskaya, L. I., and Generalov, O. N. (1974b). *Vyssh. Ucheb. Zaved. Cern. Met.* 26.

# 8

# *Analysis of Archaeological Artifacts*

**A. Kostikas**
**A. Simopoulos**

Nuclear Research Center Democritos
Athens, Greece

**N. H. Gangas**

University of Ioannina
Ioannina, Greece

## I. Introduction

The introduction of Mössbauer spectroscopy in the study of archaeological artifacts is one of the most recent additions to physical analytical techniques that have been widely used during the last twenty years in the elucidation of archaeological problems (Brothwell and Higgs, 1969). The importance of these techniques is based on the presumed elimination of subjectivity in criteria used by archaeologists to classify and extract information from ancient objects. Although macroscopic criteria such as stylistic considerations and excavation systematics have been and will undoubtedly continue to be of major importance, the use of objective physical techniques may be instrumental in removing ambiguities in interpretation. Such techniques provide information on provenance, manufacturing techniques, and dating of ancient objects that is not obtainable otherwise. Systematic work in this area has already established the field of archaeometry, with its own journal, published by the Oxford Laboratory for Archaeometry.

The study of ancient pottery wares comprises the majority of applications of physical techniques in archaeology and practically all of the applications of Mössbauer spectroscopy. This is only partly due to the abundance of samples or experimental considerations. Pottery wares are of prime importance in archaeology as cultural items, indexes of artistic and technological skill, and objects of extensive trade in the ancient world. Valuable historical information has also been obtained from inscriptions and stamps often found in decorated pottery. Among the techniques that have been used in pottery studies, the most widely applied are mineralogical analysis (Farnsworth, 1964), optical emission spectroscopy (Millet and Catling, 1967), and neutron activation analysis (Perlman and Asaro, 1971). The latter two techniques are based on the determination of the elemental composition of the sample, which characterizes with some degree of specificity the origin of the clay. These techniques have been useful in provenance studies, but they cannot contribute to questions related to manufacturing techniques or dating.

During the last four years, several studies of clays and ancient pottery have been reported, using Mössbauer techniques. The relatively high abundance of iron in clays used in ancient pottery, ranging between 5 and 10%, allows meeting the major criteria for work with archaeological objects, namely, very small damage to the artifact and reasonable measuring times. A sample of about 100 mg is sufficient for a satisfactory spectrum in a run of the order of 10 hr. The interest in the method depends, of course, on the usefulness of the information extracted from Mössbauer data. The very fact that the results pertain to only one element can be a serious disadvantage in comparison with other analytical techniques. The uniqueness of Mössbauer spectroscopy, however, rests on the detailed picture that it can provide about the iron chemistry in various clays. This information includes both the constitution of the unfired clay and the heat treatments that lead to the finished pottery item. This creates a multiparameter function that in principle could be used for provenance, classification, and/or an understanding of effects achieved by varying the firing conditions. There is evidence also that aging of the materials over several millenia may bring about alterations in the chemical and physical state of iron-containing phases in the clay. It appears, therefore, that Mössbauer spectroscopy may be an oversensitive method in pottery studies, so that the preliminary results, however promising, should be treated with a critical attitude.

In the light of these considerations, the scope of this article is to summarize the investigations reported to date and to assess on this basis the various problems that can be studied fruitfully by Mössbauer spectroscopy. The next section contains an account of the measurements on unfired clays and the transformations induced by firing. This is not only a prerequisite for the analysis of spectra of ancient pottery, but it is also valuable from the point of view of soil science and ceramic technology. Investigations related to ancient pottery are presented in Section III. Sections III,A and B describe exploratory investigations on provenance and manufacture. Finally, Section III,C contains a discussion of the effect of aging in connection with some recent results from pottery of different periods.

## II. Unfired and Fired Clays

Clay is defined in soil science as the fraction of the soil consisting of particles with diameters less than 4 $\mu$m. Its main components are inter-stratified layers of silica tetrahedra and alumina octahedra. Iron is found in the form of oxides attached on the clay particles or as free constituents of the clay. Structural iron is also present, together with other metal ions, as a substitutional ion in octahedral and tetrahedral sites of the clay minerals. The variation of chemical composition and structural forms results in a number of clay minerals, so that a clay from a particular site can be characterized by the relative abundance of each form, determined usually by x-ray diffraction techniques (Grim, 1968, pp. 126–158). As it will be shown below, additional characterization parameters can be provided by Mössbauer results, which lead to the determination of the distribution of iron in various chemical forms and the size of oxide particles.

### A. Unfired Clays

A detailed investigation of the clay fraction extracted from a typical Attic soil used for the preparation of clay by a contemporary potter in the vicinity of Athens was reported by Gangas et al. (1973). The soil was dispersed with sodium hexametaphosphate, and the clay fraction was separated by water flotation in a way similar to that adopted by ancient and modern potters. X-ray diffraction patterns characterized the clay as consisting of illite, some kaolinite and vermiculite and interstratified systems of illite–chlorite, and possibly vermiculite. There was no evidence of iron oxides in the x-ray patterns. Mössbauer spectra of the dried clay at room, liquid nitrogen, and liquid helium temperatures are shown in Fig. 1. The spectra consist generally

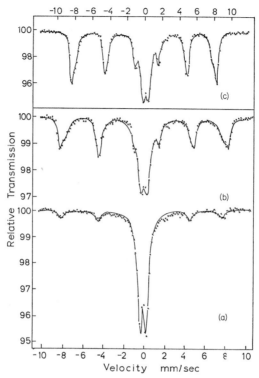

**Fig. 1.** Mössbauer spectra of unfired Attic clay at (a) 300°, (b) 77°, and (c) 4.2°K. The solid lines represent least-squares fits of the spectra.

of a central doublet and a six-line component indicating the presence of magnetic iron oxides. The major effect of the temperature variation is an enhancement of the magnetic component relative to the central doublet at the low temperatures. This is typical behavior of superparamagnetic particles (see Chapter 1), as has been shown by Mössbauer studies of fine particles of $\alpha$-$Fe_2O_3$ (Kundig et al., 1966) and $\beta$-FeOOH (Deszi et al., 1967). When the size of the oxide particles is of the order of 100 Å, they behave as an assembly of paramagnetic particles with magnetic moments of the order of $10^4$ $\mu_B$ and relaxation times for reorientation of the magnetic moment strongly dependent on temperature and size. For a given size distribution, anisotropy constant, and temperature, the Mössbauer spectrum will be separated into a paramagnetic and a magnetic component corresponding respectively to the fraction of particles with relaxation times smaller and greater than the Larmor precession time of the nucleus. The temperature dependence of the ratio of these two components can be used to determine the particle size distribution of the magnetic oxides. A quantitative treatment of the determination of particle size distribution from the temperature dependence of Mössbauer spectra is given by Gangas *et al.* (1973), where further references on superparamagnetic effects can be found. For the Attic clay, a detailed analysis of the results shown in Fig. 1 as well as others at intermediate temperatures lead to the result that the oxide particle distribution had an average diameter of 130 Å.

A complication arises in the analysis of the clay system because of the central doublet due to paramagnetic iron included in the clay mineral structure. This component may be determined either from the saturation value attained by the central doublet at low temperature or by chemical analysis after removal of the iron oxides. These two methods were applied in the clay studied and gave 33% and 38%, respectively, for the fractional abundance of structural iron. It is concluded, therefore, that in this particular clay sample, iron was present in the form of small oxide particles (65%) and as a substitutional ion in the clay mineral structure (35%). Furthermore, from the asymmetry of the lines, an analysis assuming two magnetic species identified the iron oxides as $\alpha$-$Fe_2O_3$ (Kundig *et al.*, 1966) and $\beta$-$FeOOH$ (Deszi *et al.*, 1967). This result demonstrates the uniqueness of Mössbauer spectroscopy in obtaining information not accessible by other techniques. It is clear also that the study of temperature dependence of the spectra is essential in identifying the structural and chemical form of iron oxides in the clay. A negligible magnetic component in room temperature spectra does not necessarily mean the absence of magnetic iron oxides, as concluded by Yassoglou and Peterson (1969).

The results for the Attic clay do not necessarily represent a universal compositional pattern for clays. Variations in the relative amounts of iron oxides and structural iron, as well as the possible presence of $Fe^{2+}$, are differentiating factors that have important consequences for the physical and chemical attributes of the pottery after firing. We shall return to this point in connection with studies of the effects of firing in various clays.

### B. *Transformations Induced by Firing*

The effects of firing can be studied by following the transformations of the components of Mössbauer spectra attributed to different iron-containing phases in the clay, as discussed in the previous section. A detailed investigation has been carried out on samples of the Attic clay fired in air at temperatures ranging from 200°K to 1100°K in steps of 100°K (Simopoulos *et al.*, 1975). Representative Mössbauer spectra of these samples measured at 4.2°K are shown in Fig. 2. The spectrum at 400°C shows little, if any, asymmetry in the magnetic component, indicating that a transformation of the $\beta$-$FeOOH$

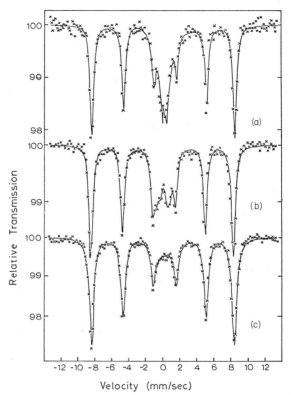

**Fig. 2.** Mössbauer spectra of clays fired at (a) 400°C, (b) 600°C, (c) 900°C measured at 4.2°K. The solid lines represent least-squares fits of the spectra.

component has taken place in this temperature range. This is confirmed by previously reported results of Deszi *et al.* (1967), who have observed that synthetically produced $\beta$-FeOOH transforms into $\alpha$-Fe$_2$O$_3$ at 673°K. For firing in the temperature range between 400° and 800°C, the major change involves the central doublet, which can be assumed to arise from structural iron, since at 4.2°K superparamagnetic effects should be very small. A second paramagnetic component with a higher quadrupole splitting appears and grows at the expense of the narrow component. This change can be attributed to the loss of hydroxyl ligands in the octahedral sites and the ensuing disorder of the lattice. Dehydroxylation is known to occur in clay minerals upon heating between 500° and 800°C, and similar effects on Mössbauer spectra have been observed in a study of kaolinite by Mackenzie (1969).

A striking change is observed in the sample fired at 900°C. The central component has almost disappeared, and the spectrum corresponds to practically pure $\alpha$-$Fe_2O_3$. A possible explanation for this result is that the transformations known to occur in the clay mineral structure in this temperature range (Grim, 1968, pp. 301–308) may lead to the migration of iron out of the lattice with subsequent oxidation to $\alpha$-$Fe_2O_3$ or inclusion in an amorphous phase with low Debye–Waller factor. It is interesting to note here that modern Attic pottery samples display similar spectra, a result that is expected, since this pottery is manufactured at the same temperature range. The central doublet is restored at higher firing temperatures with a quadrupole splitting intermediate between the values observed in the range 400°–800°C and below. This component may be associated with recrystallization and the appearance of a high-iron spinel or mullite phases, which have been found in x-ray studies of illite heated in this temperature range (Grim, 1968, pp. 332–339).

Another interesting result of this study is the increase in particle size and a broadening of the distribution, which may be inferred from the increase of the magnetic component as the firing temperature increases. This implies an aggregation process for the oxide particles and has been observed also in studies of fired Nile clay (Eissa *et al.*, 1974). An important observation in these studies is that firing periods of 5–10 hr are required before the transformations are completed.

The foregoing results, which are typical of clays with high initial oxide content, may be compared with a study of the effects of firing on a French clay (Janot and Delcroix, 1974) that apparently had most of the iron included in the clay mineral lattice. The dependence of the quadrupole splitting, isomer shift, and recoil-free fraction on the firing temperature are shown in Fig. 3. The quadrupole splitting displays a sudden increase around 550°C and a drop after 900°C to an intermediate value. The first change is accompanied by an increase in linewidth and a drop of the recoil-free fraction. It was concluded that these changes originate in a dehydroxylation process that leaves the clay mineral lattice in a disordered state and causes loosening of the iron bonds. Above 900°C, a recovery of the recoil-free fraction parallels the increase in quadrupole splitting and is taken as evidence of progressive annealing of vacancies and the establishment of a well-defined crystallographic environment in which the iron is more firmly bound. Unfortunately, the dependence of the spectra on the temperature of measurement was not studied in detail; however, the absence of a magnetic component after firing indicates that most of the iron was included as a substitutional ion.

Additional factors that influence the chemistry of iron in the final clay, and hence the Mössbauer spectra observed, are the presence of $Fe^{2+}$ in the unfired clay and the type of atmosphere (oxidizing or reducing) in which firing takes place. We shall discuss this subject further in connection with studies of ancient pottery.

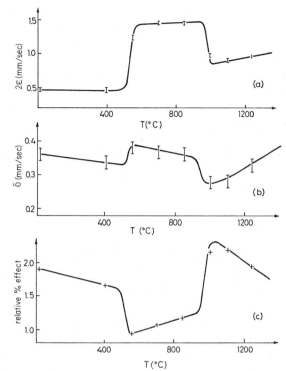

**Fig. 3.** Schematic representation of the dependence of (a) the quadrupole interaction, (b) the isomer shift, and (c) the absorption intensity of a typical French clay on the firing temperature. (Janot and Delcroix, 1974.)

## III. Investigations of Ancient Pottery

### A. Provenance Studies

As mentioned in the introduction, the analytical techniques that have been used for provenance studies of ancient pottery are based on the elemental composition of the clay material. Although these techniques have already proven their usefulness in a number of cases, in some instances they are remarkably uninformative. A typical example is the case of samples of pottery excavated from the same archaeological site that have the same elemental analysis but are distinctly different in style, color, and texture. Problems of this nature can be approached only by a technique that provides information about the effects of different manufacturing techniques, which could resolve questions related to the origin of the artifacts. This is the area where Mössbauer spectroscopy could prove most valuable, and a detailed discussion of studies related to firing techniques will be given in the next section.

Besides this application, however, the Mössbauer technique can in principle contribute to provenance studies by a phenomenological approach, i.e., by systematic classification of known groups of pottery according to their spectral parameters. The first exploratory search in this direction was a study of pottery wares from two medieval sites, Cheam and Kingston in Surrey, England, reported by Cousins and Dharmawardena (1969). They found that the hyperfine parameters of the spectra fall within a small range for samples from each site. Although the Cheam wares are superficially similar to the Kingston wares, their corresponding spectra are entirely different. Firing experiments with clay samples from Cheam reproduce the spectra of the ancient pottery for a firing temperature near 1000°C.

In the light of this application, another exploratory study was undertaken at our laboratory on ancient Greek pottery wares from the period 800–500 B.C. (Gangas *et al.*, 1971). Fourteen sherds from the excavation of a burial site in northwestern Greece (Vitsa, Epirus) were divided into four subgroups according to archaeological evidence. The classification of the sherds on the basis of Mössbauer parameters extracted from computer fits of the spectra agreed closely with the grouping established by macroscopic criteria.

The preliminary evidence of these studies on the effectiveness of Mössbauer spectroscopy for characterization of ancient pottery justifies further systematic investigations to determine whether archaeologically important groups of pottery of known provenance and homogeneity in style and clay texture exhibit Mössbauer spectra typical for each group. The successful outcome of such studies could allow differentiation between pieces genuinely belonging to a group and imitations. We have undertaken a study of this type on forty-five samples of Mycenean and thirteen samples of Minoan pottery. These groups were chosen for several reasons: (1) Minoan and, to a greater extent, Mycenean pottery were widely distributed in the area of the eastern and central Mediterranean during the Greek late bronze age and several questions of provenance are of considerable archaeological importance; (2) the shape of vases and the style of decoration of the better quality wares had reached a remarkable degree of homogeneity and sophistication; (3) detailed emission spectroscopy studies of these groups on the relation between composition and provenance have been published by Millet and Catling, (1967). Neutron activation analysis results have also been reported recently by Harbottle (1970).

Preliminary results of measurements at liquid nitrogen temperature on the Minoan samples and a subgroup of fifteen Mycenean samples have been reported by Kostikas *et al.* (1974a). Representative spectra are shown in Fig. 4. They show the usual pattern of a central doublet and a magnetic

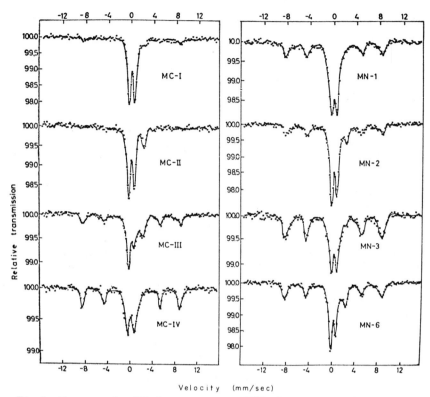

**Fig. 4.** Representative Mössbauer spectra at 77°K of two groups of Mycenean and Minoan pottery.

component of varying relative intensity. The presence of $Fe^{2+}$ is noted in some spectra of Mycenean samples. Generally speaking, there are no obvious distinctive features that would allow for differentiation between the two groups in a "fingerprint" manner. It is observed, moreover, that the spectra of the Mycenean samples do not show the uniformity that would be desired for definitive characterization. On the other hand, the texture and color of the clay of the Mycenean sherds was of considerable variety. When a sub-group of sherds with finer clay was examined, the Mössbauer results showed much greater uniformity.

In view of these results, criteria were sought to differentiate between the two groups via spectral parameter values that could be established on a statistical basis. Figure 5a shows the distribution of values for the quadrupole coupling constants of the central doublet. A small systematic difference exists between the two groups. Figure 5b shows the distribution of the values of the total area of the spectrum as determined from computer fits with one

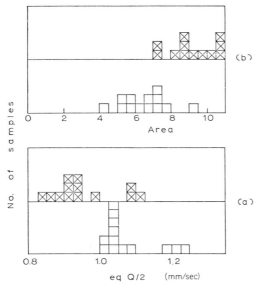

**Fig. 5.**   (a) Distribution of values of quadrupole splitting at 77°K for the Minoan (⊠) and Mycenean (□) pottery. (b) Distribution of total areas of Mössbauer absorption at 77°K for the Minoan and Mycenean pottery.

six-line component and a quadrupole doublet. The Minoan group appears with slightly larger absorption area than that of the Mycenean group, in agreement with the neutron activation results, which show larger iron content for the Minoan group (Harbottle, 1970).

Such results indicate the possibility of establishing, with more systematic and extended study, characteristic spectral parameter vectors for the Mycenean and Minoan groups. The available data are not yet sufficient to decide whether this approach could give a criterion for classification of ancient pottery even in a geographically limited area, e.g., Greece or the eastern Mediterranean, over a period of a few centuries. This is a phenomenological method in which we do not examine the role of the physical and chemical properties of the clay in determining the observed values of spectral parameters. An understanding at that level, including the effects of firing techniques and aging, could give information on provenance of a more fundamental nature. Studies of this type will be discussed in the next two sections.

## B. Firing Techniques

Among the analytical methods applied for pottery studies, Mössbauer spectroscopy is unique in providing information on manufacturing techniques. Its usefulness comes from the ability to trace the clay transformations occurring during the firing process, as has already been shown in the investigations discussed in Section II,B. From the archaeologist's point of view, the most interesting pieces of information are the firing temperature and the type of atmosphere in which the pottery was fired.

In studies aimed at the determination of the firing temperature, the variation of the quadrupole splitting with firing temperature was used as the diagnostic feature of the spectrum. This method has been used already in the work of Cousins and Dharmawardena (1969) discussed previously. It requires parallel measurements on samples of clay, from the same location as the ancient pottery, which are fired at successively higher temperatures. Results of this kind for a typical French clay (Fig. 3) have been used by Janot and Delcroix (1974) for the determination of the firing temperature of samples of French pottery dating from the tenth to the fifteenth century A.D. From the values of the quadrupole splitting the firing temperature was estimated at about 1000°C for all the samples. Unfortunately, the sensitivity to the firing temperature drops considerably near 1000°C, which is the range of interest.

We shall discuss now in detail the investigation of Bouchez *et al.* (1974) of two quite different kinds of pottery, one red and the other gray, found in the Tureng Tepe culture in Iranian Turkestan dating from the third millenium B.C. The two types coexist over five archaeological levels covering the period IIA (several centuries). An important archaeological question is whether the gray ware appeared as a result of immigration of people to northeastern Iran from elsewhere or resulted from a change in local manufacturing techniques. It is of interest, moreover, to establish relations between the Tureng Tepe gray ware and a new gray ceramic appearing in northwestern Iran and northern India shortly after the disappearance of the former by the seventeenth century B.C. Since it has been suggested that the gray ware may be typical of Indoeuropean populations, the investigation of these questions could give significant information about their spread in the area.

Mössbauer spectroscopy in combination with other analytical techniques, specifically x-ray fluorescence, spark mass spectrometry, and x-ray diffraction, was used in an extensive study of this problem. Five sherds of each type were analyzed, and they have shown no significant differences in elemental composition. The Mössbauer spectra, however, displayed pronounced differences. The iron in the gray sherds was predominantly ferrous (70%), whereas in the red sherds it was almost all ferric (70%). This suggested that the essential difference between the two ceramics was the firing process and, more specifically, that the red and gray pottery were produced under oxidizing and reducing conditions, respectively. To substantiate this conclusion, local soils and clays as well as paleosols were studied by Mössbauer spectroscopy and x-ray diffraction, and the transformations induced by firing in oxidizing and reducing atmospheres were followed by the same methods. The unfired clays were found to be of similar constitution with regard to clay minerals present, as well as relative amounts of $Fe^{3+}$, $Fe^{2+}$, and a magnetic iron component. These ratios, however, were considerably different in the samples fired in oxidizing and reducing atmospheres that reproduced the red and gray colors. It is significant that only about 15% of the iron was found in the form of oxide particles in the unfired clays. The amount of iron oxide was also small in the ancient sherds, as inferred from the absence of a magnetic component at 4.2°K. The variation of the quadrupole splitting of the central paramagnetic component for the oxidizing firings is shown in Fig. 6. From these results, in combination with x-ray data on the temperature at which illite peaks disappear from the fired clay, the firing temperature for the red wares was estimated at 1050 ± 50°C. For the gray sherds, the presence of illite x-ray diffraction peaks suggests a firing temperature of 50°–100°C lower. Thus, the application of Mössbauer spectroscopy to this problem was decisive in determining the origin of the differences in appearance of the two types of pottery and in establishing the ranges of firing temperatures used by the potters.

Along the same lines, but with significant and informative differences, is a study (Hess and Perlman, 1974) using Mössbauer techniques to relate the iron in ceramics to pottery colors. This work was stimulated by results of neutron activation analysis on pottery excavated at Tel Ashdod, Israel. It was found that ceramics that differed greatly in appearance were made from

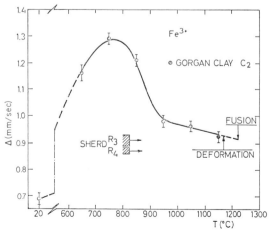

**Fig. 6.** Dependence of quadrupole splitting on firing temperature for a Gorgan clay fired under oxidizing conditions. The intervals $R_3$ and $R_4$ show the range of values for the quadrupole splitting for two sherds of the Tureng Tepe culture. (Bouchez et al., 1974.)

the same clays (Asaro *et al.*, 1971). Local clay was fired at 850°C and 1150°C in oxidizing and reducing atmospheres in order to reproduce the colors of the ancient sherds. The most important difference, in comparison to the Tureng Tepe study, is that most of the iron in the clay was found in the form of oxide particles with an estimated particle size of less than 40 Å. The spectra obtained after firing in an oxidizing atmosphere are very similar to those of the Attic clay discussed in Section II,B. The color achieved was not greatly different from that of two red ancient sherds. A third sherd, a gray-green piece of the Mycenean IIICI style, gives a spectrum that differs in important respects from those of the two red specimens, especially at 4.2°K. Although it was not possible to interpret this spectrum exactly, it was suggested that it may arise from small particles of $Fe_3O_4$ (magnetite). Similarities were found with the spectrum of the clay fired under reducing conditions, but it was obvious that $Fe^{3+}$ was more thoroughly reduced in the fired clay. It appears that with clays like the Tel Ashdod clay with a high initial content of $\alpha\text{-}Fe_2O_3$, the kiln conditions for producing colors of gray-green range involve a delicate balance in the oxygen supply. This is consistent with the fact that the Myc IIICI wares from Ashdod appear in a range of colors.

In the same study, three clays from eastern Cyprus were examined for their possible relation to ancient pottery from that region. This pottery was probably one of the styles that the Ashdod potters attempted to imitate in the gray-green ceramics. It is interesting that two of the Cyprus clays coming from the region of Encomi were gray and were found to contain most of the iron in the silicate mineral. Spectra of a clay fired at 850°C in an oxidizing atmosphere for three different measuring temperatures are shown in Fig. 7. Even at 4.2°K, very little magnetic component appears. The chemical state of the iron, which is quite different from that of the Ashdod clay, seems to preclude the possibility of making red pottery.

The results reviewed in this section point out that the chemical state of the iron in clays may occur in a rich variety of forms that are characteristic of the origin of the clay. Mössbauer spectroscopy gives definitive data on the various components and the final state of the iron in the finished pottery item. Consequently, the spectra of ancient pottery depend on both the composition and the firing history of the clay. These features, when unraveled by simulation studies, can provide unique information on manufacturing technology and provenance.

## C. Aging Effects

In studies of the chemical state of a system manufactured a few thousand years ago, the factor of aging should also be considered. A detailed investigation of samples of different periods has started in our laboratory for this

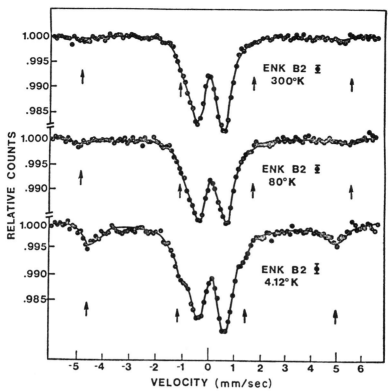

**Fig. 7.**  Temperature dependence of Mössbauer spectra for a gray clay from eastern Cyprus fired at 850°C in an oxidizing atmosphere. (Hess and Perlman, 1974.)

purpose. This study was motivated by the different temperature behavior observed in spectra of modern and ancient pottery (Kostikas *et al.*, 1974b).

The temperature dependence of the spectra of a Mycenean sample is shown in Fig. 8. A Minoan sample gave similar results. The most prominent change observed as the temperature decreases is a pronounced enhancement of the magnetic component. Its relative intensity changes from nearly 0 at 77°K to 0.5 at 4.2°K and 0.9 at 1.5°K. Such results strongly suggest an iron oxide system with small average size exhibiting superparamagnetic behavior. The ratio of the area of the magnetic component to total absorption area (hereafter called the magnetic ratio) may be employed to determine the blocking temperature of the superparamagnetic particles (Gangas *et al.*, 1973). This is the temperature at which the magnetic ratio attains the value 0.5 and is directly related to the average size of the particles. The Mycenean and Minoan pottery results imply a blocking temperature of a few degrees Kelvin corresponding to particle sizes of the order of 50 Å. It is noted that

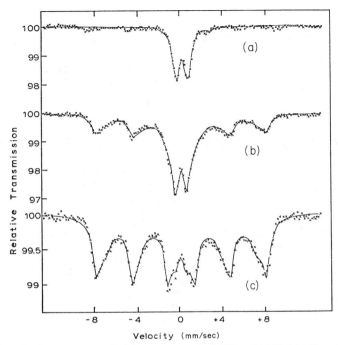

**Fig. 8.** Mössbauer spectra of a Mycenean sample (MCI) at (a) 77°, (b) 4.2°, and (c) 1.5°K. The solid lines represent computer least-squares fits. Two magnetic components have been used for the spectra at 4.2° and 1.5°K in order to approximate the effective field distribution.

the shape of the spectrum at the lowest temperature is very similar to that reported recently for iron oxide gels by Coey and Readman (1973). These authors have attributed the observed pattern to a disordered magnetic structure, which they termed spiromagnetism.

Whatever the origin of the low value of the blocking temperature, the difference from spectra of freshly fired clays as discussed in Section III,B suggests that some time-dependent process might be operative. To further examine the validity of this conjecture, measurements were made on two samples of pottery of intermediate ages, an Attic sherd dated at about 500 B.C. and a piece of Byzantine pottery from the sixth century A.D. The magnetic ratio as a function of temperature for all the samples studied is shown in Fig. 9. For comparison, the same ratio for the unfired clay sample is shown on the same figure. A difference of nearly two orders of magnitude is inferred between the blocking temperatures of the Mycenean–Minoan and modern clay samples. The data for the Byzantine sample fall in an intermediate range. The values for the Attic sherd also differentiate it from the earlier pottery, although they are somewhat lower than suggested by the

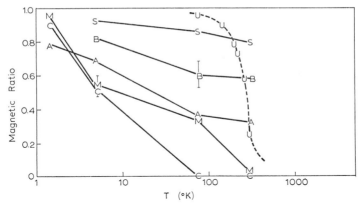

**Fig. 9.** The ratio of the magnetic component to total area as a function of temperature from Mössbauer spectra of pottery of various ages. The dashed curve represents the ratio of an unfired clay sample; M = Minoan, 1400 B.C.; C = Mycenean, 1400 B.C.; A = Attic, 500 B.C.; B = Byzantine, 600 A.D.; S = modern fired clay; U = unfired clay.

date of the sherd. All in all, however, the gross features of Fig. 9 indicate the possible existence of a correlation between time elapsed since manufacture and particle size distribution of iron oxides in the fired clay. If the initial particle size distribution in an ancient pottery item was similar to that of the modern pottery sample in Fig. 9, we would have to conclude that a disintegration process took place over a period of a few millenia.

Clearly, further work will be required in order to verify that the results described above reflect a characteristic property of the ancient wares and are not incidental to the samples studied. Other factors that may be operative in determining the final size distribution are (1) the environmental conditions, e.g., the humidity and pH of the soil in which the pottery was buried; and (2) the initial distribution of iron in various phases and the final firing temperature, as demonstrated by the investigations described in the previous section. The elucidation by Mössbauer spectroscopy of the relative importance of these factors could create a unique profile for pottery wares of a given time period and style and hence prove useful for provenance identification and authentication purposes.

## IV. Summary and Outlook for Future Work

The investigations reviewed in this article have laid the groundwork for applications of Mössbauer spectroscopy to the study of ancient pottery and have outlined the kind of problems that can be most fruitfully pursued with this technique. Summarizing the major results of this work, we shall try to point out some of the questions that are open for further research.

The studies of iron chemistry in clays, undertaken as a prerequisite to understand the Mössbauer spectra of ancient pottery, have given data on the presence of various iron compounds and their transformations during firing that would be difficult to obtain with another technique. More specifically, the presence of iron oxides in unfired clays in the form of small particles with dimensions of the order of 100 Å was first ascertained by Mössbauer methods. It has also been shown that the oxide species and the iron in the structure of the clay minerals as $Fe^{3+}$ or $Fe^{2+}$ can be identified by their spectral parameters and that a compositional profile of the clay can be established. These results have been obtained for only a few clays of widely varying origins and not always with the same degree of detail. A systematic study of clays from a given geographical area would be valuable, not only in connection with investigations of ancient pottery, but also in providing a complete picture of the variations in the iron-containing components that may exist in clays.

Mössbauer spectroscopy also has proved effective in following the transformations occurring in clays upon firing to progressively higher temperatures. Three major stages have been observed in the work reported until now, namely, an initial transformation around 400°C of $\beta$-FeOOH (if present in the initial clay) to $\alpha$-$Fe_2O_3$, followed by dehydroxylation of the clay minerals in the range of 600–800°C and disintegration, recrystallization, and vitrification processes above 900°C. Most detailed results exist for clays fired in oxidizing atmospheres and for clays initially containing the greatest part of the iron as iron oxides, but even in this area, the picture is far from complete. For example, the increase in particle size upon firing and the nature of iron-containing species appearing above 1000°C have not yet been fully investigated. On the other hand, very few results exist on the effects of firing under reducing conditions and for clays initially containing a substantial amount of $Fe^{2+}$. This question is of major importance in understanding the ways in which various pottery colors can be achieved with a given clay.

The investigations of ancient pottery can be divided into two categories: (1) an attempt is made to establish, on a statistically meaningful basis, characterization criteria for pottery of known provenance and style, e.g., the Mycenean and Minoan, by values of hyperfine parameters extracted from Mössbauer spectra; (2) (exemplified by the study of the Tureng Tepe pottery) Mössbauer spectroscopy is applied in combination with other techniques to a specific archaeological question related to changes in manufacturing technology; typically, in this type of study, parallel measurements on local clays fired at various temperatures are necessary in order to understand the conditions under which the ancient wares were produced.

The classification of pottery types by the values of hyperfine parameters requires measurements on a large number of samples and is a phenomeno-logical method at best. It appears, therefore, at least at the present stage, that information of archaeological interest could be provided by Mössbauer spectroscopy in problems where the constitution of the clay and the firing conditions have been used to vary the color of the finished object or to imitate pottery of foreign origin. An important problem of this type, which has not yet been investigated by Mössbauer spectroscopy, is the black and red figure technique for decoration of Attic vases, which reached an unsur-passed degree of technical and artistic perfection in Athens during the fifth century B.C. There is evidence (Noble, 1960) that the black and red glaze was achieved by a three-stage process of alternatively oxidizing and reducing atmosphere. This is a problem where it might be expedient to use backscatter geometry.

In a recent investigation, Longworth and Warren have examined by scattering techniques the black glaze of Greek Etruscan pottery. They suggest that it consists of $Fe_3O_4$ (magnetite) with about 30% of the iron ions in B sites substituted by aluminum (Longworth and Warren, 1975).

Further work will be required to assess the importance of an aging factor in determining the particle size distribution in samples of ancient pottery examined today. The available data are still very limited, but the significance of the problem justifies an extensive investigation.

Although this review has dealt exclusively with applications to studies of ancient pottery, a few remarks concerning potential applications to other areas might be relevant. One possibility, involving a different Mössbauer nucleus, is the study of tin alloys. The development and trade in tin alloys in the ancient world, in particular in the eastern Mediterranean during the Mycenean period, is of considerable archaeological importance. As has been demonstrated in a preliminary measurement (Kostikas *et al.*, 1974a), the spectrum of an ancient bronze can be readily decomposed into two com-ponents, one arising from tin oxide and the other from the bronze alloy. In this case, the parameter of interest could be the isomer shift of the bronze component, which could be used for characterization of the bronze alloy. Systematic work would be necessary in order to establish specific correlations between the isomer shift and the constitution of the alloy.

As another, more speculative application, we might mention the investi-gation of gold items by [197]Au Mössbauer spectroscopy. This technique might prove useful as a supplementary source of information for the study, for example, of gold–silver alloys used in ancient Greek and Roman coins studied by other physical analytical techniques (Darling and Healy, 1971). Nothing more can be said at this stage, however, since no measurements of this type have been attempted. In general, it must be emphasized that as the pottery investigations have shown, detailed background work is necessary in order to assess the archaeological relevance of information that can be provided by Mössbauer studies.

In closing, two rather general remarks should be added, that are worth emphasizing because of their importance for full evaluation of the data obtained by Mössbauer spectroscopy: (1) that active participation of an archaeologist is essential in defining a meaningful problem and avoiding trivial or uninteresting investigations; furthermore, the constant exchange of information from both sides greatly enriches the interpretation of the Mössbauer results; (2) although the information obtained by Mössbauer techniques is in some respects unique, its full value is realized when considered in conjunction with data from other analytical methods, as well as conventional archaeological criteria.

## References

Asaro, F., Perlman, I., and Dothan, M. (1971). *Archaeometry* **13**, 169–175.

Bouchez, R., Coey, J. M. D., Coussement R., Schmidt, K. P., Van Rossum, M., Aprahamian, J., and Deshayes J. (1974). *Proc. Int. Conf. Appl. Mossbauer Effect, Bendor, 1974. J. Phys. Colloq.* **35**, C6-541–546.

Brothwell, D., and Higgs, E. S. (eds.) (1969). "Science in Archaeology," 2nd ed. Thames and Hudson, London.

Coey, J. M. D., and Readman, P. W. (1973). *Nature (London)* **246**, 476–477.

Cousins, D. R., and Dharmawardena, K. G. (1969). *Nature (London)* **223**, 733–734.

Darling, A. S., and Healy, S. F. (1971). *Nature (London)* **231**, 443–444.

Deszi, I., Keszthelyi, L., Kulgawczuk, D., Molnar, B., and Eissa, N.A. (1967). *Phys. Status Solidi* **22**, 617–629.

Farnsworth, M. (1964). *Amer. J. Archaeol.* **68**, 221–228.

Eissa, N. A., Sallam, H. A., and Keszthelyi, L. (1974). *Proc. Int. Conf. Appl. Mossbauer Effect, Bendor, 1974. J. Phys. Colloq.* **35**, C6-569–570.

Gangas, N. H., Kostikas, A., Simopoulos, A., and Vokotopoulou, J. (1971). *Nature (London)* **229**, 485–486.

Gangas, N. H., Simopoulos, A., Kostikas, A., Yassoglou, N. J., and Filippakis, S. (1973). *Clays Clay Min.* **21**, 151–160.

Grim, R. E. (1968). "Clay Mineralogy." McGraw-Hill, New York.

Harbottle, G. (1970). *Archaeometry* **12**, 23–34.

Hess J., and Perlman I. (1974). *Archaeometry* **16**, 137–152.

Janot, C., and Delcroix, P. (1974). *Proc. Int. Conf. Appl. Mossbauer Effect, Bendor, 1974 J. Phys. Colloq.* **35**, C6-557–561.

Kostikas, A., Simopoulos, A., and Gangas, N. H. (1974a). *J. Phys. Colloq.* **35**, C1-107–115.

Kostikas, A., Simopoulos, A., and Gangas, N. H. (1974b). *Proc. Int. Conf. Appl. Mossbauer Effect, Bendor, 1974 J. Phys. Colloq.* **35**, C6-537–539.

Kundig, W., Boemmel, H., Konstabaris, G., and Lindquist, R. H. (1966). *Phys. Rev.* **142**. 327–333.

Longworth, G., and Warren, S. E. (1975). *Nature* **255**, 625–627.

Mackenzie, K. J. D. (1969). *Clay Min.* **8**, 151–160.

Millet, A., and Catling, H. (1967). *Atchaeometry* **10**, 70–77.

Noble, J. V. (1960). *Amer. J. Archaeol.* **64**, 306–318.

Perlman, I., and Asaro, F. (1971). *In* "Science and Archaeology" (R. H. Brill, ed.). M. I. T. Press, Cambridge, Massachusetts.

Simopoulos, A., Kostikas, A., Sigala, I., Gangas, N. H., and Moukarika, A. (1975). *Clays Clay Min.* **23**, 393–399.

Yassoglou, N. J., and Peterson, J. B. (1969). *Proc. Soil. Sc. Soc. Amer.* **33**, 967–970.

# 9

# *Analysis of Works of Art*

## B. Keisch

National Gallery of Art Research Project
Carnegie-Mellon University
Pittsburgh, Pennsylvania

## I. Introduction

Because of the common occurrence of iron compounds in rocks of the earth's crust and the variety of color exhibited by these compounds, iron is an important constituent, by design and by accident, in many works of art. In particular, the iron-bearing pigments used by artists, historically as well as in the present, are numerous and represent a variety of colors. Table I gives a list of such pigments, the approximate percentage of iron in them, and the colors or color ranges exhibited by them. The exact identification of many of these, by means usually available to those working in museum laboratories or in association with restorers, is difficult. An important cause of this difficulty is the small size of samples that can be removed from valuable or potentially

**TABLE I**

Artists' Pigments Containing Iron

| Name and some alternate names | Chemical form of iron | Colors | Range of iron content (%) |
|---|---|---|---|
| Red iron oxide—Spanish red, Venetian red, Indian red, metallic brown, Mars orange, burnt ochre, burnt sienna, burnt umber | $\alpha$-$Fe_2O_3$ (hematite) | Red, maroon, brown, orange | 28–70 |
| Black iron oxide | $Fe_3O_4$ (magnetite) | Black | 72 |
| Yellow iron oxide—lemon yellow, yellow ochre, raw sienna, raw umber | $\alpha$-FeOOH (goethite) | Yellow to brown | 10–60 |
| Brown oxide | $\alpha$-$Fe_2O_3$ + $\alpha$-FeOOH + $Fe_3O_4$ (sometimes) | Shades of brown | 50–70 |
| Van Dyke brown—Cassel brown | Unknown | Dark brown | 1–2 |
| Prussian blue, Berlin blue, Paris blue | Iron ferrocyanides | Blue | 35–45 |
| Green earth—terre verte, burnt terre verte | Ferric and ferrous: substituted potassium aluminum magnesium calcium silicates (glauconite, celadonite) | Green, blue-green, yellow-green | 15 or less |

valuable paintings and hence the restriction of analytical techniques to microscopy and microchemical techniques.

While many elements have yielded Mössbauer effect spectra, the system represented by iron–57 (natural abundance approximately 2%) is the easiest to observe. Hence our interest in the method as a means of studying iron compounds in art. While most Mössbauer effect data have been collected with transmission geometry, which requires either thin samples or some sample preparation to achieve "thinness," the collection of data by means of scattering allows one to achieve the same results with no sample preparation whatsoever. That is, provided the compound to be studied lies at or very near the surface of the material in which the compound occurs.

In the study of works of art, the latter proviso is usually satisfied, since iron compounds occur as pigments at the esthetic surface of a painting (often covered by only a thin layer of varnish) or are nearly homogeneously distributed throughout the matrix of statuary material, such as terracotta. We have now surveyed numerous samples of all the important iron-bearing pigments, using both transmission and scattering geometry. We have also studied terracotta statuary, which usually contains from 5 to 10% iron. The results provide information of an extraordinary richness and great potential value for identification purposes and for the study of the chemistry and physics of the various iron-bearing materials.

What follows is an account of our research in this field to date. It begins with the results of our survey of pigments in which we have analyzed approximately sixty samples of pigments that fairly represent all the available types of iron-bearing materials. Thus, we have now built up a kind of catalog of spectra for iron-bearing pigments, the contents of which are summarized and discussed below. The next part of the account is a description of the application of the technique to paintings and other objects. This part describes the development of a back-scattering detector so that Mössbauer effect spectra can be collected nondestructively in an efficient manner. Some examples are given of such applications to paintings and also to terracotta statuary.

## II. Mössbauer Effect by Transmission

### A. Experimental Arrangement

Since we wished to begin our studies of pigments using samples of the "pure" material, mainly transmission geometry was used with no particularly novel techniques. The source employed was a cobaltous oxide source (New England Nuclear Corporation, Boston, Massachusetts) containing approximately 10 mCi of $^{57}$Co. The spectrometer was a Nuclear Science Instruments (Duquesne, Pennsylvania) model MM–60 coupled to a Northern Scientific (Middletown, Wisconsin) model NS–600, 1024-channel pulse height analyzer. An ORTEC (Oak Ridge, Tennessee) model 485 amplifier and model 109PC preamplifier were used in conjunction with a home-made proportional counter filled with a mixture of 90% krypton and 10% methane. The high-voltage supply was a Fluke (Seattle, Washington) model 412B. The velocity calibration of the system was established with natural iron foil as an absorber. All spectra were obtained at room temperature.

Powdered samples were mounted between two 22 mm diameter acrylic disks with a total thickness of 2 mm. The sample, typically weighing approximately 50–100 mg, was lightly compressed to a thickness of 1 mm between the disks and confined by a rim on one of them. The assembly was held together by self-adhesive tape applied around its perimeter.

The data obtained for all of the spectra were processed by computer (IBM 360/67), and the fitted plots were also generated by the computer. The particular program used (Chow *et. al.*, 1969) locates peaks and adjusts peak height and width in an iterative process to obtain a least-squares fit.

## B. Survey of Materials by Mössbauer Effect Spectroscopy

### 1. Red Iron Oxides

Red iron oxide pigments, chemically $\alpha$-$Fe_2O_3$, may be natural or produced synthetically. Natural material may be merely ground and washed or may be heated in addition. Synthetic material may be produced by direct precipitation, by calcination of ferrous sulfate—with or without the presence of other materials such as $Ca(OH)_2$—or by calcination of synthetic or natural yellow oxides. The last may be termed "burnt" ochre or sienna and are discussed in Section II,B,5. Mixtures of synthetic and natural material may also occur.

We obtained Mössbauer spectra for fifteen different samples as listed in Table II with their source and/or method of production as well as the manufacturers claimed $Fe_2O_3$ content where known. All of these exhibit very similar six-peak spectra (magnetic hyperfine splitting) of which Fig. 1 is an example. Figure 1 is almost an ideal spectrum for the compound $\alpha$-$Fe_2O_3$,

**TABLE II**

Red Iron Oxides Analyzed

| Sample no. | Name[a] | Manufacturer or source[b] | Description | $Fe_2O_3$ (%) |
|---|---|---|---|---|
| B–52 | Spanish red oxide | RC | Natural, possibly calcined | 88 |
| B–43 | Red oxide | RC | "Chemical by-product" | 95 |
| B–49 | Red iron oxide | W | Natural | 74–78 |
| B–47 | Venetian red | RC | Calcined mixture of ferrous sulfate and calcium hydroxide yielding $Fe_2O_3 + CaSO_4$ | 40 |
| B–50 | Kroma® red oxide | W | Direct precipitation | 96 |
| B–55 | Pure Indian red | RC | Calcined ferrous sulfate | 97 |
| B–45 | Super Aetna® crimson red | RC | Natural possibly calcined | 82 |
| B–48 | Red iron oxide | W | Calcined ferrous sulfate | 99 |
| B–342 | Natural oxide | Italy | — | — |
| B–326 | Terra Pozzuoli | Italy | Natural | — |
| B–39 | Red oxide | Unknown | Unknown | — |
| B–368 | Red iron oxide (maroon shade) | RC | Natural, possibly calcined and/or mixed with synthetic material | 89 |
| B–374 | Red iron oxide | RC | Chemically precipitated | 98 |
| B–376 | Metallic brown iron oxide | RC | Calcined natural | 97 |
| B–375 | Red iron oxide | RC | Calcined ferrous sulfate | 98 |

[a] Names in italics are manufacturer's designations.
[b] RC = Reichard-Coulston, Inc.; W = C. K. Williams (Minerals, Pigments, and Metals Division, Chas. Pfizer & Company).

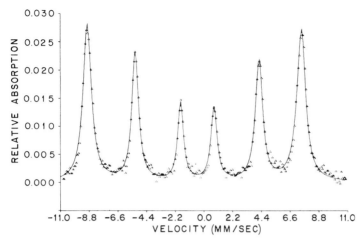

**Fig. 1.**  Mössbauer effect spectrum of hematite, $\alpha$-Fe$_2$O$_3$ (sample B–374).

but many of the others show peak broadening and peak areas that are not in the theoretical ratios of $3:2:1:1:2:3$. These "nonideal" spectra may be due to a particle-size phenomenon (see Section II,C). Samples prepared by processing natural material (B–52, B–49, B–45, B–342, B–326, and B–368) are somewhat different than the natural reds included in Section II,B,5.

### 2. Black Iron Oxide

Black iron oxide, a ferro–ferric oxide having an empirical formula Fe$_3$O$_4$, is not commonly used as an artist's pigment and is not listed in various compendia (Harley, 1970; Gettens and Stout, 1966). Nowadays, however, it is sometimes used alone, mixed with carbon black, or in the preparation of synthetic brown oxides.

Figure 2 is the Mössbauer spectra obtained on a sample obtained from Reichard-Coulston containing 93% iron oxide (sample number B–367). There are actually twelve peaks, but only nine are resolved here. The relative intensities of the two superimposed six-peak spectra are reversed from published data (Sawatzky *et al.*, 1969), perhaps indicating that this sample is partially oxidized and not pure Fe$_3$O$_4$.

### 3. Yellow Iron Oxides—Synthetic

Yellow iron oxide is the hydrate of Fe$_2$O$_3$. The preferred empirical formula is FeOOH. This compound has been reported in four structural modifications (Bernal *et al.*, 1959), the most common of which is $\alpha$-FeOOH, or goethite. Synthetic yellow oxide is produced by chemical precipitation and is goethite. Figure 3 is the Mössbauer spectrum we obtained from sample number B–57,

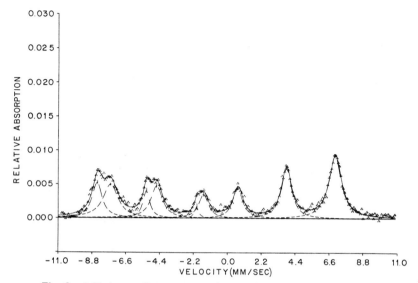

**Fig. 2.**    Mössbauer effect spectrum of magnetite, $Fe_3O_4$ (sample B–367).

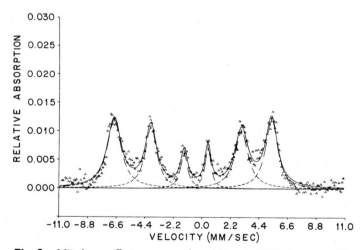

**Fig. 3.**    Mössbauer effect spectrum of goethite, α-FeOOH (sample B–57).

containing 87% $Fe_2O_3$ (theoretical amount for pure α-FeOOH is 89.9%). A synthetic lemon yellow oxide, obtained from sample number B–103, yielded a similar spectrum. A spectrum obtained from a sample of a synthetic oxide called Mars orange (sample number B–106), once known as an artificial ochre, showed that this material was not FeOOH but probably very finely divided $Fe_2O_3$.

### 4. Yellow Iron Oxides—Natural (Ochre, Sienna, and Umber)

Yellow ochre, raw sienna, and raw umber are naturally occurring yellow iron-bearing earths containing varying amounts of silica and alumina that owe their color principally to $\alpha$-FeOOH. Umber also contains varying amounts of manganese oxides as well, which imparts a greenish brown color to the pigment. Ochres are considered to contain between approximately 15 and 40% $Fe_2O_3$ (based on dry weight), while siennas contain more than 40% $Fe_2O_3$ (ASTM Standards, 1958), but the ready distinction between ochre and sienna is not easy. It has been said that sienna is a special kind of ochre, the best of which occurs near Sienna, Italy, with a greater transparency than most ochres (Gettens and Stout, 1966, p. 156).

Table III is a descriptive list of the natural yellow earths from which ME spectra were obtained. There are two general types of Mössbauer spectra obtained. Type I, shown in Fig. 4 (sample number B–19), appears to exhibit

**TABLE III**

Natural Yellow Earth Oxides Analyzed

| Sample no. | Name | Manufacturer or source[a] | Description | Mössbauer type[b] |
|---|---|---|---|---|
| B–87 | Yellow ochre | N | — | I |
| B–19 | Yellow ochre | SOF | Screened | I |
| B–22 | Yellow ochre | SOF | "Micronized" | I |
| B–21 | Yellow ochre | SOF | "Micronized" | I |
| B–84 | Yellow ochre | France | — | I |
| B–415 | Yellow ochre | W | — | II |
| B–82 | Ochre | FM | Late nineteenth century | II |
| B–26 | Raw sienna | RC | Italian import, 70% $Fe_2O_3$ | I |
| B–27 | Raw sienna | RC | Domestic | II |
| B–414 | Raw sienna | W | — | |
| B–416 | Raw sienna | W | — | |
| B–411 | Raw sienna | W | 68–73% $Fe_2O_3$ | II |
| B–17 | Raw sienna | — | — | II |
| B–51 | Raw Turkey umber | RC | Cyprus import, 49% $Fe_2O_3$, 8% $MnO_2$ | I |
| B–28 | Raw Turkey umber | L | Cyprus import | I |
| B–65 | Cappagh brown[c] | — | — | I |

[a] N = Northampton Co., Pennsylvania; SOF = Société des Ochres de France (Auxerre); W = C. K. Williams (Minerals, Pigments, and Metals Division, Chas. Pfizer & Company; FM = Farmers Museum, Cooperstown, New York; RC = Richard-Coulston, Inc.; L = Lowe Bros., Dayton, Ohio.

[b] See text.

[c] Cappagh brown is an umber-like material (containing manganese) that comes from a particular place in Cork County, Ireland. Samples courtesy of Winsor & Newton, Ltd.

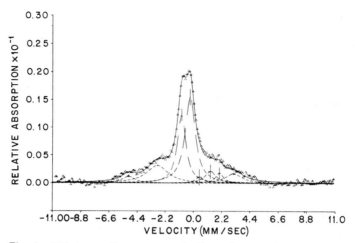

**Fig. 4.** Mössbauer effect spectrum of yellow ochre, type I (sample B–19).

superparamagnetism. This phenomenon has been previously described (Shinjo, 1966; Van der Krann and Van Loef, 1966), and the degree to which it occurs is a function of particle size (see Section II,C). Type II, shown in Fig. 5 (sample B–27), is a complicated variation of the six-peak spectrum for $\alpha$-FeOOH (Fig. 3) probably exhibiting some superparamagnetism. Some Type II samples, B–26 for instance, also show the presence of a small amount of $\alpha$-Fe$_2$O$_3$ as two additional small peaks at velocities of approximately $-9.0$ and $7.4$ mm/sec. Two Type I samples of this group, B–28 and B–65, show complete superparamagnetic relaxation of the hyperfine structure, with only two central peaks observed in the spectrum.

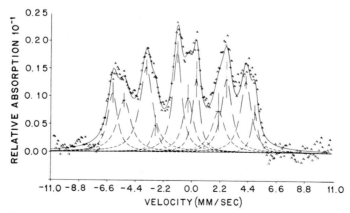

**Fig. 5.** Mössbauer effect spectrum of raw sienna, type II (sample B–27).

## 5. Red Ochre, Burnt Sienna, and Burnt Umber

These natural red or reddish brown materials are either produced by heating the corresponding yellow, or raw, form or may be found naturally, as in the case of red ochre. In any case, the iron oxide content is mainly in the form of $\alpha$-$Fe_2O_3$. Note that in Section II,B,1, a number of natural or processed natural red oxides are listed. Chemically, the state of the iron is the same. However, those shown in Table IV do not yield the clean six-peak spectrum (magnetic hyperfine splitting) of the pure red oxide.

### TABLE IV

Red Ochres and Burnt Siennas and Umbers Analyzed

| Sample no. | Name | Manufacturer or source[a] | Description | Mössbauer type |
|---|---|---|---|---|
| B–23 | Red ochre | SOF | Screened | IV |
| B–24 | Red ochre | SOF | "Micronized" | IV |
| B–18 | Red ochre | SOF | "Micronized" | IV |
| B–20 | Red ochre | SOF | "Micronized" | IV |
| B–356 | Armenian bole[b] | Venice | c. 1906 | IV |
| B–350 | Light red | WN | — | IV |
| B–44 | Burnt sienna | RC | Italian import, 71% $Fe_2O_3$ | III |
| B–42 | Burnt sienna | LC | Italian import | IV |
| B–53 | Burnt sienna | RC | Domestic | IV |
| B–41 | Burnt sienna | RC | — | IV |
| B–413 | Burnt sienna | W | — | IV |
| B–412 | Burnt sienna | W | — | IV |
| B–40 | Burnt sienna | W | — | IV |
| B–373 | Burnt Turkey umber | RC | Cyprus import, 47% $Fe_2O_3$, 18% $MnO_2$ | III |
| B–33 | Burnt umber | Unknown | — | III |
| B–54 | Burnt Turkey umber | RC | Cyprus import, 53% $Fe_2O_3$, 9% $MnO_2$ | III |

[a] SOF = Société des Ochres de France (Auxerre); WN = Winsor & Newton, RC = Reichard-Coulston, Inc.; LC = Lucas Company; W = C. K. Williams (Minerals, Pigments, and Metals Division, Chas. Pfizer & Company.

[b] Bole is a claylike material similar to an ochre but softer. It was originally found in Armenia but now elsewhere.

Two general types of spectra are obtained. Type III shows a prominent central group flanked by a six-peak spectrum. Figure 6 (sample B–54) is an example. The outer peaks seem to correspond to the black oxide ($Fe_3O_4$) rather than the red, and because of the very dark color of this pigment, a burnt umber, one might be inclined to presume that the black oxide predominates, with the central peaks indicating superparamagnetism as in the

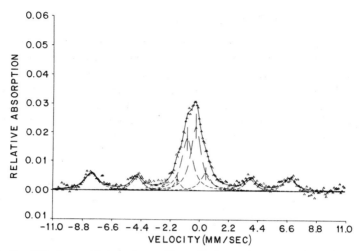

**Fig. 6.** Mössbauer effect spectrum of burnt Turkey umber, type III (sample B–54).

case of the yellow oxides. However, sample B–44, a definite red color, yields the same type of spectrum. It was also found that there is a similarity between the spectra for sample B–54, a burnt umber, and that for the synthetic Mars orange mentioned in Section II,B,3.

The second type, IV, consists mainly of a six-peak spectrum corresponding to that obtained for $\alpha$-$Fe_2O_3$, except that (a) the outermost peaks are not quite so far apart and (b) there are smaller peaks in the central part of the spectrum. Figure 7 (sample B–23) is a typical example of this type of spectrum.

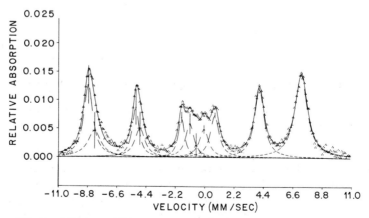

**Fig. 7.** Mössbauer effect spectrum of red ochre, type IV (sample B–53).

Again, while the possibility exists for the presence of some $Fe_3O_4$ or $\gamma$-$Fe_2O_3$ (maghemite) (Kelly *et al.*, 1961) in these materials, it is likely that we are observing the presence of some fraction of a superparamagnetic material. It is quite possible that the two peaks at the extreme left are really only one broadened peak. Resolution of these problems of composition could probably be aided by measurements at low temperatures, which have not been done as yet. The prominence of the central peaks varies considerably among the samples in this group. In some cases, for example, B–41, these "extra" peaks are barely discernible.

### 6. Brown Pigments Containing Iron

There are several iron-bearing pigments that are brown in color. First, there are calcined naturally occurring materials known in the industry as metallic brown oxides. Sample B–376, an example of this, has been classified among the red oxides in Section II,B,1, and the color is actually a reddish brown. The pigment industry also prepares brown iron oxide synthetically by mixing various proportions of $\alpha$-$Fe_2O_3$ (red), $\alpha$-FeOOH (yellow), and either carbon black or $Fe_3O_4$ (black).

When the yellow oxide of the sienna, umber, or ochre type consists of extremely fine particles, it sometimes appears darker and approaches brown in color. Umbers are particularly good examples of this, but they also contain $MnO_2$, which is black, and/or $Mn_3O_4$, which is red.

Finally, Van Dyke brown (also called Cassel earth or Cologne earth) is a brown peat-like material, the color of which is mainly derived from the organic material present. It also contains approximately 1% iron and is therefore included here. Mössbauer spectra for these were obtained on lightly packed powder approximately $\frac{1}{4}$ inch thick. Table V lists brown pigments for which Mössbauer spectra were obtained.

The spectra obtained for the synthetic brown mixtures showed, as expected, the presence of the separate pigments making up the mixtures. Also, as expected, the varying proportions of the oxides were evident as the shade of brown varied. A typical spectrum, that of sample B–370, is shown in Fig. 8. The difference between sample B–46 and the others is evident in that the width and proportions of the outermost peaks in the spectrum from B–46 (not shown) is effected by the presence of $Fe_3O_4$ instead of carbon black.

The spectra obtained from the three Van Dyke brown samples were similar to one another. Two peaks were obtained. For two of the samples, B–419 and B–421, these peaks were located at $-0.92$ and $-0.20$ mm/sec. For the other sample, B–420, the peaks were located at $-0.86$ and $-0.03$ mm/sec. No attempt was made to determine the source of this difference or the nature of the iron compounds involved.

**TABLE V**

Brown Iron-Bearing Pigments Analyzed

| Sample no. | Name | Manufacturer or source | Description |
|---|---|---|---|
| B–46 | Brown iron oxide | Williams | Synthetic mixture of yellow, red, and black oxides |
| B–369 | Brown iron oxide | Reichard-Coulston | Synthetic mixture of yellow and red oxides and carbon black |
| B–371 | Brown iron oxide | Reichard-Coulston | Synthetic mixture of yellow and red oxides and carbon black |
| B–372 | Brown iron oxide | Reichard-Coulston | Synthetic mixture of yellow and red oxides and carbon black |
| B–370 | Brown iron oxide | Reichard-Coulston | Synthetic mixture of yellow and red oxides and carbon black |
| B–420 | Van Dyke brown | Habich | Cassel Region, Germany (0.8% Fe) |
| B–421 | Van Dyke brown | Urban | Cassel Region, Germany (2% Fe) |
| B–419 | Van Dyke brown | Hawley | Czechoslovakian import (0.8% Fe) |

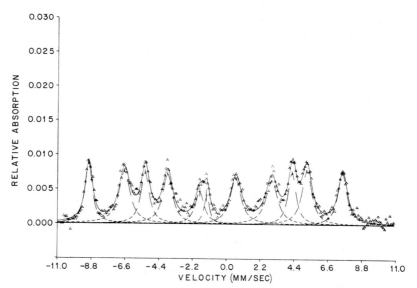

**Fig. 8.** Mössbauer effect spectrum of a synthetic brown pigment (sample B–370).

### 7. Prussian Blue

Prussian blue is the well-known iron ferrocyanide that was first produced early in the eighteenth century. Our interest in this pigment is not for identification purposes as it is readily identified by other means, but for conservation purposes. This pigment is not completely lightfast, and the mechanism by which the color fades is not understood. As a part of our investigation into the use of Mössbauer effect spectroscopy, we have begun a study of the fading phenomenon. Panels painted with a preparation containing Prussian blue synthesized with enriched $^{57}Fe$ in the two different positions were prepared and are being exposed to intense light for a prolonged period. Using a scattering configuration for Mössbauer effect spectroscopy, it is hoped that changes in the spectrum will be observed that can be related to the change in color and that information thus obtained will help elucidate the mechanism of fading.

### 8. Green Earth (Terre Verte)

Green earth consists of either glauconite or celadonite or both, which are mica-like clay minerals containing interstitial ferrous and ferric ions. The color varies from blue-green to yellow-green. "Burning" results in an olive green color. Mössbauer spectroscopy is being used to help characterize samples of this material and to classify sources. Typical spectra of green earth and burnt green earth are shown in Fig. 9. (These were obtained in a scattering configuration, see Sections III,A and B). The area of the most positive peak, which is half of a ferrous doublet, seems to be directly related to the color of the material. Further studies are under way, including the effect of heating green earth to various temperatures. Preliminary results of the latter experiment indicate that a change occurs at about $300°-350°C$ in which the ferrous component disappears, the color changes, and an additional ferric component appears.

## C. Effect of Particle Size on Mössbauer Spectra

As mentioned above, there is evidence that one of the important properties effecting the color as well as the Mössbauer spectrum obtained from a sample of pigment is the size of the particles. This phenomenon, the change of Mössbauer spectrum with very small particle size, has been explained (Shinjo, 1966; Van der Krann and Van Loef, 1966) as a collapse of the internal magnetic field due to superparamagnetism.

For four of our samples—B–103 (a synthetic lemon yellow oxide, Fig. 10), B–22 (a yellow ochre, Fig. 11), B–28 (a raw turkey umber), and B–65 (cappagh

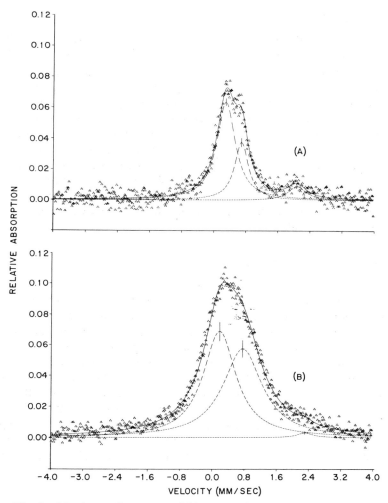

**Fig. 9.** Mössbauer effect spectra of (A) green earth and (B) burnt green earth.

brown, Fig. 12)—x-ray diffraction showed no evidence of the presence of anything but $\alpha$-FeOOH, even though the Mössbauer spectra of these four, in the order given here, showed progressively less of the six-peak spectrum normally expected for $\alpha$-FeOOH. Electron micrographs of these samples and one other (B–411, a raw sienna) were made, and the corresponding Mössbauer spectra obtained showed an obvious correlation with particle size. We therefore conclude that the spectra obtained for natural $\alpha$-FeOOH-bearing pigments usually show superparamagnetism, while those for synthetic $\alpha$-FeOOH usually do not. This is due mainly to the smaller particle size occurring in the

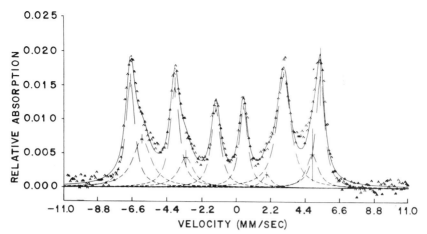

**Fig. 10.**  Mössbauer effect spectrum of lemon yellow oxide (sample B–103).

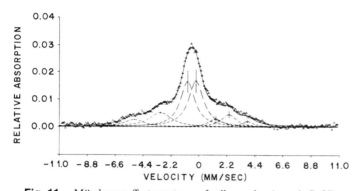

**Fig. 11.**  Mössbauer effect spectrum of yellow ochre (sample B–22).

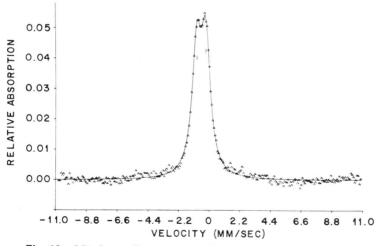

**Fig. 12.**  Mössbauer effect spectrum of Cappagh brown (sample B–65).

natural materials, but it is likely that some of this phenomenon may also be due to the presence of clay in the natural material (Yassaglou and Peterson, 1969). Such spectra would show slightly greater quadrupole splitting than we observe, but while these peaks may contribute to the total spectrum, the small differences in peak location and the low intensity defies their resolution.

One may also explain the unusual spectra obtained for some natural $\alpha$-$Fe_2O_3$ on the basis of particle size (Kundig and Bommel, 1966) and/or the presence of clay in a similar manner. A series of simple experiments confirmed this for the materials with which we are working. These experiments consisted of calcining synthetic $\alpha$-FeOOH samples of known particle size and observing the spectrum obtained from the $\alpha$-$Fe_2O_3$ thus produced. We were able to conclude that natural red oxides are, as in the case of the yellow oxides, more finely divided than their synthetic counterparts and may contain some clay-bound iron, because in a few samples, one could distinguish the wider spacing of the central peaks typical of the spectra obtained for clay–iron preparations.

### D. Quantitative Analysis

Thus far we have been mainly concerned with the qualitative assessment of the properties of pigments that might be encountered. While this kind of information can be of use in identifying the sources of certain artists' works, in many cases it would also be useful to obtain an estimate of iron content. Generally, the magnitude of an observed Mössbauer effect is related to the iron content in a sample undergoing analysis.

We have only just begun to explore the possibilities of making direct determinations of iron content in typical samples. Already we have found that, in binary mixtures of red, black, and yellow oxides (see Fig. 8), we can determine the approximate proportions of these components. Thus far, our limits of detection are roughly one part of one oxide in the presence of ten parts of another. Difficulties arise, however, when the effects of media or admixtures of other pigments are considered.

### III. Mössbauer Effect by Scattering

#### A. Apparatus and Experimental Arrangement

The key element in any attempt to obtain Mössbauer spectra in a scattering configuration is the radiation detector. A number of workers (Koch et al., 1967; Terrell and Spijkerman, 1968; Ord, 1969) have published results obtained with various techniques. Our detector is based on a design that originated with Chow (Chow *et al.*, 1969), with improvements suggested by Flinn (Flinn and O'Connell, 1972), and certain modifications of our own. The basic detector is described elsewhere (Keisch, 1972), and with some additional refinements has performed remarkably well. Throughout our experiments, the electronic equipment used was the same as that described above for our transmission work. The cobalt-57 source used was diffused into a chromium matrix. The spectrometer drive was calibrated with natural iron.

## B. Bulk Pigments by Scattering

Several typical pigments were analyzed by scattering for comparison with the data obtained for the same materials by transmission. These samples were loosely packed in a small acrylic holder and covered with a 0.05 mm polyethylene film. Some of the following results were obtained while the detector was still under development. Results of several comparisons not shown here demonstrated, as expected, the equivalence of the two methods of measurement.

## C. Nondestructive Mössbauer Spectrum of Pigments in Paintings

Figure 13 shows the set-up for obtaining a Mössbauer spectrum of the iron-bearing pigments in an eighteenth century painting. At the time the photograph was taken, we were limited to an area of the painting within 3.5 inches of the edge of the work. However, we have rearranged the base of the system so that any point on the surface of a painting up to 45 cm in its

**Fig. 13.** Analyzing a painting by backscattering.

narrowest dimension or of any iron-bearing object of comparable size is accessible for this analysis. In the figure, the part of the painting undergoing analysis, a circular area approximately 2.5 cm in diameter, is a light brown decorative detail on the subject's cape. The resulting spectrum, shown in Figure 14, is typical for a natural yellow ochre. Figures 15 and 16 show the dark backgrounds of two other paintings from the eighteenth century. The former appears to be typical of natural red ochres, while the latter is typical of a raw umber. Each of these spectra was obtained over a period of 2–3 days, although with the detector optimized and a more active source, much shorter times would probably be sufficient.

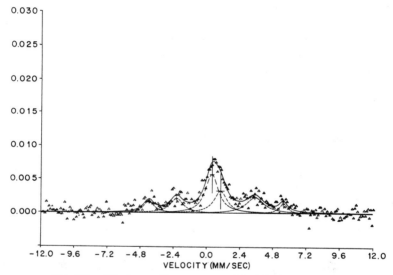

**Fig. 14.**   Mössbauer effect spectrum of the painting in Fig. 13.

The use of this technique as an aid to solving problems in identification would involve the study of several works by an artist (or school) to which an unknown work is attributed. While this might not permit one to make a positive identification, it could provide strong corroboration with absolutely no harm to the works involved.

## D. Mössbauer Spectra of Iron in Terracotta Statuary

### 1. General Background

The color of terracotta is mainly due to the presence of iron in the material. Using neutron activation analysis, we found weight concentrations of iron ranging from 2.3 to 21% with a mean of approximately 6%. We undertook

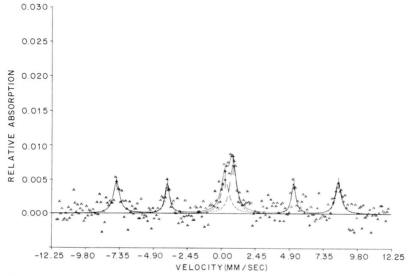

**Fig. 15.** Mössbauer effect spectrum of the dark background of an eighteenth century painting.

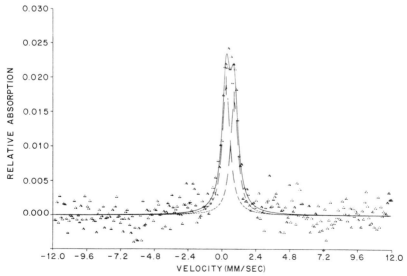

**Fig. 16.** Mössbauer effect spectrum of the dark background of an eighteenth century painting.

to obtain Mössbauer spectra by scattering to determine if there were differences that might lead to a means of identification. We also investigated the effect of heating samples of some of these to temperatures high enough to change their color and the chemical state of the iron in them.

While these measurements could have been made *in situ*, thirty-two samples removed from twenty-four terracotta works were available to us. Not all of the works were believed genuine. Among those that were beyond suspicion we generally found that one of two "types" of Mössbauer spectra were obtained. These are illustrated in Figs. 17 (TC–3) and 18 (TC–19).

The spectrum shown for sample TC–3 shows virtually no magnetic hyperfine splitting and is not unlike that obtained for a clay–iron complex by other workers (Yassaglou and Peterson, 1969). Sample TC–19 appears to be a

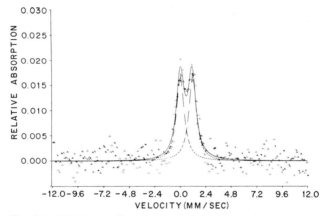

**Fig. 17.** Mössbauer effect spectrum of a terracotta (sample TC–3).

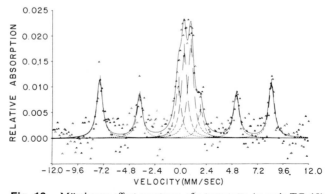

**Fig. 18.** Mössbauer effect spectrum of a terracotta (sample TC–19).

mixture of at least two components, $\alpha$-$Fe_2O_3$ and another nonmagnetic material. In nearly every well-authenticated case, one or the other of these spectra appeared (i.e., two peaks separated by $\sim 0.8$–1 mm/sec or superimposed spectra of $\alpha$-$Fe_2O_3$ and a component yielding two peaks with a separation of $\sim 0.5$–0.8 mm/sec). In a few cases, a four-peak spectrum was obtained that was predominantly like that obtained for sample TC–3. In a few instances, where two or three works were known to be by the same artist, very similar spectra were obtained. The use of this technique appears to offer useful corroboration of authenticity pending the acquisition of a larger collection of data.

### 2. Effect of Baking Temperature

We suspected that the TC–3 type of spectrum represented clay that had not been heated to the highest temperatures used for baking terracotta. Therefore, an experiment was initiated using sample TC–4, which consisted of heating the sample to successively higher temperatures with Mössbauer spectra taken after each heat treatment. Figure 19 shows some of the results obtained. Indicated temperatures are only approximate, as the furnace controls were not calibrated.

The spectrum obtained at room temperature was not changed by heating to a temperature of 770°C (the curve for 770°C is not shown). After heating at 1000°C, we noted a deepening of the color as well as the change of the

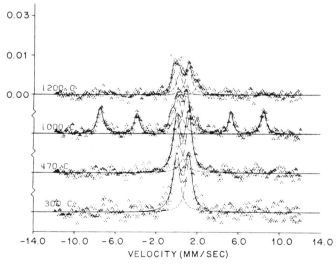

**Fig. 19** Mössbauer effect spectra of a terracotta heated to successively higher temperatures as indicated. Spectra were taken at room temperature after cooling.

Mössbauer spectrum from a typical TC–3 type to a TC–19 type. Finally, at 1200°C, the sample melted to a brownish, glassy solid, which, of course, would not ever be observed in a museum piece.

We have now been able to obtain a supply of clay that when baked, results in a product similar in chemical composition and in physical properties to renaissance terracotta. This clay is being used in an ongoing study to determine the conditions of firing that will result in Mössbauer spectra similar to those observed in museum specimens. Thus far, it appears that the transition noted above takes place between 900° and 1000°C, depending somewhat upon the time of heating.

With the tentative assumption that the observance of the TC–19 type spectrum indicates a higher firing temperature, we noted several relationships in our data that lead us to the tentative conclusion that certain artists and craftsmen were more skilled than others in the management of their furnances and thus produced a more evenly fired work at a higher temperature. Evaluation of whether or not this results in a higher quality product is beyond the present scope of this project; we merely indicate that these differences occur.

### 3. Identification of Fakes

A small number of works of highly dubious authenticity were also included in our survey. We have been able to make some tentative conclusions that may help classify some of them. While all superficially appear to be terracotta, a closer look at the chemical and physical properties of the material indicated that some are not. Deceptively good imitations of terracotta are produced by a cementlike material homogeneously colored to look genuine. Mössbauer spectra obtained from such materials appear to be quite different than those obtained from genuine works.

The detection of mismatched materials of works that presumably should be identical is also possible. For example, out of one unified group of three figures, two were analyzed by Mössbauer Effect scattering. One of these yielded a TC–3 type spectrum. The second yielded a TC–19 type spectrum. The spectra obtained, smoothed by computer but not fitted, are shown in Fig. 20. The distinction is quite clear and suggests that they were not produced at the same time by the same artist.

A suspect work yielding a TC–3 type spectrum (Fig. 17) was attributed to an artist of whom two genuine samples yielded TC–19 type spectra (Fig. 18). Tentatively, one is tempted to rule out the attribution of the suspect work. This conclusion also was reinforced by data obtained by neutron activation analysis.

Finally, x-rays showed that one arm of a small statue of David had been repaired sometime in the past. The joint was essentially invisible, and outwardly it appeared that the broken arm had been merely reattached. How-

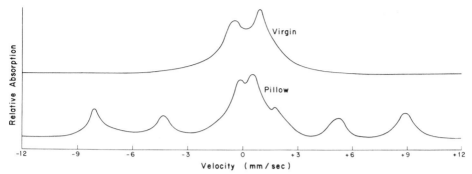

**Fig. 20.**  Mössbauer effect spectra for two objects in a three-part group. The pillow is suspected of being a relatively modern addition to the group.

ever, Mössbauer spectroscopy by scattering clearly showed, without the need of sampling, that the arm and the remainder of the piece were of two different materials and that the attached arm was not original.

At this point in our survey of terracottas, one can safely conclude that a valuable tool is at hand. However, as in much of this type of work, a much larger body of data is necessary to ascertain the reliability of determinations such as those tentatively made here.

## IV.  Summary and Conclusions

Mössbauer effect spectroscopy can be used to characterize iron oxide pigments on the basis of such properties as oxidation state, crystallography, degree of hydration, and particle size. This information is useful for distinguishing between certain natural and synthetic pigments and can be obtained, using a scattering geometry, with no sampling required. The method can also be used as an aid to characterizing and classifying terracotta objects. In any case, the experimental arrangement for analysis of a presumably valuable object would be similar to that shown in Fig. 13. Various parts of the object could be brought into the area of analysis and much information obtained in this way. For the present, however, the routine use of such information for purposes of identification awaits the outcome of additional efforts at standardization of the techniques involved and classification of the spectra observed.

With further development, the method may also provide a tool for investigating undesirable transformations, such as the fading of Prussian blue. It is also possible that semiquantitative analyses for iron may be obtained, which would add yet another dimension to the utility of the method.

The nondestructive character of this method is one of its chief attributes. Typically, less harm will be done to a work than that done by the usual x-radiography. The estimated absorbed dose of radiation for a typical Mössbauer analysis is somewhat less than 1 rad in a scattering geometry.

### Acknowledgments

The work represented here would not have been possible without the help of the following persons and institutions: The United States Atomic Energy Commission, the National Gallery of Art (Washington, D.C.), Dr. P. A. Flinn and Mr. T. O'Connell of the Carnegie-Mellon University Physics and Metallurgy Departments, Dr. Robert L. Feller and the Pigment Bank of the Art Research Project at Carnegie-Mellon University, and various pigment manufacturers indicated in the tables throughout this work.

### References

ASTM Standards (1958). Part 8, pp. 58–63. *Amer. Soc. Testing Mater.*, Philadelphia, Pennsylvania.

Bernal, J. D., Dasgupta, D. R., and Mackay, A. L. (1959). *Clay Minerals Bull.* **4**, 15–30.

Chow, H. K., Weiss, R. F., and Flinn, P. A. (1969). Mössbauer Effect Spectrometry for Analysis of Iron Compounds, U.S. At. Energy Comm. Rep., No. NSEC–4023–1 (October).

Flinn, P. A., and O'Connell, T. (1972). U.S. At. Energy Comm. Project, No. AT(11–1)–3033.

Gettens, R. J., and Stout, G. L. (1966). "Painting Materials." Dover, New York.

Harley, R. D. (1970). "Artists' Pigments, c. 1600–1835." Butterworths, London and Washington, D.C.

Keisch, B. (1972). *Nucl. Instrum. Methods* **104**, 237–240.

Kelly, W. H., Folen, V. J., Hass, M., Schreiner, W. N., and Beard, G. B. (1961). *Phys. Rev.* **124**, 80.

Koch, R. C., Chow, H. K., and Bogner, R. L. (1967). Mössbauer Spectrometry for Analysis of Iron Compounds. U.S. At. Energy Comm. Rep., No. NSEC–4010–1.

Kündig, W., and Bömmel, H. (1966). *Phys. Rev.* **142**, 327–333.

Ord, R. N. (1969). *Appl. Phys. Lett.* **15**, 279–281.

Sawatzky, G. A., Van der Woude, F., and Morrish, A. H. (1969). *Phys. Rev.* **183**, 383.

Shinjo, T. J. (1966). *Phys. Soc. Japan* **21**, 917–922.

Terrel, J. H., and Spijkerman, J. J. (1968). *Appl. Phys. Lett.* **13**, 11–13.

Van der Krann, A. M., and Van Loef, J. J. (1966). *Phys. Lett.* **20**, 614–616.

Yassoglou, N. J., and Peterson, J. B. (1969). *Soil Sci. Soc. Amer. Proc.* **33**, 967–970.

# 10

# Some Applications to Polymer Science

**V. I. Goldanskii**
**L. A. Korytko**

Institute of Chemical Physics
USSR Academy of Science
Moscow, USSR

## I. Introduction

Since the Mössbauer effect was discovered, it has time and again shown its effectiveness in solving complicated problems in various fields of science and technology. Gamma resonance investigations have found the most intensive use in studying chemical and physicochemical subjects. The gamma resonance method has proved highly effective in studying a number of problems encountered in the physical chemistry of polymers, examples of which will be treated in this chapter.

Most of these gamma resonance investigations have been carried out with tin and iron compounds. This is a result of the relative simplicity of experimental techniques with these Mössbauer isotopes and a keen interest in organometallic polymers (their structure, properties, new methods of production), polymer stabilizers, and the effect of polymerization catalysts.

The Mössbauer effect was first observed in a polymer by Bryukhanov *et al.* (1962). The compound studied was the copolymer of methyl methacrylate and ethyl stannyl methacrylate, $CH_2=C(CH_3)-COOSn(C_2H_5)_3$, with 1:1 composition and weight content of tin $\sim 30\%$. From x-ray analysis, this polymer was known to be amorphous. A doublet was observed in the Mössbauer spectrum at 77° and 195°K. It is noteworthy that, as exemplified by this polymer, structures of the $R_iSnX_{4-i}$ type (where R = alkyl or aryl groups and X = substituents with a higher degree of electronegativity) were for the first time shown to be characterized by an electric field gradient at the tin nuclei and by quadrupole splitting appearing in the gamma resonance spectra. This observation of the Mössbauer effect in a polymer was of fundamental significance for the further development of Mössbauer spectroscopy (GRS), as the previously unrecognized possibility of conducting Mössbauer experiments with organometallic systems where light (hydrogen, carbon, oxygen) atoms predominate was demonstrated conclusively.

This work marked the beginning of GRS application to the chemistry of macromolecular compounds. At first, which is quite natural, the studies concentrated on the changes in the parameters of Mössbauer spectra that occur when passing from monomer to polymer compounds. In the investigations of Aleksandrov *et al.* (1964a), during the polymerization of tin butyl methacrylic derivatives, $(C_4H_9)_nSn[OCOC(CH_3)CH_2]_{4-n}$, ($n = 1, 2, 3$), quadrupole splitting was observed to decrease by 20% in comparison with the initial monomers, the isomer shift remaining practically unchanged.

A relatively large number of works have been devoted to investigating the structure of various classes of polymer systems. Based on the established classification of isomer shift values characteristic of tetra- and divalent tin compounds (Goldanskii, 1963), the compounds of nominally divalent tin, $Bu_2Sn$, and $Ph_2Sn$ were actually found (Goldanskii, *et al.*, 1964) to have a polymer structure of the polyethylene type with a tetravalent tin atom in the main chain $-SnR_2-SnR_2-$.

As another example of such investigations we shall cite the work of Harrison and Zuckerman (1969), who studied tin dicyclopentadienyl. The value of the isomer shift in freshly prepared $(C_5H_5)_2Sn$ suggests that the tin is divalent, but changes with aging to tetravalent tin, with the formation of the above-mentioned polymer structure. In this transition, the nominal formula of the compound is preserved.

In the investigations carried out by Goldanskii *et al.* (1963), it was found that $(R_2SnO)_n$ polymer oxides have a network structure, the coordination number of the tin atoms being equal to 5. The same study showed that the nature of the R radical significantly affected the recoil-free fraction $f$ (see Chapter 1). The asymmetry of the intensities of quadrupole doublet lines was observed and assumed to be caused by the anisotropy of tin atom oscillations in such compounds—the phenomenon known at present in literature as Goldanskii–Karyagin effect (Goldanskii *et al.*, 1962; Karyagin, 1963).

Intensive studies of the structure of both organic and inorganic polymers with tin coordination bonds have been started lately. Methods and procedures of these studies are close to those used in structural investigations of conventional nonpolymeric compounds, described in many monographs and reviews on Mössbauer effect (Greenwood, 1971; Goldanskii and Herber, 1968; Zuckerman, 1970). They are therefore not considered here, except for some special cases.

Among the early works on GRS, let us once again note (Aleksandrov *et al.*, 1964a) that $f$ was observed to rise during the polymerization of monomeric tin derivatives of methyl methacrylate. The magnitude of this parameter is known (Greenwood, 1971; Goldanskii and Herber, 1968) to be related to the mean square of the amplitude of Mössbauer atom oscillations. This amplitude depends on the bonds of the atom with its neighbours and on the structure of the matrix in which it is situated. It was of interest to ascertain the sensitivity of $f$ to the changes in molecular weight, geometry of molecules, their structure, etc. As it was shown by Aleksandrov *et al.* (1964a), the value of $f$ increases from 0.072 to 0.18 during the polymerization of dibutyltin dimethacrylate, and from 0.085 to 0.25 for tributyltin methacrylate.

Aleksandrov *et al.* (1962) were the first to study the effect of $\gamma$ irradiation on the spectral parameters of the $[(C_4H_9)_2Sn(OCOCCH_3CH_2)_2]_n$ polymer. Irradiation doses reached 250 Mrad. One of the lines of the doublet spectrum was observed to broaden, with a simultaneous decrease in the intensity of the other one. At larger doses, the Mössbauer spectrum split into several poorly resolved components. The authors noted, however, that the experimental data did not make it possible to reach unambiguous conclusions on the source of the changes observed in the spectra. This was not surprising, for at that time Mössbauer spectroscopy was at an early stage of its development. It has taken a long period of time to systematize and critically analyze the numerous experimental data and apply this knowledge to the relatively complicated investigations of structural and chemical transformations.

As Mössbauer spectroscopy data on polymers accumulated, it became clear that useful information is conveyed by practically all the parameters of the spectra. Since each problem requires an individual approach, there is hardly any sense in making general recommendations as applied to the problems of macromolecular chemistry. In the present review, we shall therefore consider some investigations that are, as we see it, especially typical. We will use these as examples to demonstrate the concrete approach to solving specific problems of the physical chemistry of polymers with the help of GRS methods.

As a rule, when interpreting experimental results in these investigations, no quantitative calculations are presented. Useful information is revealed on the basis of detailed qualitative analysis of Mössbauer spectra, obtained from samples designed to test certain specific assumptions. The most conclusive studies combine such analysis with data obtained using other methods. It should also be noted that gamma resonance investigations in the field of the physical chemistry of polymers have been performed in many diverse fields. This has presented some difficulties when writing this review. To clarify the role of gamma resonance investigations in solving these problems, we thought it useful to begin some of the sections with a brief description of "pre-Mössbauer" studies and the formulation of the problem for Mössbauer spectroscopy.

Mössbauer spectroscopy has demonstrated its effectiveness in solving a broad range of questions posed by the physical chemistry of macromolecular compounds. Within this field, a number of subfields have been found where such investigations have proved to be most productive. These are primarily studies of the mechanisms of polymer stabilization with tin and iron compounds, kinetics of polymerization processes, reactivity of some polymers and their catalytic properties, and the dynamic parameters of polymer systems (e.g., molecule oscillations, mobility in the region of phase transitions). These questions will be considered here. The investigations included in this review are mainly those pertaining to the chemistry of macromolecular compounds. Some of that research was previously dealt with by Goldanskii (1971).

We do not consider here the structural investigations of inorganic and organic coordination polymers and the effect of their properties on various parameters of Mössbauer spectra, the investigations connected with studying ion-exchange resins, biopolymers, and certain other topics.

## II. Effects of Some Stabilizers of Polyethylene

### A. Problems of Stabilization of Radiation-Modified Polyethylene

The wide use of polymer materials and the need to preserve their physico-mechanical properties for a long period of time have led to active search for the methods and ways of protecting polymers from the effect of various aggressive factors (e.g., high temperature, ionizing radiation, light, oxygen). Work in this field proceeds in two ways—by developing new modified polymers and by selecting appropriate stabilizing additives. An enormous number of original papers, monographs, and patents is devoted to these questions. It is, however, necessary to note that though the most widespread stabilizers are grouped in the literature according to the principle of their effect, investigating the protective mechanism of compounds, even within one group, sometimes requires distinct approaches, depending on the actual conditions of the polymer usage. Many organotin compounds are known to be effective stabilizers of polymers, in particular polyethylene, against thermooxidative aging (Ingham et al., 1960; Neiman, 1964; Gordon, 1963; Voigt, 1966). In spite of the wide practical use of stabilizers, however, the mechanism of their protective action remains in many respects obscure. Until recently, the empirical approach predominated in selecting effective stabilizers. This was also true of organotin compounds. Some of them were found to serve as excellent antioxidants for radiation-modified polyethylene as well. It is to be noted that recently the methods of radiochemical modification of polymers, including polyethylene, have developed rapidly, aimed at effecting preassigned changes in their structure and properties (Kargin, 1973). Polyethylene, under the effect of $\gamma$ irradiation, is known to undergo irreversible changes (Kargin, 1973; Charlesby, 1960) resulting in the formation of a new material with electrical, physicomechanical, and chemical properties superior to those of the initial polymer. Such modified polyethylene, for example, can operate at temperatures of up to $\sim 200°-250°C$, substantially exceeding the melting point of unirradiated polyethylene. Irradiated polyethylene has, however, one significant disadvantage. It is succeptible to thermooxidative aging to a greater degree than nonirradiated polyethylene. Thus, the problem of preserving its properties at high temperatures when exposed to air (conditions when oxidation reactions proceed at high rates) acquires vital significance. The task of finding effective antioxidants for radiation-modified polyethylene is complicated by the fact that additives must be introduced into the polymer before it has been irradiated. Therefore, they are themselves to some extent subjected to radiolysis, which can radically affect their properties as inhibitors of oxidation processes. This imposes an additional requirement on such additives—the inhibiting properties must survive doses necessary for the radiation modification of polyethylene. It should be noted that the properties of modified compositions (polyethylene + additive) in many respects depend on the polymer and additive interaction with reactive formations created during the irradiation. To understand the processes taking place in stabilized polyethylene during its radiation modification and subsequent service at elevated temperature, a detailed study of the mechanism of interaction between the stabilizing additive and the polymer is necessary.

Certain organotin salts of fatty acids proved to be among the most promising stabilizing additives for radiation-modified polyethylene. These are, in particular, dibutyltin maleate, dibutyltin distearate, dibutyltin dilaurate, etc., which, having been introduced into the irradiated polyethylene, improved manyfold its subsequent properties (Finkel' et al., 1968; Gladkova et al., 1965; Gashnikova et al., 1963). From the analysis of infrared spectra of the mixtures and the data of some other investigations, Gashnikova et al. (1963) attributed this improvement to the inhibition by the above compounds of chain reaction polymer oxidation. These compounds were also observed to produce improved radiation resistance. The data obtained, however, proved to be insufficient to describe the mechanism of their action during the radiation modification and the thermooxidative aging of the compositions. It should be said, incidentally, that even for ordinary unmodified polyethylene, there is no experimental evidence in the literature reliably supporting the existence of specific mechanisms by which thermooxidative processes are inhibited by organotin compounds of the given class, though they are widely used in industry. This problem has been solved to a considerable extent with the help of Mössbauer spectroscopy.

### B. Mechanism of Polyethylene Stabilization with Organotin Compounds

A. Y. Aleksandrov et al. (1964b) were the first to make an attempt to investigate the mechanism of polymer stabilization using the Mössbauer effect. They studied changes in the spectra resulting from the radiation modification of the mixture of polyethylene with dibutyltin maleate,

$$(C_4H_9)_2Sn\begin{array}{c} O-CO-CH \\ \phantom{O-CO-}\| \\ O-CO-CH \end{array}$$

Samples were irradiated in vacuum with doses of 25–600 Mrad of $^{60}Co$ $\gamma$ rays. Introducing dibutyltin maleate into polyethylene by means of hot rolling was shown not to affect the Mössbauer spectrum, as compared with the initial organotin compound, for which the isomer shift was $\delta = 1.67^{\dagger}$ mm/sec and quadrupole splitting was $\Delta = 3.35$ mm/sec. During irradiation, with the increase of the dose, a gradual decrease in the intensity of

---

† This and other isomer shifts of tin compounds are given with respect to $SnO_2$, and for iron compounds, with respect to sodium nitroprusside.

the initial dibutyltin maleate lines is observed in the spectra of the mixtures, and a new doublet, which has been found to belong to the maleate of divalent tin appears. The intensity of this doublet increases while the dose rises up to $\sim 300$ Mrad at which the Mössbauer spectrum of dibutyltin maleate has practically disappeared. The spectrum of the sample that was irradiated with this dose consists of a doublet belonging to divalent tin and a singlet of considerably smaller intensity in the vicinity of zero velocity. The changes observed in the spectra are indicative of tetravalent tin being transformed into divalent tin as a result of the disruption of $Sn(—C—)$ bonds. Butyl radicals formed in the process are assumed to block unsaturated sections of the polymer chain, which are most susceptible to oxidation, and, in this way, inhibit the development of oxidation chains. Based on this assumption, Aleksandrov *et al.* (1964b) maintain that the protective action of the additive is exhausted in the dose interval of 100–300 Mrad, since, as it has been established from the intensity of the lines in gamma resonance spectra, for a sample that received a dose of 100 Mrad, the initial additive is still present in an appreciable amount, whereas at the dose of 300 Mrad it is already completely absent. It is necessary to emphasize the value of this work, which for the first time showed GRS to be a promising method for solving such problems. The authors, however, restricted themselves to investigating the process of modifying stabilized polyethylene by radiation, without in any way touching upon the processes that take place in irradiated compositions in different service conditions and, in particular, under heating in air. The assumption of the inhibition of oxidation reactions thus remained unproved.

Aleksandrov *et al.* (1968a) worked to clarify these questions. They attempted to establish the mechanism of the action of organotin stabilizers during the thermal oxidation of irradiated and unirradiated mixtures of polyethylene with dibutylin distearate, $(C_4H_9)_2Sn[OCO(CH_2)_{16}CH_3]_2$; dibutyltin dilaurate, $(C_4H_9)_2Sn[OCO(CH_2)_{10}CH_3]_2$; and the ethyl ester of dibutyltin dimaleate, $(C_4H_9)_2Sn[OCOCH=CHCOOC_2H_5]_2$. These compounds belong to the same class of polyethylene stabilizers as dibutyltin maleate, and it was of interest, therefore, to confirm the common nature of the processes taking place during the radiation modification of various compositions. Aleksandrov *et al.* (1968a) irradiated samples of stabilized polyethylene by a $\gamma$ source of $^{60}Co$ in helium atmosphere at room temperature simultaneously with control samples of organotin compounds. Irradiation doses varied from 5 to 300 Mrad. Comparative analysis of Mössbauer spectra changes was performed for stabilized polymer samples and those of initial compounds. As an example, Figs. 1 and 2 show the spectra of polyethylene mixtures with dibutyltin dilaurate and the ethyl ester of dibutyltin dimaleate, and the spectra of pure organotin compounds that have been

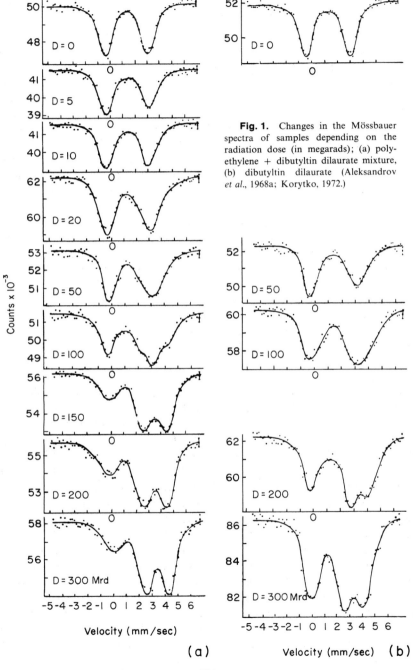

**Fig. 1.** Changes in the Mössbauer spectra of samples depending on the radiation dose (in megarads); (a) polyethylene + dibutyltin dilaurate mixture, (b) dibutyltin dilaurate (Aleksandrov *et al.*, 1968a; Korytko, 1972.)

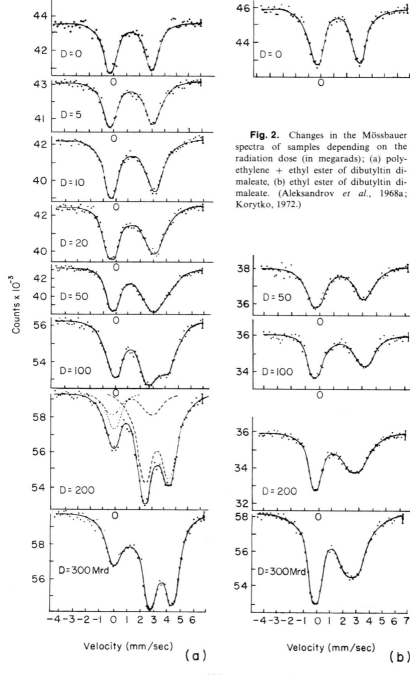

**Fig. 2.** Changes in the Mössbauer spectra of samples depending on the radiation dose (in megarads); (a) polyethylene + ethyl ester of dibutyltin dimaleate, (b) ethyl ester of dibutyltin dimaleate. (Aleksandrov *et al.*, 1968a; Korytko, 1972.)

irradiated under the same conditions. The spectra of polyethylene mixtures with dibutyltin stearate are similar to those shown in Fig. 1. The doublet spectra of the pure organotin compounds and of the samples of polyethylene stabilized by them are, as seen from the figures, identical and have the parameters $\delta = 1.35 \pm 0.07$ mm/sec and $\Delta = 3.3 \pm 0.15$ mm/sec. These values coincide with the data from literature on related compounds (Goldanskii and Herber, 1968). As the irradiation dose increases, radical changes take place in the Mössbauer spectra of both the mixtures and the organotin compounds themselves (obviously due to the radiochemical transformations) eventually showing the disappearance of the primary organotin compound that had been introduced into polyethylene. As a result, as was also shown by Aleksandrov (1964b), in Mössbauer spectra of those samples that have received doses of about $\sim 200-300$ Mrad, a doublet is observed whose parameters correspond to Sn(II) compounds (tin distearate, tin dilaurate, ethyl ester of tin dimaleate), and a singlet with isomer shift close to zero. It can thus be inferred from these data that the transformations of such organotin compounds during $\gamma$-irradiation proceed in two main directions:

$$\text{Organotin compounds} \underset{k_2}{\overset{k_1 \,(-2C_4H_9)}{\rightleftarrows}} \begin{array}{l} \text{Sn(II)} \\ \text{Sn(IV) with } \delta = 0, \end{array}$$

and, as will be seen below, $k_1 \gg k_2$ in the polymers. By correlating the Mössbauer spectra of simultaneously irradiated mixtures and pure organotin compounds, it was possible to draw new conclusions on the mechanism of radiation-induced transformations of organotin compounds in the compositions studied.

Figures 3 and 4 show the relative changes in the concentrations of the initial stabilizer and the compounds of Sn(II) and Sn(IV) with $\delta = 0$, depending on the irradiation dose. The curves were obtained by decomposing the Mössbauer spectra presented in Figs. 1 and 2 into their components (Korytko, 1972). From these graphs, it follows that the maximum accumulation of divalent tin compounds in stabilized polyethylene corresponds to doses of 250–300 Mrad, with the consumption of primary organotin compounds and Sn(II) formation in stabilized polyethylene going much faster than the radiolysis of these same compounds in the absence of the polymer. The relative content of Sn(II) in the polymer is also higher. The natural explanation of these facts is that when organotin compounds themselves are irradiated, a significant role is played by recombination (cellular) effects, whereas in stabilized polyethylene, butyl radicals leave the cell due to additional reactions with the polymer.

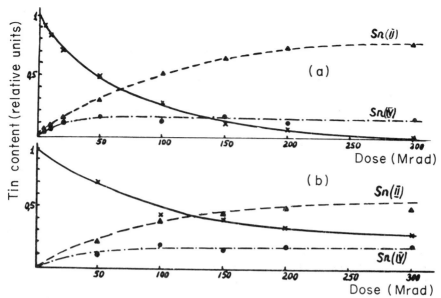

**Fig. 3.** Variations of the relative concentrations of the initial stabilizer and Sn(II) and Sn(IV) versus radiation dose; (a) polyethylene + dibutyltin dilaurate mixture, (b) dibutyltin dilaurate. (Korytko, 1972.)

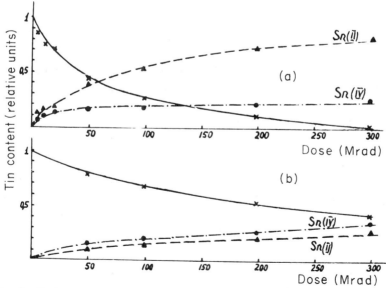

**Fig. 4.** Variations of the relative concentrations of the initial stabilizer and Sn(II) and Sn(IV) versus radiation dose; (a) polyethylene + ethyl ester of dibutyltin dimaleate, (b) ethyl ester of dibutyltin dimaleate. (Korytko, 1972.)

Below it will be made clear that the Sn(IV) compound with $\delta = 0$ has nothing to do with the thermal stabilization effect in radiation-modified polyethylene, and therefore, the mechanism of its formation was not studied in detail. It is to be noted, however, that the proximity of the isomer shift to zero gives grounds for assuming that the tin atom is symmetrically surrounded by oxygen in this compound. One of the possible formation reactions is the cross-linkage of the stabilizing addition molecules via the oxygen of carbonyl groups. When polyethylene mixtures with organotin compounds are irradiated, the polymer matrix prevents the formation of such bonds. In these conditions, the reaction whereby butyl radicals formed during radiolysis interact with polymer radicals becomes predominant, resulting in the rapid formation of Sn(II) in relatively large amounts. The interaction of butyl radicals with the polymer radicals, rather than the protection from thermal oxidative processes, explains the radiation-protective properties of organotin compounds in the polyethylene mixtures studied. It appears likely that, as it is assumed by Aleksandrov et al. (1964b), this reaction results in partially preventing the appearance in the polymer bonds and structure of radiation damage that subsequently, when radiation-modified polyethylene is heated in air, could become the centers of the chain oxidation reaction. It is difficult, however, to explain by this fact the increased stability observed in high-temperature oxidation in the compositions irradiated with doses greater than 150–200 Mrad, i.e., doses at which stabilizers in their initial form are already absent in the polymer (see, e.g., Fig. 1). This suggests that the radiolysis products of initial stabilizers play a role in inhibiting the thermal oxidative processes.

To clarify this question, transformations of organotin compounds and the products of their radiolysis were studied when the irradiated and non-irradiated stabilized polyethylene were heated in air at 200°C. The samples studied had received a dose of ∼150 Mrad, and the stabilizing effects in them could only be caused by the products of the radiolysis of the additives. By direct measurements of the oxidation induction periods for these samples, radiolysis products of the initial organotin compounds were found under these conditions to act as good antioxidants. From the experiments with the samples heated in air at 200°C, it was ascertained that as a result of thermo-oxidative processes, the area under the Sn(II) spectrum gradually decreases and, simultaneously, the intensity of the line of Sn(IV) with $\delta = 0$ increases (Fig. 5b).

Correlating this fact with the existence of a prolonged induction period, Aleksandrov et al. (1968a) concluded that chain oxidation reactions are disrupted when Sn(II) interacts with the RO˙ and ROO˙ polymer radicals, and an inactive product of Sn(IV) with $\delta = 0$ is formed.

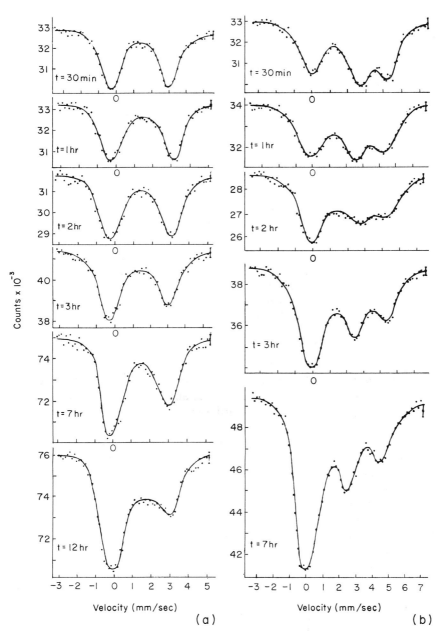

**Fig. 5.**    Mössbauer spectra of polyethylene + dibutyltin distearate mixture oxidized in air at 200°C; (a) unirradiated sample, (b) sample irradiated with a dose 150 Mrad. (Aleksandrov *et al.*, 1968a; Korytko, 1972.)

Similar studies were performed for samples of nonirradiated stabilized polyethylene. It has been established that in this case, too, the final product of the transformation of organotin stabilizers when heated in air is a compound of Sn(IV) with $\delta = 0$ (Fig. 5a). Analysis of the Mössbauer spectra and the data on the oxidation induction periods made it possible to hypothesize that in unirradiated polyethylene, when it is heated in air, thermooxidative reactions are inhibited as a result of the breakaway of butyl radicals and subsequent interaction of Sn(II) compounds with the oxidized hydrocarbon chains. Aleksandrov *et al.* (1968a) note that the compounds of Sn(IV) with $\delta = 0$ formed during the oxidation of both the irradiated and the unirradiated compositions cannot be identified as $SnO_2$, as the Mössbauer effect for them already disappears at 200°K, whereas for $SnO_2$ it is observed at room temperature. This is an argument in favor of the above assumptions. An additional confirmation is the fact that Sn(IV) with $\delta = 0$ was detected in gel fractions isolated from the oxidized samples.

The spectra shown in Fig. 5 have been obtained from thick stabilized polyethylene samples. For these samples, the diffusion delay of penetrating of oxygen into the polymers distorted the true consumption rates of Sn(II) compounds. In experiments on thin stabilized polyethylene films, a direct confirmation of Sn(II) compounds inhibiting thermooxidative processes in irradiated polyethylene was obtained. Experiments were performed with three samples. The first was a mixture of polyethylene with dibutyltin distearate, which was irradiated with a dose of 300 Mrad prior to being introduced into the polymer. Into the second sample, a stabilizer irradiated with a dose of 150 Mrad, was introduced, and the mixture was irradiated with a dose of 150 Mrad. The third sample contained an unirradiated stabilizer and was irradiated with a dose of 300 Mrad. The total dose received by the stabilizer in all of the three samples thus amounted to ~300 Mrad, and it could be said with assurance that the initial organotin compound had completely decomposed into Sn(II) and Sn(IV) with $\delta = 0$. At the same time, the dose received by the polymer in the first sample was equal to zero, in the second 150 Mrad and in the third 300 Mrad. The chain oxidation reaction in irradiated polyethylene is known to proceed much faster than in unirradiated polyethylene. Therefore, if Sn(II) effectively interacts with oxygen-containing radicals that develop the oxidation chains, one should have expected a more rapid decrease of Sn(II) concentration in irradiated polyethylene samples. Figure 6 taken from Korytko (1972) shows the change of the areas under the Sn(II) Mössbauer spectra produced by heating in air at 200°C, for the three samples of stabilized polyethylene mentioned above. As seen from the figure, Sn(II) in the first (unirradiated sample) is

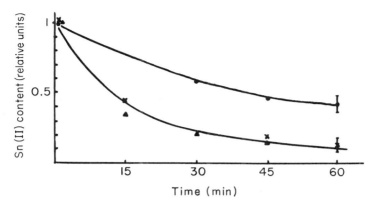

**Fig. 6.**   Variation in the content of Sn(II) in stabilized polyethylene samples upon heating in air; ● = composition made up of unirradiated polyethylene + dibutyltin distearate irradiated with a dose 300 Mrad; × = composition of polyethylene + dibutyltin distearate irradiated with a dose of 150 Mrad and additionally irradiated with a dose of 150 Mrad; ▲ = polyethylene + dibutyltin distearate composition irradiated with a dose of 300 Mrad.

consumed more slowly than in the other two. Similar results have also been obtained for other mixtures of polyethylene with organotin compounds.

From these experiments, it follows that Sn(II) compounds formed during the irradiation of stabilized polyethylene inhibit thermal oxidation reactions at high temperatures by interacting with oxygen-containing macroradicals that develop the oxidation chains, Sn(IV) (with $\delta = 0$) being formed in the process.

The experiments described here made it possible to understand the efficacy of the given class of organotin compounds as stabilizers of un-irradiated and irradiated polyethylene when articles made of them operate under severe service conditions—in radiation fields and at elevated temperatures in the presence of oxygen. It is the formation of Sn(II) carboxylates from the initial organotin compounds during the radiomodification of stabilized polyethylene that promotes the preservation of its properties in air and at high temperatures. It is necessary to note one characteristic feature of the investigated Sn(II) carboxylates. They slowly oxidize in air even at room temperature, whereas the initial organotin compounds are stable under these conditions. At high temperatures, however, when the interaction of Sn(II) compounds with the oxidized polymer becomes the predominant reaction, they behave as effective inhibitors of chain thermal oxidative reactions.

## C. Effect of Tin and Iron Chlorides on Thermooxidative Processes in Polyethylene

The application of Mössbauer spectroscopy also proved fruitful in solving another problem of radiation-modified polyethylene stabilization connected with the effect of tin and iron chloride additives on thermooxidative processes in irradiated materials. We shall briefly describe the studies performed prior to Mössbauer experiments and touch upon the questions that arose when interpreting them.

Berlyant et al. (1968) showed that additions of tin and iron chlorides improve the physicomechanical characteristics of irradiated polyethylene (as compared with samples without additives) for operation at elevated temperature in air. It was a rather surprising result, since most previous research has shown that admixtures of metals with variable valency are known to produce an adverse effect on polymer stability (Neiman, 1964; Gordon, 1963; Voigt, 1966). In particular, their interaction with organic hydroperoxides results in the initiation of thermal oxidative processes. Nevertheless, as was shown by Berlyant et al. (1968), additions of $SnCl_2 \cdot 2H_2O$, $FeCl_2 \cdot 4H_2O$, and $FeCl_3 \cdot 6H_2O$ substantially improve the properties of irradiated compositions—the temperature interval where elastic deformation is observed increases and the temperature at which thermomechanically tested samples rupture rises. Samples of irradiated polyethylene with additives were shown to preserve their physicomechanical properties at high temperatures longer than pure polyethylene samples. With the help of DTA, the rate of polymer oxidation in irradiated samples was established to be much slower than in the absence of additives, reaching its maximum at higher temperatures. Being restricted by the size of this review, we shall not consider any other experimental data obtained by Berlyant et al. (1968). Let us only note that when analyzing the properties and characteristics of the mixtures studied, they came to the conclusion that the above chlorides interact somehow with the polymer and inhibit thermooxidative processes at high temperatures ($\sim 150°$–$250°C$). They considered various possible reactions of additives with the polymer when the samples were first irradiated and then oxidized. In spite of extensive experimentation, however, the data then available did not make it possible to arrive at any definite conclusion on the mechanism of the action of additives.

With the help of Mössbauer spectroscopy (Aleksandrov, 1969), it became possible to answer the questions arising and to define the processes in which tin and iron chlorides participate during the irradiation and oxidation of stabilized polyethylene.

Parallel investigations of Mössbauer spectra for the tin and the iron chlorides and their compositions with polyethylene were performed at varying conditions of exposure to radiation and thermal effect. Doses received by the samples were 50, 100, and 300 Mrad from a $^{60}$Co $\gamma$-ray source. Irradiation was performed in vacuum, and for some of the samples, in air.

Let us first consider the mixture of polyethylene with $SnCl_2 \cdot 2H_2O$. The doublet Mössbauer spectrum of $SnCl_2 \cdot 2H_2O$ ($\delta = 3.7 \pm 0.1$ mm/sec, $\Delta = 1.15 \pm 0.1$ mm/sec), after the introduction of this additive into the polymer using the method of hot rolling, was found to turn into a singlet with $\delta = 4.4 \pm 0.1$ mm/sec, which coincides with the parameters of the dehydrated tin chloride spectrum. This was found to be accompanied by the additive being partially oxidized, which is revealed by the appearance of a line in the vicinity of zero velocity. It has been shown that during irradiation in vacuum, the additive does not practically interact with the polymer and is not subjected to radiolysis. In the case when irradiation is performed in air, the line belonging to $Sn^{2+}$ is broadened and the relative intensity of the line at zero velocity increases. This result was to be expected, since during the irradiation in air the radiation oxidation of the polymer is taking place and the oxygen-containing radicals interact with the additive. It should be noted that when pure $SnCl_2 \cdot 2H_2O$ was irradiated both in vacuum and air, no appreciable changes were observed in the parameters of Mössbauer spectra.

The study of the Mössbauer spectra for samples of hydrous and anhydrous tin chloride and for their mixtures with polyethylene, all oxidized in air at the temperatures of 150° and 250°C, made it possible to establish that the oxidation of the first two compounds proceeds until $SnO_2$ is formed. But in the polymer, the additive interacts with oxygen-containing macroradicals, resulting in the inclusion of tin into the polymer net via oxygen bonds with the formation of the $Sn(—O—)_n$ type structure. Simultaneously, the additive is partially oxidized by the oxygen of air to $SnO_2$.

Let us now turn to polyethylene compositions with iron chlorides. It was previously shown, that heating $FeCl_2 \cdot 4H_2O$ ($\delta = 1.55 \pm 0.1$ mm/sec, $\Delta = 3.0 \pm 0.1$ mm/sec) and $FeCl_3 \cdot 6H_2O$ ($\delta = 0.80 \pm 0.1$ mm/sec, $\Delta = 0.44 \pm 0.1$ mm/sec) in vacuum at the temperatures at which the compositions are rolled and pressed results in a partial dehydration of initial compounds, observed from the changes in the parameters of their Mössbauer spectra. For $FeCl_2 \cdot xH_2O$, $\delta = 1.60 \pm 0.1$ mm/sec, $\Delta = 2.5 \pm 0.1$ mm/sec, and for $FeCl_3 \cdot xH_2O$, $\delta = 0.80 \pm 0.1$ mm/sec, $\Delta = 0.6 \pm 0.1$ mm/sec. When introducing additives into polyethylene by hot rolling, significant changes are observed in the Mössbauer spectra. Figures 7a, b show the spectra of the samples of polyethylene mixed with $FeCl_2 \cdot 4H_2O$ and $FeCl_3 \cdot 6H_2O$.

As seen from the figures, in the spectrum of polyethylene mixed with $FeCl_3 \cdot 6H_2O$, a line appears whose position corresponds to one of the lines of the $FeCl_2 \cdot xH_2O$ doublet, and in the spectrum of polyethylene combined

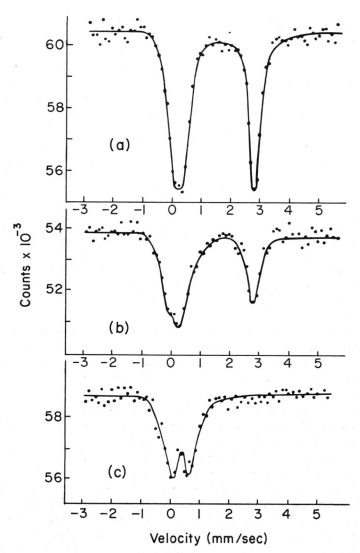

**Fig. 7.** Mössbauer spectra of polyethylene compositions with iron chlorides; (a) poly-ethylene + $FeCl_2 \cdot 4H_2O$ mixture, (b) polyethylene + $FeCl_3 \cdot 6H_2O$ mixture, (c) sample of the polyethylene + $FeCl_3 \cdot 6H_2O$ mixture after storage at room temperature for 2 months. (Aleksandrov *et al.*, 1969.)

with $FeCl_2 \cdot 4H_2O$, the left line of the doublet is markedly broadened due to the presence of $Fe^{3+}$. Such changes in the Mössbauer spectra of iron chlorides in polyethylene are explained by the capability of the salts of variable valency metals to enter oxidation–reduction reactions with hydroperoxides (Neiman, 1964; Gordon, 1963; Voigt, 1966):

$$ROOH + Me^{3+} \rightarrow RO_2^{\cdot} + Me^{2+} + H^+$$

$$ROOH + Me^{2+} \rightarrow RO^{\cdot} + Me^{3+} + OH^-$$

When samples are kept in air for a long time, the intensity of the lines belonging to $Fe^{2+}$ decreases (Fig. 7b). Short-duration heating of such samples both in vacuum and air restores the spectra shown in Figs. 7a, b.

It is to be noted that when these samples are irradiated in vacuum, their Mössbauer spectra do not change as compared with the spectra taken immediately after rolling. When samples are irradiated in air, the relationship between the intensities of the $Fe^{2+}$ and $Fe^{3+}$ lines changes somewhat, which must be caused by the radiation-induced oxidation processes.

Mössbauer spectra undergo the most radical changes when samples are heated in air at temperatures of $150°–250°C$. During such heating, the hydrous and the partially anhydrous iron chlorides are oxidized to $Fe_2O_3$, which was identified by the appearance of a corresponding magnetic structure in the spectra.

When polyethylene compositions with $FeCl_2 \cdot 4H_2O$ and $FeCl_3 \cdot 6H_2O$ are heated under the same conditions, the changes in the spectra indicate intensive interaction of the additives with the oxidized polymer. Figure 8 shows, as an example, the Mössbauer spectra of compositions irradiated with a dose of 100 Mrad and heated at $250°C$.

Without dwelling on the detailed analysis of these spectra performed by Aleksandrov et al. (1969), let us note the most interesting facts. Irrespective of the duration of heating, the magnetic structure does not manifest itself in the spectra of the polymer mixtures, whereas, when the additives themselves are oxidized for $1–1.5$ hours, a substantial amount of $Fe_2O_3$ is formed. It should also be noted that in the spectra of the mixtures, the lines of iron chloride starting materials practically disappear after as little as 1 hr of heating. This is also supported by the data on chloride evolution (Berlyant et al., 1968) from such samples. Attention must be drawn to the Mössbauer spectrum shown in Fig. 8a, which is of great importance for understanding the stabilizing effect of the investigated additives. This is the spectrum of the gel fraction isolated from the composition of polyethylene with $FeCl_3 \cdot 6H_2O$ irradiated with a dose of 100 Mrad and heated at $250°C$. From the observed spectrum, only $Fe^{3+}$ is present in the gel fraction of this sample, whereas the spectrum of the gel fraction isolated from the oxidized composition of

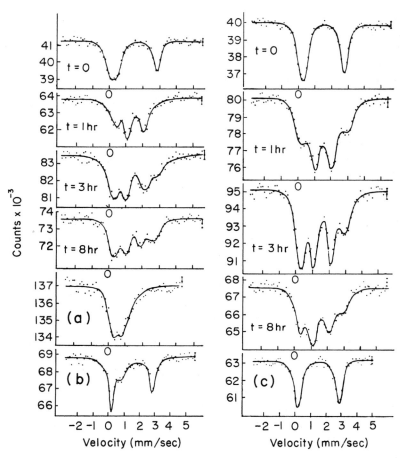

**Fig. 8.** (Left) Mössbauer spectra of the polyethylene + $FeCl_3 \cdot 6H_2O$ mixture irradiated in a vacuum with a dose of 100 Mrad and heated in air at 250°C for $t$ hr. (Aleksandrov *et al.*, 1969; Korytko, 1972.) (a) Spectrum of a gel fraction separated from a heated sample whose initial Mössbauer spectrum is given in Fig. 7c. (Right) Mössbauer spectra of the polyethylene + $FeCl_2 \cdot 4H_2O$ mixture irradiated in a vacuum with a dose of 100 Mrad and heated in air at 200°C for $t$ hr. (b, c) Mössbauer spectra of samples heated for 8 hr and subsequently stored for 3 months at room temperature. (Gol'danskii *et al.* 1968b.)

polyethylene with $FeCl_2 \cdot 4H_2O$ does not differ from the one shown in Fig. 8c. What follows from these data is (1) the conclusion that iron is being incorporated into the polymer net; (2) the iron in gel fractions isolated from samples oxidized at 250°C appears, for the most part, in the same valency state it was in when heating began. These facts are explainable, assuming that under high-temperature heating, in addition to the oxidation–reduction reactions of the additions with the hydroperoxides, chlorine substitution by oxygen-containing polymer radicals is also taking place. Thus, iron atoms would be incorporated into the polymer net via oxygen bonds, with nominally di- and trivalent structures of the $Fe(-O-)_2$ and $Fe(-O-)_3$ types being formed. It may be noted here that when polyethylene mixtures with ferric and ferrous chlorides are heated in air at 150°C, their Mössbauer spectra, as well as the spectra of the gel fractions, are practically the same and are similar to the one shown in Fig. 7a. This difference in the spectra of samples heated at 150° and 250°C seems to be connected with a considerable increase of the contribution of the above-mentioned substitution reactions, as com-pared with the oxidation–reduction reactions.

From the data presented by Aleksandrov *et al.* (1969), it follows that the investigated iron chlorides can participate in reactions with hydroperoxides, with the oxygen of air, with the secondary products of polyethylene oxidation, and with oxygen-containing radicals as shown in scheme (**A**). Depending on

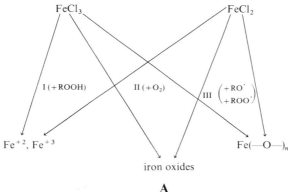

**A**

the thermal conditions of treatment, one or the other reaction can predomi-nate. The contribution of reaction II during the oxidation of the compositions in air at 150° and 250°C is insignificant in comparison with reactions III and I. At 250°C, the relative contribution of reaction III increases, and the additional cross-links of the $Fe(-O-)_n$ type rapidly forming in such condi-tions can result in improvement in the physicomechanical properties of modified mixtures. It is possible that, in such reactions of chlorine substitu-tion by oxygen-containing macroradicals, the chain oxidation process is disrupted, producing thermooxidation protection.

Concluding this section, let us note that the interaction of tin and iron chlorides with polyethylene does not depend on the dose and takes place both in irradiated and unirradiated samples. The process of irradiation in vacuum does not by itself cause any significant reaction of the additives with the polymer radicals. Radiation modification of samples made it possible to extend the temperature range of thermomechanical tests (Berlyant *et al.*,

1968), beyond the melting point of the starting polyethylene. It is this very fact that provided the possibility of observing the above-mentioned unusual features in the behavior of metallic chlorides during high-temperature heating of mixtures. This process is characterized by the increase of the role of such reactions, which are insignificant at low temperatures.

### D. Effect of Ferrocene and Its Derivatives on Thermal Degradation of Polyethylene

It is commonly known that certain ferrocene-containing compounds behave as inhibitors in the reactions of thermal decomposition and oxidation of polymers. Karakozova *et al.* (1972) studied the effect of ferrocene and its derivative *n*-ferrocenylaniline on the course of thermal degradation of high-pressure polyethylene at a temperature of 300°–400°C in a vacuum. It was found that, added to polyethylene to the extent of 0.5% by weight, these compounds markedly inhibited the thermal decomposition.

Mössbauer spectroscopy was used to study the transformations of the inhibitors caused by various conditions of thermal processing of stabilized polyethylene. Using *n*-ferrocenylaniline as an example, it was shown that destruction of polyethylene entailed severe decomposition of the inhibitor. After heating a mixture of polyethylene with this compound for 1 hr at a temperature of 350°C, the Mössbauer spectrum showed, alongside the initial quadrupole doublet ($\delta = 0.80$ mm/sec; $\Delta = 2.24$ mm/sec), a new singlet line with a 0.40 mm/sec isomer shift. This was ascribed by the authors to the nonferromagnetic finely divided iron formed in the decomposition of the inhibitor. The fraction of *n*-ferrocenylaniline decomposed was estimated at roughly 15% of the initial amount. It has been assumed that the decomposition of the inhibitor is triggered by the macroradicals emerging in the system at the early stages of polyethylene destruction.

Comparing the variations of the Mössbauer spectra with the results of other studies of this system, Karakozova *et al.* (1972) concluded that the polyethylene thermal decomposition inhibition effect must be attributed to the cyclopentadienyl radical that emerges as a result of decomposition of the initial additives.

### III. Stabilization of Polyvinyl Chloride and Some of Its Organotin Derivatives

Problems of stabilization are extremely important for polyvinyl chloride (PVC) and vinyl chloride copolymers, just as they are for polyethylene. Dehydrochlorination, oxidation, chain breakage, and cross-linking are known to be the chief processes that occur as PVC ages (Neiman, 1964; Gordon, 1963; Voigt, 1966; Minsker *et al.*, 1972). However, simultaneous investigation of all these processes and mechanisms of PVC stabilization in the presence of aggressive factors (oxygen, heat, radiation, light, etc.) is a task of formidable difficulties. Thus, the effects of these factors are usually studied separately. The present chapter deals with just one problem, the ways certain organotin compounds affect PVC stability in the course of heat treatment. It is currently believed that PVC decomposition proceeds by a conjugation mechanism involving ion–molecule and radical reactions, with particular decomposition routes prevailing, depending on the type and potency of energy and chemical factors as well as on the degree of decomposition. The best approach to studying the PVC stabilization problem in thermal decomposition is to inhibit chain reactions and stress the dehydrochlorination. Use is made of numerous substances, including organotin compounds, as PVC stabilizing agents against thermal and thermooxidative decomposition. The high effectiveness of organotin stabilizers in various aggressive environments gives them a leading position among PVC stabilizers and has brought forth a range of hypotheses on the mechanism of their effect. Yet, in spite of the many experimental studies with the use of various physicochemical methods, the proposed models still arouse a great deal of controversy.

Alkyltin salts such as $R_nSnX_{4-n}$ are widely used as PVC stabilizers. It is assumed that their effectiveness derives both from their ability to suppress free-radical processes, for instance by

$$R^{\cdot} + (C_4H_9)_2Sn(OCOCH_3)_2 \rightarrow RC_4H_9 + C_4H_9\overset{\cdot}{Sn}(OCOCH_3)_2,$$

and from the ability to substitute labile chlorine atoms in the polymeric molecule; furthermore, they are capable of accepting hydrogen chloride to form alkyltin chlorides:

$$Bu_nSnX_{4-n} + (4-n)HCl \rightarrow Bu_nSnCl_{4-n} + (4-n)HX$$

or

$$Bu_2SnX_2 + 2RCl \rightarrow Bu_2SnCl_2 + 2RX.$$

Some other theories attempting to explain the mechanism of these stabilizing agents have been advanced (Neiman, 1964; Gordon, 1963; Voigt, 1966; Minsker *et al.*, 1972). Very frequently symmetrical $SnR_4$ and mixed $R_{4-n}SnR'_n$ organotin compounds, such as tetraphenyltin and dibutyldiphenyltin, are employed as PVC stabilizers. Special mention should be made of the organotin derivatives of PVC (Purinson *et al.*, 1968; Plate *et al.*, 1968; Minsker *et al.*, 1968), which apart from being PVC stabilizers, are modified polymers featuring high intrinsic thermal stability. Substantial work has been done on various classes of organotin PVC stabilizers by non-Mossbauer methods; the results are presented in a review (Minsker *et al.*, 1971) and a monograph (Minsker *et al.*, 1972). These investigations revealed some specific effects of the organotin stabilizers in question and led to a number of models of the mechanism of their reactions with the polymer and with hydrogen chloride. The latter models required direct experimental confirmation, and so Aleksandrov *et al.* (1971a) performed gamma resonance experiments to study the mechanism of PVC stabilization by tributyltin laurate, dibutyltin dicaprylate, and tetraphenyltin. The information obtained from the Mössbauer spectra regarding the immediate environment of the tin atom gives an insight into the relative importance of the various reactions of the organotin compounds in the course of PVC decomposition and also enables the reaction products to be identified.

Aleksandrov *et al.* (1971a) studied the Mössbauer spectra of PVC mixtures with organotin compounds as a function of the time of dehydrochlorination in an atmosphere of nitrogen at a temperature of 175°C. The spectra of the polymer mixed with tributyltin laurate and dibutyltin dicaprylate presented symmetrical doublets of the parameters $\delta = 1.45 \pm 0.07$ mm/sec, $\Delta = 3.60 \pm 0.1$ mm/sec and $\delta = 1.50 \pm 0.07$ mm/sec, $\Delta = 3.8 \pm 0.1$ mm/sec, respectively. Heating of these compositions for 1 hr caused a 0.6–0.7 mm/sec reduction of quadrupole splitting. Further heating (up to 6 hr) caused minor variations (within 0.1–0.2 mm/sec) in isomer shifts and quadrupole splittings. Since the organotin chlorides formed as a result of dehydrochlorination of the mixtures have similar parameters of the Mössbauer spectra, the following technique was employed for identifying them. The samples were hydrolyzed in an aqueous solution of ammonia, the organotin mono-, di-, and trichlorides forming organotin oxides, hydroxides, and acids with substantially differing Mössbauer spectrum parameters (Goldanskii and Herber, 1968). Analysis of these spectra permitted clearly identifying the products of stabilizer transformation in the polymer. The data obtained indicated that the initial stage of dehydrochlorination of the PVC compositions with tributyltin laurate and dibutyltin dicaprylate proceeds through the reactions

$$(C_4H_9)_3SnR + HCl \rightarrow (C_4H_9)_3SnCl + HR$$

$$(C_4H_9)_2SnR'_2 + 2HCl \rightarrow (C_4H_9)_2SnCl_2 + 2HR',$$

where R and R' are lauric and caprylic acid radicals. The reaction of substitution of labile chlorine atoms mentioned above may likewise play a certain part in the formation of organotin chlorides. For the PVC mixture with tributyltin laurate, partial substitution of chlorine for one butyl group was observed; this phenomenon may either be associated with the addition of this group to the polymeric macroradical by the route

$$R^{\cdot} + (C_4H_9)_3SnCl \rightarrow RC_4H_9 + (C_4H_9)_2SnCl^{\cdot}$$

or else ascribed to the reaction of tributyltin chloride with hydrogen chloride by the route

$$(C_4H_9)_3SnCl + HCl \rightarrow (C_4H_9)_2SnCl_2 + C_4H_{10}.$$

Apparently the former reaction is less probable than the latter one.

The Mössbauer spectra of frozen gaseous dehydrochlorination products showed presence of negligible $SnCl_4$, suggesting that under such conditions the butyl radicals retained their bond with the tin atom. It will be noted that heating of the compositions studied in air at 175°C for 1 hr eliminates all traces of organotin compounds, in contrast to the polyethylene mixtures with similar stabilizers. Comparison of the spectral parameters of the organotin chlorides left in the polymer and present in the gaseous dehydrochlorination products revealed that the former are characterized by an abnormally small value of quadrupole splitting (approximately 2.9–3.2 mm/sec), whereas for the latter this quantity is within the norm (approximately 3.5 mm/sec). This decrease of quadrupole splitting is assumed to result from the emergence of dative bonds between the tin atom and the structures formed in the course of dehydrochlorination of the mixtures. Indirect evidence is available confirming the existence of such complexes (Minsker et al., 1971). On the basis of the foregoing assumption, it is easy to explain some peculiarities noted in the course of PVC dehydrochlorination in the presence of organotin compounds (Minsker et al., 1971).

Figure 9 presents the Mössbauer spectra of the PVC mixtures with tetraphenyltin as a function of heat treatment time.

Analysis of these spectra as well as the spectra of hydrolyzed samples revealed that during the induction period, chlorine atoms are substituted for the phenyl groups, but without the formation of triphenyltin chloride. The reason for this is believed to be associated with the simultaneous pairwise break-off of the phenyl groups following the formation of intermediate complexes with hydrogen chloride, as noted by Gol'danskii et al. (1968) and as will be discussed below. The gaseous products of dehydrochlorination were a mixture of organotin chlorides and $SnCl_4$.

It has been mentioned earlier that the trialkyl- and triarylstannyl derivatives of PVC, which are self-stabilized polymers, are effective PVC stabilizers. Fruitful gamma resonance experiments on the structure and mechanism of interaction with hydrogen chloride of the tributyl-, triphenyl-, and triethylstannyl derivatives of PVC have been performed by Goldanskii et al. (1968) and Aleksandrov et al. (1971b). The methods of synthesizing these compounds are described by Purinson et al. (1968). Their Mössbauer spectra exhibit singlet lines with an isomer shift of roughly 1.4 mm/sec, typical of tin compounds such as $R_4SnR'_{4-n}$, where R and R' are organic radicals. This is direct evidence in support of the following structure of these PVC

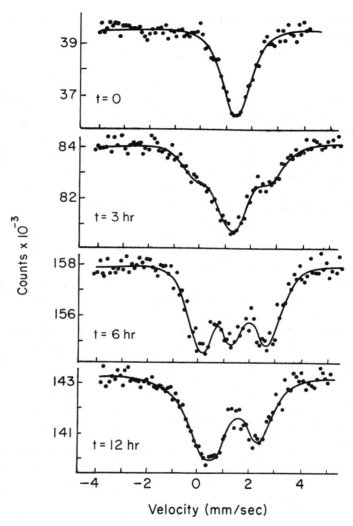

**Fig. 9.** Mössbauer spectra of the polyvinyl chloride + tetraphenyltin composition after heating for $t$ hr in an atmosphere of nitrogen at 175°C. (Aleksandrov *et al.*, 1971a.)

derivatives:

$$\sim CH_2-CH-CH_2-CH\sim$$

$$SnR_3 \qquad Cl \qquad\qquad R = n\text{-}C_4H_9, C_2H_5, C_6H_5.$$

The reaction of these polymers with hydrogen chloride was studied in a model system by passing HCl through their solutions in tetrahydrofuran.

These conditions ruled out all possibility of polymer destruction affecting the reaction mechanism. As HCl reacted with the triphenylstannyl derivative of PVC, the initial singlet line turned to a doublet (as is illustrated in Fig. 10) that seems to be associated with the formation of organotin chlorides in the side chains of the polymer. Identification of the reaction products by the hydrolysis technique mentioned above revealed total lack of mono- and trichlorostannyl groups. Immediately following the emergence of organotin dichloride, the content of tin in the polymer decreases. All these facts led Gol'danskii *et al.* (1968b) to assume that the reaction proceeds through pairwise splitting of organic radicals from the tin atom to form a phenyl-dichlorostannyl group at the first stage, followed by a simultaneous bond cleavage of the tin atom with the phenyl group and with the polymeric chain to form $SnCl_4$. According to Gol'danskii *et al.* (1968b), such a reaction mechanism is determined by the formation of intermediate hexacoordinate complexes by the route shown in scheme (**B**). Reaction of the tri-*n*-butyl-

$$\sim CH_2\!-\!CH\!-\!CH_2\!-\!CH \sim \xrightarrow{2HCl} \sim CH_2\!-\!CH\!-\!CH_2\!-\!CH \sim \xrightarrow{-2PhH}$$

with $SnPh_3$ and $Cl$ substituents on the left, and $HCl$, $Cl$ on the right with a hexacoordinate Sn center bearing Ph, HCl, Ph, Ph groups.

$$\sim CH_2\!-\!CH\!-\!CH_2\!-\!CH \sim \xrightarrow{2HCl} \sim CH_2\!-\!CH\!-\!CH_2\!-\!CH \sim \longrightarrow$$

with $Ph\!-\!Sn\!-\!Cl$ (and $Cl$ below) and $Cl$ substituents on the left, and $Cl$ on the right with a hexacoordinate Sn center bearing Ph, HCl, Cl, Cl groups.

$$\sim CH_2\!-\!CH_2\!-\!CH_2\!-\!CH \sim + PhH + SnCl_4.$$

with $Cl$ substituent.

**B**

stannyl derivative of PVC with hydrogen chloride brought about no additional lines in the Mössbauer spectra, the sole alteration being a reduction in intensity of the initial singlet line, suggesting that the tri-*n*-butylstannyl group broke off the polymer, bypassing intermediate stages by the route

$$\sim CH_2\!-\!CH\!-\!CH_2\!-\!CH \sim \xrightarrow{HCl} \sim CH_2\!-\!CH_2\!-\!CH_2\!-\!CH \sim + Bu_3SnCl$$

with $SnBu_3$ and $Cl$ substituents on the left, and $Cl$ on the right.

For the triethylstannyl derivative of PVC, the changes in the Mössbauer spectra following its reaction with hydrogen chloride are accounted for by

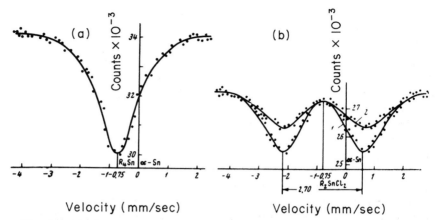

**Fig. 10.** Mössbauer spectra of the triphenylstannyl derivative of polyvinyl chloride; (a) initial sample; (b) sample treated with HCl: 1. for 30 min, 2. for 90 min.

the two-stage reaction discussed by Alexsandrov *et al.* (1971b). According to this reaction, at the first stage a chloride atom is substituted for one ethyl group,

$$\sim CH_2\!-\!CH\!-\!CH_2\!-\!CH \sim \xrightarrow[-EtH]{HCl} \sim CH_2\!-\!CH\!-\!CH_2\!-\!CH \sim,$$
$$\qquad\;\; | \qquad\qquad | \qquad\qquad\qquad\qquad \backslash \qquad\qquad |$$
$$\quad SnEt_3 \qquad Cl \qquad\qquad\qquad (Et)_2SnCl \qquad Cl$$

so that an appropriate doublet appears in the Mössbauer spectrum. At the second reaction stage, the tin–polymer bond breaks off and diethyltin dichloride is formed,

$$\sim CH_2\!-\!CH\!-\!CH_2\!-\!CH \sim \xrightarrow{HCl} \sim CH_2\!-\!CH_2\!-\!CH_2\!-\!CH \sim \; + \; (Et)_2SnCl_2$$
$$\qquad | \qquad\qquad | \qquad\qquad\qquad\qquad\qquad\qquad\qquad\qquad |$$
$$(Et)_2SnCl \qquad Cl \qquad\qquad\qquad\qquad\qquad\qquad\qquad\qquad Cl$$

Thus, the mechanism of this reaction is intermediate with regard to the two discussed above. This agrees to a certain extent with the data on the increasing strength of the bond of tin with the radicals in the series Ph < Et < Bu. In this series, the polymer chain occupies a position close to the ethyl group.

For the triethylstannyl derivative of PVC, the Mössbauer spectra were studied, with the compound subjected to dehydrochlorination at 175°C in an atmosphere of nitrogen (Aleksandrov *et al.*, 1971b). Under these conditions, the mechanism of its reaction with hydrogen chloride does not differ from the one described above. It will be noted that by the end of the HCl elimination induction period, which for this polymer was about 90 min, the initial singlet line had disappeared from the spectrum, with only the doublet of the diethylchlorostannyl group left. It follows that the process of thermo-stabilization of this polymer occurs in the first stage of the previously discussed reaction of the polymer with hydrogen chloride.

Special mention should be made of the work by Morris and Rockett (1972), who investigated the Mössbauer spectra of a number of asymmetrically substituted ferrocenes. This work merits attention, as we see it, on account of the fact that two of the compounds studied, dibutyltin $SS'$-bis(ferrocenyl thioglyconate) and dibutyltin bis(ferrocenyl carboxylate), are classified by the authors as PVC stabilizers. Unfortunately, Morris *et al.* (1972) give no information on the mechanism of stabilization, although the problem seems all the more interesting in view of the fact that the compounds in question contain two Mössbauer isotopes, permitting comparative evaluation of the data offered by the Mössbauer spectra of each isotope.

### IV. Polymerization Processes and Structure of Catalytically Active Centers

This section is concerned with several examples of successful application of Mössbauer spectroscopy for studying the specific features of polymerization of certain systems. Aleksandrov *et al.* (1968b) first demonstrated the effectiveness of the gamma resonance method for studying the polymerization processes and the structure of the catalytically active centers in the reaction of *n*-butyllithium with triethylvinyltin. Researchers were attracted to this system by some extraordinary features that manifested themselves in the course of anionic polymerization of triethylvinyltin and its copolymerization with styrene (Platé *et al.*, 1966). It was found that these reactions catalyzed with *n*-BuLi are characterized by surprisingly low rates and small yields, and the reaction products have a low molecular weight, although *n*-BuLi is known to be an effective catalyst of anionic polymerization of vinyl monomers. Moreover, according to Platé *et al.* (1966), addition of triethylvinyltin to styrene (which effectively polymerizes on *n*-BuLi) brought about a sharp reduction in the rate of polymerization. These and some other studies all indicate that *n*-BuLi loses its activity as a polymerization initiator in such systems, but the causes and mechanism of this phenomenon were obscure prior to this research.

The reaction of this catalyst with triethylvinyltin in the course of polymerization was studied by Mössbauer spectroscopy (Aleksandrov, 1968b). The experiments with the above-mentioned systems were conducted by use of frozen solutions in benzene, a solvent that does not affect the Mössbauer spectra of the compounds under study. Addition of *n*-BuLi to the solution of $(CH_2{=}CH)Sn(C_2H_5)_3$ in benzene in the range of ratios of from 1:20

to 10:1 was found not to affect the shape and halfwidth of the singlet line of triethylvinyltin having an isomer shift of 1.4 ± 0.07 mm/sec. The probability of the Mössbauer effect (recoil-free fraction) proved to be the sole parameter carrying useful information. Figure 11 illustrates the magnitude of the Mössbauer effect versus the molar ratio of *n*-BuLi to triethylvinyltin. A higher probability of the Mössbauer effect is one of the characteristic symptoms of an increase in the coordination number of tin atoms (Gol'danskii and Herber, 1968). The concentration relationship given in Fig. 11 led to the conclusion that a $(CH_2=CH)Sn(C_2H_5)_3 \cdot 2(n\text{-BuLi})$ complex emerged in the system in question. It must be noted that complexes of the same composition are formed by reaction of *n*-BuLi with tetraethyltin and polytriethylvinyltin in the absence of the double bond. These facts were used as a basis for a hypothesis accounting for the low activity of triethylvinyltin in the process of anionic polymerization initiated with catalytic quantities of *n*-BuLi. According to this hypothesis, part of the catalyst is tied down in the complex and hence takes no part in the polymerization process, and the formation of an active polymerization center at the initiation stage is limited by the instability of the $(CH_2=CH)Sn(C_2H_5)_3 \cdot 2(n\text{-BuLi})$ complex. Since it was ascertained that *n*-BuLi reacted with polytriethylvinyltin as well, the hypothesis considers it probable that complexing may obstruct polymerization at the chain-growth stage, too. On the assumptions of Aleksandrov *et al.* (1968b), Platé and Mal'tsev (1970) continued a detailed study into the mechanism of polymerization and the products of reaction of the triethylvinyltin–*n*-BuLi system. And although the latter authors conducted no gamma resonance experiments, we shall still relate in brief their principal results, for they offer graphic proof of the assumptions of Aleksandrov *et al.* (1968b) and are in fact an extension of them. Chromatographic analysis of the gases liberated by the polymerization system being decomposed with

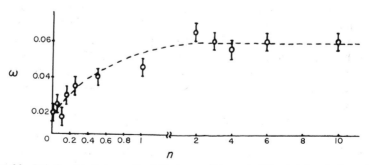

**Fig. 11.** Relationship between the Mössbauer effect probability ($\omega$) (in relative units) and the molar ratio of *n*-BuLi to triethylvinyltin (*n*). (Aleksandrov *et al.*, 1968b.)

methanol revealed that the bound moiety of *n*-BuLi, taking no part in the polymerization process, remained chemically invariable. The authors further investigated the kinetics of styrene polymerization on *n*-BuLi in the presence of tetraethyltin (a saturated analog of triethylvinyltin), which also forms a complex with *n*-BuLi, as is shown by Aleksandrov *et al.* (1968b). Comparing the data obtained with the laws governing the *n*-BuLi–triethylvinyltin system, Platé *et al.* (1970) concluded that the growing macroanions of triethylvinyltin were the primary factor behind the bonding and deactivation of *n*-BuLi. As a result of this reaction, their growth is terminated and products having a low molecular weight are formed. Platé *et al.* (1970) succeeded in identifying a complex of *n*-BuLi with polytriethylvinyltin that was insoluble in the monomer or in benzene, though both polytriethylvinyltin and *n*-BuLi separately are soluble in them. Thus, the investigations reported have done much to elucidate the mechanism whereby the monomer and polymer alkyl derivatives of tin affect the process of anionic polymerization of vinyl monomers initiated with the aid of *n*-BuLi.

Sanaya *et al.* (1971) discuss the problem of formaldehyde polymerization under the effect of $SnCl_4$. It had been found earlier that the reaction could be boosted by adding formaldehyde oligomers having an acetal-type chain structure $CH_3O(CH_2O)_nCH_3$, e.g., methylal (M). The effect had been tentatively ascribed to the formation of some complexes by $SnCl_4$ and the additive. To verify this hypothesis, the authors used Mössbauer spectroscopy in an attempt to trace the chemical alterations of the catalyst state at the various stages of formaldehyde polymerization. The process was carried out at room temperature, with gaseous formaldehyde continuously fed into a solution of $SnCl_4$ in toluene. The results of preliminary experiments in which Mössbauer spectra showed new lines differing from the $SnCl_4$ singlet indicated that $SnCl_4$ formed hexacoordinate complexes with formaldehyde with an isomer shift equal to 0.1 mm/sec. Addition of methylal to the $SnCl_4$ solution in toluene also gave rise to $SnCl_4 \cdot 2M$ complexes detectable by the conversion of the $SnCl_4$ singlet to a doublet with a quadrupole splitting of approximately 1.20 mm/sec. The full conversion corresponds to a ratio of $SnCl_4$ to methylal of 1 to 2. The $SnCl_4$ complexes with formaldehyde and methylal were found to have substantially differing isomer shifts and quadrupole splittings, enabling the structure of the catalytically active centers to be successfully studied in detail. Figure 12 presents the Mössbauer spectra corresponding to various stages of formaldehyde polymerization in the presence of $SnCl_4$.

One can see from these spectra that as polymerization goes on, the line with $\delta = 0.1$ mm/sec belonging to the above-described complex increases in intensity, with up to 70% of $SnCl_4$ being involved in the complexing

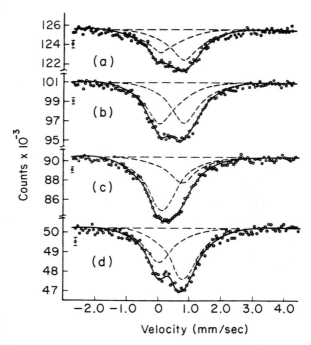

**Fig. 12.** Mössbauer spectra corresponding to various stages of formaldehyde polymerization. $[SnCl_4] = 9.1 \times 10^{-2}$ moles/liter. Polymerization time: (a) 1 min 40 sec, (b) 3 min, (c) 6 min 15 sec. (d) Spectrum of a solution of $[SnCl_4] = 9.4 \times 10^{-2}$ moles/liter in toluene with an addition of polyformaldehyde (MW = 50,000). (Sanaya *et al.*, 1971.)

phenomenon. The spectrum of the $SnCl_4$ solution in toluene with poly-formaldehyde (Fig. 12d) likewise shows a characteristic line with $\delta = 0.1$ mm/sec. The spectral analysis described led Sanaya *et al.* (1971) to conclude that $SnCl_4$ tended to form a complex with polyformaldehyde rather than with the monomer of the polymerization system. The changes in the Mössbauer spectra in the course of polymerization caused by methylal added to the system are shown in Fig. 13.

It was found that alongside the quadrupole doublet characteristic of the $SnCl_4 \cdot 2M$ complex, a new broad line with $\delta = 0.4$ mm/sec appeared, with intensity increasing with the degree of polymerization. The isomer shift of this line was found to lie between $\delta = 0.55$ mm/sec for the $SnCl_4 \cdot 2M$ complex and $\delta = 0.1$ mm/sec for the $SnCl_4$ complex with polyformaldehyde. It is assumed that this new line represents a mixed hexacoordinate complex of $SnCl_4$ with methylal and polyformaldehyde, and analysis of quadrupole splitting data bears this assumption out. On the basis of their experiments, the authors concluded that the specific course of polymerization of the

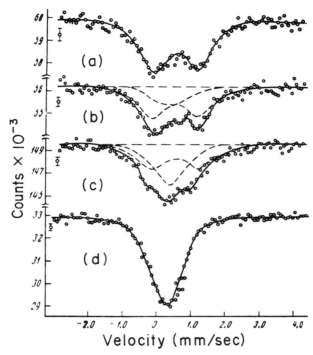

**Fig. 13.**  Mössbauer spectra corresponding to various stages of formaldehyde polymerization in the presence of methylal (M). $[SnCl_4] = 9.2 \times 10^{-2}$ moles/liter; $SnCl_4 : M = 1 : 10$. Polymerization time: (a) 1 min, (b) 2 min 20 sec, (c) 5 min 15 sec, (d) 30 min (dry polymer separated from the polymerization system). (Sanaya *et al.*, 1971.)

systems in question was associated with the change in the nature of the active center caused by methylal. Subsequently, the conclusions of Sanaya *et al.* (1971) as to the formation of complexes by $SnCl_4$ and the methylal additive were corroborated by NQR findings for $^{35}Cl$ (Morozova *et al.*, 1971).

Formaldehyde polymerization in organic media in the presence of catalysts, e.g., salts of Sn(II), is discussed by Kessler *et al.* (1973). It had been previously known that systems of this kind were less sensitive to the impurities brought by the gaseous monomer flow. In order to elucidate the attendant processes, the authors studied the changes in the Mössbauer spectra of the catalysts, i.e., tin dicaproate and diacetate. Polymerization and sample preparation were carried out both in an atmosphere of argon and by a conventional method in air. Figure 14 presents the spectra of the starting catalyst $Sn(C_5H_{11}COO)_2$, the polymer, and tin tetracaproate. As can be seen in the figure, the spectrum of the catalyst shows a doublet with $\delta = 3.55$ mm/sec and $\Delta = 1.85$ mm/sec representing tin dicaproate, and also

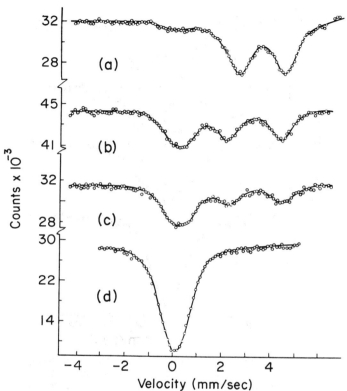

**Fig. 14.** Mössbauer spectra of (a) Sn(II) caproate, (b) polyformaldehyde produced in air, (c) polyformaldehyde produced in argon, (d) Sn(IV) tetracaproate. (Kessler *et al.*, 1973.)

a weak singlet close to zero velocity, which represents compounds formed through partial oxidation of the catalyst in the process of its production. A conclusion was drawn that the oxygen environment of the tin atom in this compound had an octahedral configuration. The polyformaldehyde spectra (Fig. 14b and c) showed a variation in the ratio of the oxidized form of the catalyst to the nonoxidized form; this ratio was equal to 1.46 for the sample polymerized in air and 0.90 for the case of polymerization in argon. The authors assume that in the former case oxidation of Sn(II) to Sn(IV) is promoted by air oxygen, while in the latter case additives introduced into the polymerization system in the gaseous monomer flow are responsible. The following routes of tin dicaproate oxidation by air oxygen were proposed:

$$2Sn(C_5H_{11}COO)_2 + O_2 \rightarrow (C_5H_{11}COO)_2Sn\underset{O}{\overset{O}{\diagup\diagdown}}Sn(C_5H_{11}COO)_2$$

and

$$(C_5H_{11}COO)_2Sn \overset{O}{\underset{O}{\diamondsuit}} Sn(C_5H_{11}COO)_2 \rightarrow (C_5H_{11}COO)_4Sn + SnO_2.$$

It was shown that the catalytic power of the Sn(II) salts in the polymerization reaction derived solely from the content of the nonoxidized form, and a higher concentration of the oxidized phase in the catalyst would result in a lower yield and viscosity of the polymer. The parameters of the catalytically active nonoxidized form of $Sn(C_5H_{11}COO)_2$ were found to vary in the course of polymerization. Thus, the isomer shift dropped from 3.54 to 3.16 mm/sec, while the quadrupole splitting rose from 1.84 to 2.07–2.19 mm/sec. These alterations were ascribed to a complex of the catalyst with formaldehyde formed in the course of polymerization, $Sn(C_5H_{11}COO)_2 \cdot nCH_2O$, in which the coordination number of tin was found in additional experiments to be 3. It was assumed that this complex could be an active polymerization center. Thus, the data of Kessler *et al.* (1973) indicate that the system under study undergoes partial oxidation of the catalyst resulting in the emergence of nonactive hexacoordinate tin compounds, but the nonoxidized moiety of Sn(II) retains its catalytic activity.

Here is another example of the application of the gamma resonance technique to the study of polymerization processes. George and Hayes (1973) noted the difference between the kinetics of free-radical polymerization of vinylferrocene and the common mechanism of polymerization characteristic of vinyl compounds. These authors found the chain termination step to be monomolecular, though the current theory stipulates its bimolecular nature. The reaction proceeded by a strange route—intramolecular interactions. The causes of such pecularities were brought to light after the polymerization system had been investigated by the Mössbauer spectroscopy, NMR, and ESR methods. We will discuss the Mössbauer data in some detail. As distinct from the usual Mössbauer spectra of ferrocene and ferrocene-base polymers, which present well-resolved doublets (Gol'danskii and Herber, 1968), in this case, the spectrum showed a third peak accounting for 3–14% of the total spectrum area, depending on the molecular weight and yield of the polymer. Analysis of this additional peak made it clear that it could not represent the ferricinium cation, and so it was tentatively ascribed to a complex of the high-spin Fe(III) with ionic binding. On the basis of these data it was assumed that the chain termination phenomenon occurred as a result of electron transfer from the iron atom to the growing macroradical, inactivating it and giving rise to an unpaired electron in the ferrocene nucleus. The iron atom, having lost an electron, fails to retain the low-spin configuration $d^5$. Its electronic environment undergoes a change resulting in the emergence of a polymer molecule incorporating the high-spin Fe(III) complex. These transformations are described by reaction equation **(C)**. The above mechanism of the chain termination step is based

**C**

on the assumption that the polymeric molecules formed in the reaction contain unpaired electrons. This assumption was confirmed by the NMR and ESR findings. According to the ESR data, the spin concentration is equal to unity per molecule at a mean molecular weight of the polymer of around 7000.

Vértes *et al.* (1972) discuss the mechanism of the catalytic effect of $Fe_2O_3$ in the condensation of chlorosilanes with alkoxysilanes used to synthesize siloxanes. The reaction was carried out with a solution of dimethylethoxychlorosilane in toluene in the presence of $Fe_2O_3$ enriched with $^{57}Fe$ to 80%. The reaction was assumed to take the route

$$n\left(\begin{array}{c} CH_3 \\ | \\ Cl-Si-OC_2H_5 \\ | \\ CH_3 \end{array}\right) \xrightarrow[\text{toluene}]{Fe_2O_3} Cl\left[\begin{array}{c} CH_3 \\ | \\ -Si-O \\ | \\ CH_3 \end{array}\right]_n -C_2H_5 + (n-1)C_2H_5Cl.$$

Figure 15a shows the spectra of a freshly prepared reaction mixture and of a sample taken at the time of the highest reaction rate (30 min after the reaction was started) (Fig. 15b). One can see how the clear-cut magnetic structure of the initial spectrum loses its definition in the course of the catalytic process. Analysis of the Mössbauer spectra led the authors to attribute this fact to a mononuclear iron complex arising in this reaction, in which, they believe, the iron is involved in coordination interaction with the oxygen atoms bonded with the silicon. Such a complex was also found in the reaction mixture that had been stored for 24 hr at room temperature prior to polymerization.

Holzinger *et al.* (1972, 1973) used Mössbauer spectroscopy to study tin-containing polymer films obtained in a glow-discharge tube. In the first work, benzene vapor and $SnCl_4$ were passed through the discharge tube. Analysis of the Mössbauer spectra of the polymers produced indicated the presence of tin-containing groups and also the formation of compounds such

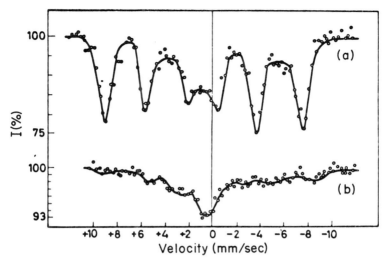

**Fig. 15.** Mössbauer spectra of the reaction mixture (dimethylethoxychlorosilane + $Fe_2O_3$ + toluene) at 77°K; (a) freshly prepared reaction mixture, (b) sample taken at the highest rate of the reaction ($t = 30$ min). (Vértes *et al.*, 1972.)

as $Sn(—R)_4$ ($\delta = 1.66$ mm/sec) at a ratio of the reaction mixture components of 1:1. If more $SnCl_4$ was present than the other constituent, the Mössbauer spectrum of the polymer showed an asymmetrical doublet whose parameters suggested that the Sn—Cl bonds were partially retained. Aging of the films in air caused the tin-containing groups to be oxidized to $SnO_2$.

Holzinger *et al.* (1973) introduced tin into the polymer chain by passing benzene vapor over molten metallic tin under glow-discharge conditions. They were able to prove the formation of Sn—C bonds in the polymer.

Khrapov *et al.* (1970) established the emergence of the carboxylate grouping $(C_4H_9)_3Sn—O—C{=}O$ in a number of organotin compounds derived from cellulose by reacting a graft copolymer of cellulose and polyacrylic acid with hexabutyldistannooxane $[(C_4H_9)_3Sn]_2O$ or by reacting sodium salts of these substances with tributylchlorostannane.

All the works reviewed in this section convincingly demonstrate the advantages of Mössbauer spectroscopy for studying polymerization processes and also the growing popularity of such studies.

## V. Photochemical Processes

Matalygina and Moshkovskii (1972) studied the photochemical processes in polymethylmetacrylate containing iron salts (introduced in the polymer in the form of $FeCl_3$ enriched with the [57]Fe isotope) and obtained intriguing results. The study was performed on samples containing 0.5 and 2.5% by weight of iron. Figure 16 presents the Mössbauer spectra of the former sample. The initial spectrum (Fig. 16a) presents a broad line ($\Gamma \approx 6$ mm/sec.) with a poorly resolved magnetic hyperfine structure, apparently due to paramagnetic relaxation. After the sample had been irradiated with ultraviolet light, this line was replaced by a doublet (Fig. 16b) having parameters characteristic of high-spin divalent iron ($\delta = 1.53 \pm 0.09$ mm/sec and $\Delta = 2.75 \pm 0.09$ mm/sec). Exposure to air for several days resulted in a new doublet with $\delta = 0.75 \pm 0.04$ mm/sec and $\Delta = 0.75 \pm 0.04$ mm/sec (Fig. 16c). This alteration was caused by the transformation of the high-spin

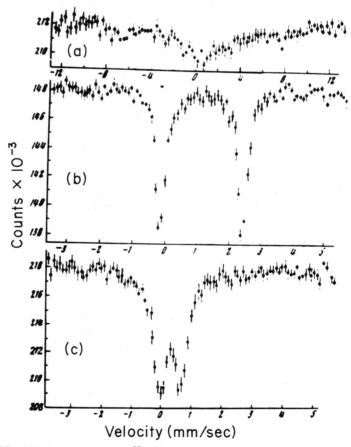

**Fig. 16.** Mössbauer spectra of $^{57}FeCl_3$ in polymethylmethacrylate; (a) initial sample, (b) sample irradiated with ultraviolet light for 1 hr, (c) sample exposed to air for 4 days. (Matalygina *et al.*, 1972.)

divalent iron either into a trivalent form or into a low-spin state Fe(II). The authors point out that a similar Mössbauer spectrum is obtained without ultraviolet irradiation and exposure to air if the starting composition is given an addition of antoxidants such as diphenylamine or carbazole. Irradiation of such samples failed to affect their spectra in any way.

Studies on compositions richer in $FeCl_3$ (2.5%) revealed similar transformations of the additive; however, exposure of the irradiated sample to air for 2 weeks resulted in the emergence in the Mössbauer spectrum of all the three forms of iron shown in Fig. 16. The transformations resulting from exposure to ultraviolet light are explained by Matalygina *et al.* (1972) by a phototransfer of electrons to iron orbitals. The atomic chlorine appearing in the process may react with the polymer molecule to form a cation radical; or else it may escape into the atmosphere as gaseous $Cl_2$. The changes in the Mössbauer spectra caused by the exposure of the irradiated samples to air were ascribed by the authors to the slow reaction of the cation radical with $Fe^{2+}$, although oxidation by air oxygen, less probable as it is, cannot be ruled out entirely as a possible mechanism. It was also noted that the paramagnetic hyperfine structure in the Mössbauer spectrum of the high-iron sample irradiated and then kept in air could be accounted for by a dark transfer of an electron from the $Fe^{2+}$ ion to the chlorine atom.

Unfortunately, no further research along this line has been reported in spite of the obvious fact that Mössbauer spectroscopy is a powerful tool in the hands of a scientist attempting to gain an insight into the kinetics and mechanism of photochemical processes and reactions of photostabilizers of polymers.

## VI. Structure and Action Mechanisms of Catalytically Active Polyferrocene Derivatives and Pyropolymers

The structure and properties of ferrocene and its derivatives and polymers intrigue many researchers. The interest in these compounds centers on their structure, reactivity, catalytic activity, and their possible use for developing organic semiconductors. Bypassing the results of work concerned with ferrocene proper reviewed by Goldanskii and Herber (1968), we are going to discuss only the problems having a direct bearing on the subject of this chapter.

It will be recalled that the Mössbauer spectrum of ferrocene presents a symmetrical doublet with $\delta \approx 0.78$ mm/sec and $\Delta \approx 2.4$ mm/sec. A number of authors have established that introduction of various substituents into the cyclopentadienyl ring affects the values of $\delta$ and $\Delta$ relatively little. The reason for this should be sought in the peculiar structure of ferrocene, where the $\pi$ electrons of the cyclopentadienyl ring are involved in the bonding of the iron atoms so that the effect of substituents is not felt directly but only to the extent they are capable of affecting the structure of the $\pi$-electron cloud.

Belov *et al.* (1964) obtained interesting gamma resonance data on polyferrocene, polyvinylferrocene, and various copolymers of the general formula

$$\left[ -\underset{\text{Fe}}{\underset{\bigcirc}{\bigcirc}}\!\!-A- \right]_n,$$

where A represents copolymer components acetone, naphthalene, $\alpha$-bromonaphthalene, *n*-dichlorobenzene, salicylaldehyde, benzaldehyde, or phthalaldehyde. The work was concerned with the soluble and insoluble fractions of the copolymers. It was found that the isomer shift and quadrupole splitting values for all soluble copolymers were close to those of monomeric ferrocene and its derivatives. This is obvious from the linear structure of these polymers given above, where the iron atom is not involved in the conjugated chain, and the electronic structure of iron in ferrocene is preserved in these compounds.

The Mössbauer spectra of insoluble copolymers show a second doublet with a lower value of quadrupole splitting characteristic of ferricinium salts (Fig. 17) alongside the doublet of the same parameters that characterize soluble fractions. The authors believe that the second doublet is due to the peculiar structure of the insoluble fraction and assume that the space network structure is formed by two cyclopentadienyl rings. In this case, the iron atoms get involved in the conjugated chain, and one of the electrons of this atom may be found in the group orbit of the conjugated chain. For this reason, the electronic structure of iron approaches that characteristic of the ferricinium cation as the quadrupole splitting value diminishes. Studies of samples of soluble and insoluble ferrocene copolymers revealed some interesting features regarding the temperature dependence of the parameter $f$. This point will be discussed in greater detail in Section VII.

The work by Belov *et al.* (1964) triggered an extensive research program on the structure and reduction–oxidation properties of ferrocene-based polymers and pyrolysis products.

Aliyev *et al.* (1970a) studied the oxidized states in the copolymers of ferrocene with phthalaldehyde and methyl ethyl ketone, an important problem related to the field of new redoxites. They concentrated primarily on the ratio of the oxidized and reduced forms of ferrocenylene groups, determined from calculating the areas under the respective spectral lines. One of the

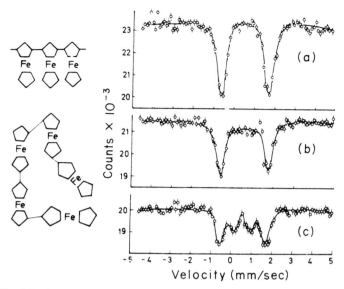

**Fig. 17.**   Mössbauer spectra of (a) ferrocene, (b) soluble polyferrocene, (c) insoluble poly-ferrocene. (Belov *et al.*, 1964.)

processes studied was polymer oxidation by ferric chloride; the observed changes of the Mössbauer spectra were studied against the quantities of the oxidant used. The spectra of the oxidized samples showed three peaks—two peaks corresponding to the doublets of the initial copolymer samples, and a third broad peak, assumed by the authors to be an unresolved ferricinium doublet. This assumption agrees with the theory that oxidative–reductive processes in such systems are associated with the ferrocene–ferricinium pair. It was further found that as the consumption of $FeCl_3$ increased, the fraction of the ferricinium form rose to a certain maximum value approximately equal to 65%, after which no changes in the Mössbauer spectra were observed. It was noted that the oxidation number determined chemically by the oxidant consumption proved to exceed that according to Mössbauer data. The apparent reason for this discrepancy is the effect of undesirable parallel oxidation of the side groups of copolymers and destructive oxidation.

Aliyev *et al.* (1970a) point out that, if the best redox systems on the basis of ferrocene polymers are to be chosen, the reaction is to continue to a certain degree of oxidation, avoiding destruction of the macromolecule, and monitoring the reaction by Mössbauer spectroscopy. They also worked out recommendations as to the set of conditions for the production of a ferrocene copolymer with phthalaldehyde showing the highest degree of ferrocene oxidation and a minimum of side reactions.

An impressive body of experimental material on the catalytic behavior of ferrocene copolymers and the products of their pyrolysis was obtained by Paushkin *et al.* (1969), Voronina *et al.* (1969), and Aliyev *et al.* (1970b), who dealt with the state of the iron atom in the course of some catalytic processes. This particular application of Mössbauer spectroscopy proved to be very promising, for not only ferricinium but also other iron compounds (carbides, oxides, and metallic iron) may arise in the catalytically active pyropolymers, and the role they play in catalytic processes is a problem of enormous theoretical and practical interest.

Paushkin *et al.* (1969) discuss the properties and catalytic activity of a number of ferrocene-containing polymers employed in the dehydration and dehydrogenation of alcohols. These reactions occur only very slowly with heating alone, so their realization is due to the electronic properties of the catalysts. Without dwelling on the non-Mössbauer investigations, it will be simply noted that Paushkin *et al.* (1969) concluded that the catalytic properties are largely dependent on the structure of the polymer and the valence of the iron atom.

This assumption prompted a Mössbauer study on a ferrocene copolymer with phthalaldehyde. The spectrum of the initial polymer presented a doublet characteristic of soluble polyferrocenes, with the ferricinium form accounting for not more than 5%. The polymer was brought to a catalytically active state by heating to a temperature of around 300°C. After this heat treatment, the spectrum turned into a quadrupole doublet due to a higher contribution of the ferricinium state, whose relative level rose to 20%. After catalytic dehydrogenation of isopropyl alcohol, the fraction of the ferricinium form dropped to the initial level. All inorganic forms of iron, such as oxides that could furnish at least some grounds for the catalytic process were conspicuously absent. The authors explain the observed changes in the Mössbauer spectra on the theory that alcohol dehydrogenation is a donor reaction resulting in the transfer of an electron from the alcohol to the catalyst, whereas in the course of alcohol dehydration the electron is transferred back from the iron atom of the ferrocene structure to the alcohol.

Voronina *et al.* (1969) and Aliyev *et al.* (1970b) continued this work with detailed studies into the catalytic properties of a number of ferrocene copolymers and the products of their pyrolysis obtained in varying conditions in the course of dehydration of alcohols, in particular, dimethylvinylcarbinol. Detailed description of these works is beyond the scope of this review, and we shall just list the results obtained.

As in the previously cited work, the Mössbauer spectra of the initial polymers showed both ferrocene and ferricinium. Under heat treatment, however, as the temperature and time of heating increased, the polymers would break down to form oxides of iron (apparently $Fe_3O_4$) in a finely divided state. The spectra presented a characteristic magnetic hyperfine structure. The mechanism of this destruction was ascribed by the authors to the oxygen (approximately 5–10%) present in the initial compounds, both that involved in the structural formula of the polymer and that of extraneous origin. Aliyev (1970b) demonstrated that for some of the ferrocene polymers studied, the catalytic activity was associated with the formation of such oxides. For others, just as shown by Paushkin *et al.* (1969), the dominant factor was the ratio of the ferrocene form to the ferricinium state. In yet another case, where the catalyst was a ferrocene copolymer with *p*-dichlorobenzene, the inorganic oxides accounting for the magnetic hyperfine structure in the spectra were formed in the course of the catalytic reaction. According to the authors, this was due to the destruction by the reaction products of the ferricinium form, whose Mössbauer spectrum decreases in area as the reaction proceeded, accompanied by a rise in intensity of the oxide lines.

The Mössbauer spectrum changes observed while some pyropolymers progressively lost their catalytic activity suggest a possibility of electron exchange between the ferrocene and ferricinium forms, as well as between these forms and the reaction products in the systems studied.

Aliyev *et al.* (1970b) also undertook to discover the cause behind the rise in catalytic activity of the ferrocene copolymer with *p*-dichlorobenzene exposed to *γ* radiation from a $^{60}$Co source. Analysis of the Mössbauer spectra showed that irradiation did not affect the ferrocene–ferricinium ratio and brought forth no new forms of iron. However, in the course of catalytic decomposition of alcohols, iron oxides (detected by the emergence of a magnetic hyperfine structure in the spectrum) were formed faster in the irradiated polymer than in the nonirradiated one. The catalytic activity of this sample was assumed to be associated not only with various forms of iron but also with the state of the matrix of the irradiated polymer. The catalytic process proved to be more intricate than expected, and its mechanism is still to be elucidated. The main conclusion of Aliyev *et al.* (1970b), in which the investigations described above are summed up, is that the catalytic activity of the pyropolymers studied either derives from the fine particles of iron oxides evolving in the course of pyrolysis or is due to the reduction–oxidation reactions of the ferrocene and ferricinium states, whose ratio varies in pyrolysis. The catalyst activity depends on the kind of initial polymer, on the conditions of its activation, and on the set of conditions under which the catalytic process is effected. The research summed up here has thrown much light on the mechanism of the catalysts prepared on the basis of ferrocene copolymers as well as on the nature of the catalytic processes under study. This formidable task would have hardly been accomplished with the same measure of success by any other physicochemical method.

Seishi and Mamoru (1972, 1974) studied the products of pyrolysis of the polymers produced by the polycondensation of acetylferrocene with furfural and acetone. They were able to show that pyrolysis in a vacuum at 400°C brought about free atoms of iron, $Fe^+$ ions, and iron clusters, as well as superparamagnetic and ferromagnetic iron particles in the matrix of glasslike carbon. The ratio of these forms depended on the ratio of the components making up the polymer. Although they make no mention of any problem of macromolecular chemistry, the work still merits attention here, for as we see it, the results presented by these authors may be of value for gamma resonance investigations of other pyropolymers, too.

Problems of radiation transformations of polydinitrile–ferrocene are discussed by Markova *et al.* (1969). It was found that, under $\gamma$ irradiation, the polymer underwent various structural changes, partial destruction, and cross-linking (around 35%), whereas equivalent doses of the same irradiation failed to affect ferrocene in any way.

## VII. Molecular Motions in Polymers and Monomers

A number of the parameters of Mössbauer spectra (linewidth, peak intensity asymmetry, and $f$, the probability of the Mössbauer effect[†]) are related to the mobility and orientation of the molecules and to the mean-square displacement of the Mössbauer atom. This offered a promise of valuable information about the specific features of molecular motions in polymers and polymerization systems. Several works (Kargin *et al.*, 1959, 1961a,b,c; 1964; Papisov *et al.*, 1965; Kabanov *et al.*, 1965; Kritskaya *et al.*, 1968) demonstrated that in thin frozen layers of monomers in the presence of finely divided catalysts, polymerization proceeded at fantastically high rates in the phase transition region. The reactions of solid-phase radiation-induced polymerization (Kritskaya *et al.*, 1968) and postpolymerization (Gusakovskaya *et al.*, 1968) of acrylonitrile were also speeded up in the vicinity of phase transitions. Omitting the details of the hypotheses advanced to explain the mechanism of these processes, we will just point out that preliminary ordering of the monomer molecules and their enhanced mobility are considered as the major prerequisites for fast solid-phase polymerization, and these are precisely the factors that can be verified by Mössbauer spectroscopy. Unfortunately, the most interesting compounds contain no Mössbauer isotopes, but this difficulty was obviated by introducing a Mössbauer "mark" into the systems under study. Indeed, the collective nature of the resonance absorption process permits studying the dynamic properties of the host matrix by the spectral parameters of the tracer. The technique of Mössbauer tracer compound introduction was successfully employed by Dézsi *et al.* (1965) for phase transition studies. Their work provided a foundation for this line of research and triggered intensive

---

[†] $f'$ is used below to represent the absorber recoil-free fraction.

exploration of this field. The method likewise proved fruitful for studying phase transitions and dynamics of molecular motions in polymerizable systems and polymers. Gusakovskaya *et al.* (1972a) studied the structural transformations and radiation-chemical processes in solid butanediol dimethacrylate. Earlier it had been found by calorimetric analysis (Gusakovskaya *et al.*, 1967) that defrosting of a vitreous monomer irradiated at 77°K was accompanied by a vigorous postpolymerization reaction in the β-transition region. It seemed interesting, therefore, to compare the data on the phase transformations in such a matrix with the Mössbauer evidence. The Mössbauer spectra of 3% solutions of anhydrous $FeCl_3$ and $SnCl_4$ in butanediol dimethacrylate, nonirradiated and irradiated with a dose of 2.2 Mrads, were studied in the temperature range from 110° to 250°K. Figure 18 shows the calorimetric thermograms for these systems and the temperature dependence of the Mössbauer effect probability. Introduction of Mössbauer tracers was found to have no effect on the thermograms. In the thermogram, the γ-transition corresponds to the freeing of limited molecular mobility, possibly connected with the rotation of the side methyl groups; the β-transition is accompanied by vitrification or glass softening; peak I corresponds to glass crystallization and minimum II to the butanediol dimethacrylate melting region. The correlation between the Mössbauer and

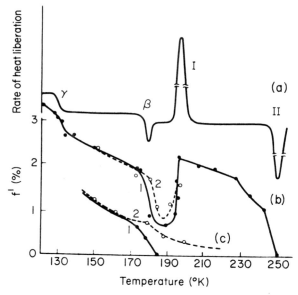

**Fig. 18.**   (a) Calorimetric thermograms for pure butanediol dimethacrylate glass and for the monomer with an addition of $FeCl_3$. (b) Resonance absorption probability ($f'$) versus temperature for the monomer with an addition of $FeCl_3$. (c) $f'$ versus temperature for the monomer with an addition of $SnCl_4$: 1. unirradiated samples, 2. irradiated samples. (Gusakovskaya *et al.*, 1972a.)

calorimetric data is obvious, particularly so where $FeCl_3$ is used as the tracer. The temperature dependence $f'$ reflects the changes in the dynamics of molecular motions in the system under study. The freeing of methyl group rotation, as well as the motion of individual segments or molecules as a whole in the $\beta$-transition region, reduce the probability of resonance absorption. Restriction of molecular mobility during crystallization causes a rise in the $f'$ value, whereas in the premelting region the area of the spectrum again diminishes and a diffusion line broadening is observed. Gusakovskaya *et al.* (1972a) also point out that in the $\beta$-transition region the doublet spectrum of the tracer acquires marked asymmetry brought about, as they show very convincingly, by orientation of the monomer molecules. Thus the experiments discussed above furnish ample evidence supporting the theory that higher molecular mobility in the monomer goes hand in hand with molecular orientation in the temperature range characterized by intensive postpolymerization. This confirms the conditions favoring solid-phase polymerization mentioned at the beginning of this section.

Gusakovskaya *et al.* (1972a) discovered one more interesting phenomenon; as distinct from the pure monomer and the system containing $SnCl_4$, irradiated samples of butanediol dimethacrylate having an addition of $FeCl_3$, exposed to heating conditions, failed to polymerize. The Mössbauer spectrum of such a sample showed lines typical of $Fe^{2+}$ in the $\beta$-transition region. According to that research, the reason for this phenomenon should be sought in the fact that the polymerization reaction is inhibited due to the capture by the $Fe^{3+}$ ions of the irradiation-produced electrons that would otherwise serve as polymerization initiators. As a result of such capture, $Fe^{2+}$ ions are formed, and the growth of the polymer chain is totally inhibited. Studies of a frozen solution of $SnCl_4$ in butanediol dimethacrylate showed the additive to partially precipitate as a separate phase. Thus its Mössbauer spectrum cannot fully represent the changes occurring in the monomer matrix, and the temperature dependence of $f'$ is by and large determined by the structure of the Mössbauer tracer itself. And yet, the $\beta$-transition was observed in this system (Fig. 16), and it was more pronounced in the irradiated samples. The fact is that the evolving polymer network (in case of a $\beta$-transition the polymer yield is around 40%) imposes certain constraints on the mobility of $SnCl_4$ and thus reduces the temperature dependence of $f'$ as compared with the unirradiated sample.

It is noteworthy that a similar phenomenon was observed while studying the irradiated samples of polyethylene stabilized by organotin compounds (Aleksandrov *et al.*, 1968a), which also qualify as Mössbauer tracers. The emergence of a cross-linked polymer as a result of irradiation of the initial mixture increased the probability of the Mössbauer effect and reduced its temperature dependence as against the unirradiated samples. The abrupt diminution of the $f'$ value was observed at the glass transition point at $T$ approximately equal to $-70°C$.

It should be noted at this juncture that for the mixtures of polyethylene with chlorides of tin and iron (Aleksandrov et al., 1969), the $f'$ value for the irradiated and unirradiated samples remained the same; that is to say, the Mössbauer additives that did not react with the polymer matrix upon irradiation proved insensitive to the formation of cross-links. These observations call for a cautious approach to the interpretation of data on the motions in polymer matrices obtained by use of various Mössbauer additives.

Studies into the system of butanediol dimethyacrylate with $FeCl_3$ by Gusakovskaya et al. (1972b) that investigated the effect of the glass–crystal phase transition on the spin–lattice relaxation time were continued. It had been found earlier that such a transition not only caused a sharp rise in the $f'$ value but also brought about marked variations in the magnetic hyperfine structure (Gusakovskaya et al., 1972a). Gusakovskaya et al. (1972b) studied this phenomenon in detail, showing that the spin–lattice relaxation time is halved sharply upon transition to the crystalline phase. Thorough analysis of this work is beyond the scope of the present review; we will only observe that the authors ascribe this effect to the changes in the intra- and intermolecular parameters of the butanediol dimethacrylate matrix upon transition from the amorphous to the crystalline state.

The same problems are discussed by Gusakovskaya et al. (1973), who studied the phase transitions in methacrylonitrile, acrylonitrile, and acetonitrile with the use of the same $FeCl_3$ tracer. The authors present strong evidence to substantiate their claim that the Mössbauer tracer forms complexes with the monomers in these systems, raising the sensitivity of the Mössbauer parameters to changes in the state of the monomer matrices. It was demonstrated by calorimetric analysis that $FeCl_3$ introduced into the monomers studied failed to affect the phase transitions experienced by the monomers. At these points, gamma resonance investigations revealed sharp fluctuations in the Mössbauer effect probability, as well as in the shape and width of the lines. These changes are analyzed in detail on the basis of the known data as to the phase transformations in the systems under study.

The examples cited, above all, point to the great future of the Mössbauer tracer technique for studying various parameters of polymerizing systems. However, several remarks on this problem are in order. It is clear now that the Mössbauer tracer is not always adequate to transmit the information about the changes occurring in the host matrix. The most favorable conditions arise in those cases, as Gusakovskaya et al. (1972a) point out, where the additive forms a single-phase system with the compound studied. The possibility of the tracer forming an independent phase imposes limitations of the choice of tracer and in fact on the applicability of the technique itself, for in such a case the additive is likely to show its own temperature dependence of $f'$, masking the changes in the system under study. Another disadvantage of the Mössbauer tracer method consists in the ever-present danger of the additive interfering with the processes in the host matrix, especially in the microregions adjoining the additive. One example illustrating such an adverse effect of the additive is the above-mentioned inhibition of polymerization in the butanediol dimethacrylate system with $FeCl_3$ (Gusakovskaya et al., 1972a). Those authors also observed some alterations in the isomer shift and quadrupole splitting values in the Mössbauer spectra of $FeCl_3$ and $SnCl_4$ dissolved in the monomer, compared to the pure compounds. This is definitely an indication that the Mössbauer tracer was involved in chemical bonding with the matrix studied, and in spite of the fact that this phenomenon failed to affect the main results, still, the risk of the matrix properties changing in the vicinity of the additive must be reckoned with. A method of studying systems containing no Mössbauer atoms, which is free of this disadvantage, is discussed later in this section.

Interesting results were obtained by Belov et al. (1964), who studied the behavior of the Mössbauer effect probability in soluble and insoluble ferrocene copolymers. In soluble copolymers at room temperature, $f'$ sharply decreased as against ferrocene. Belov et al. (1964) account for this by the assumption that higher temperatures release the oscillatory or rotational degrees of freedom of the ferrocenylene units of the polymeric molecule. At liquid nitrogen temperature, these motions are hindered, and the $f'$ parameter for soluble copolymers is close to that for crystalline ferrocene. For the insoluble fractions of the copolymers, it was found that the intensity of the intrinsic doublet of the cross-linked polymer was less dependent on temperature than the intensity of the extrinsic doublet (Fig. 17). According to the authors, this was attributed to the hindered mobility of the ferrocenylene units in the cross-linked structures. On the basis of the shape of the temperature curve $f'$ for the soluble and insoluble fractions of the ferrocene copolymers, the authors suggested a method for determining the degree of cross-linking of these copolymers.

Reich and Michaeli (1972) found the Mössbauer effect a suitable means of studying molecular motions in polymeric systems. The authors studied polyacrylonitrile produced by polymerization with the aid of $Fe(ClO_4)_2 \cdot 6H_2O$. As distinct from the work of Gusakovskaya et al. (1972a), the Mössbauer tracer was naturally incorporated into the polymer in the course of its production. Hence, in this case, the effect of the tracer on the properties of the polymer matrix is almost no problem. The following points were investigated: (1) the possibility of observing the Mössbauer effect above the glass transition temperature ($T_c$); (2) the effect of rotation freezing at $T$ less than $T_c$ on the rise in $f'$; and (3) the possibility of describing the temperature function $f'$ in a glass state of the polymer by means of the effective Debye temperature by analogy with the crystalline state. Reich and Michaeli (1972) made an important observation—that the isomer shift and quadrupole splitting values of $Fe(ClO_4)_2 \cdot 6H_2O$ present in the polymer matrix differed from those of the pure salt. Mössbauer studies in a broad range of temperatures showed that at $T = T_c$, the $f'(T)$ function experienced a break, but for $T > T_c$, the Mössbauer effect probability was still considerable and its temperature dependence became weaker. In the $80°K < T < T_c$ region, freezing of the rotational degrees of freedom in the glass state caused a sharp rise in $f'$, after which this parameter stabilized. The temperature dependence of $f'$ reflects the properties of the polymer matrix, as the authors correctly held. They also drew an interesting analogy with the temperature dependence of the elasticity modulus, which in the general case is inversely proportional to the mean square of the oscillation amplitudes of the atoms $\langle x^2 \rangle$. It will be remembered that the relationship between $f'$ and this parameter is expressed by the formula $\ln f' = -\langle x^2 \rangle / \lambdabar^2$, where $\lambdabar$ is the $\gamma$-quantum wavelength over $2\pi$ (see Chapter 1). Finally, this same work demonstrated that the effective Debye temperature concept applies to the system studied in an interval of at least $50°C$ in the $T < T_c$ range. Undoubtedly, the work of Reich and Michaeli (1972) is of interest as bearing on the feasibility and effectiveness of Mössbauer spectroscopy for investigating the dynamic properties of polymers and studying phase transitions and structural transformations in them.

Special consideration should be given to the highly original and promising investigations of molecular motions in a number of organic glasses, supercooled liquids, and polymers performed by Champeney et al. (1968, 1971, 1972a, b). None of the systems studied contained Mössbauer nuclei. The specific feature of the method used consisted in their application of the Mössbauer technique to investigate the elastic Rayleigh scattering of resonance $\gamma$-quanta.

Without dwelling on the theory and experimental subtleties of this method fully described in the original papers, it will be just noted that the resonance component of the Rayleigh scattering is given by the equation

$$f = A \exp(-K^2 \langle x^2 \rangle),$$

where $A$ is the fraction of Rayleigh scattering in the total scattering cross section; and

$$K = (4\pi \sin \theta)/\lambdabar,$$

where $2\theta$ is the scattering angle of quanta; and $\lambdabar$ is the wavelength. As follows from the above formulas, the Rayleigh-scattered $\gamma$-quanta bear the same information about the molecular motions in a scattering medium as the Mössbauer effect probability in conventional Mössbauer experiments. The work of Champeney and Sedgwick (1972b) deserves special notice. This work dealt with three polymers—polyisobutene, polytetrafluoroethylene, and poly(ethylene glycol). The temperature dependence of $f$ and the resonance line broadening were the parameters studied. The first of these three polymers is an amorphous substance with an average molecular weight of around 1000. For this polymer, the authors observed an abrupt variation of the parameter $f$ in the glass transition temperature region ($T_c \cong 200°K$) apparently associated with a freeing of the molecular motions. At higher temperatures (about 283°K), the resonance line broadened, presumably by the diffusion mechanism. As distinct from the first polymer, polytetrafluoroethylene has both crystalline and amorphous phases and shows two first-order transitions at 292° and 303°K. The peculiar features of the temperature dependence of $f$ seem to convey overall information about the pattern of molecular motion in the crystalline and amorphous states. It was noted that the resonance line did not broaden in the phase transition region ($T \cong 295°K$). The third compound studied, poly(ethylene glycol), is crystalline at room temperature and has a melting point range between approximately 303° and 313°K. At these temperatures were approached, no resonance line broadening was observed.

The work of Champeney et al. has so far been unique in employing the Rayleigh scattering principle for studying polymers with the help of Mössbauer spectroscopy. It must be pointed out that this technique is still in its infancy, and there can be no doubt that in the future it will find wide application, yielding valuable information about the specific patterns of molecular motion in systems possessing no Mössbauer atoms.

At the beginning of this review, a reservation was made that polymeric systems with a coordination structure fall beyond the scope of the review. However, the list of methods for investigating the motions of Mössbauer atoms in polymers would be incomplete without mention of the works of Stöckler and Sano (1967, 1968, 1969) Herber and Chandra (1970), and Herber et al. (1970), which deal with the temperature dependence of the asymmetry of the quadrupole doublet intensities—Goldanskii–Karyagin effect (Gol'danskii et al., 1962; Karyagin, 1963)—for a number of tin-containing polymers. Convincing proof was presented to show how observations of this effect can furnish information about the structural properties of the systems under study in various temperature regions.

## VIII. Chemical Processes in Butyl Rubber Vulcanization

A group of Bulgarian scientists have been systematically studying the mechanism of the processes accompanying butyl rubber vulcanization with phenolic resins, with Mössbauer spectroscopy featuring prominently in these investigations. The compounds studied are complex compositions where the molecules of rubber, resin, vulcanization accelerators, and other ingredients may react one with another. Studies into the mechanism of these reactions, synthesis of new resins, and the quest for effective vulcanization accelerators, are all of enormous practical importance. Christov and co-workers (1966, 1970a, b, c, 1971a), remark that investigation of the reactions involving the various components of complex mixtures is a task of formidable proportions, and that only Mössbauer spectroscopy could give hope of accomplishing it. The Mössbauer experiments centered on the chemical transformations undergone by the vulcanization accelerators. Salts of various metals, including tin and iron, are often employed as vulcanization accelerators, but the mechanism of acceleration has so far remained a mystery, with different authors advancing conflicting views as to the role of their interaction with the rubber, resins, and other ingredients. So, research into this problem is important not only from the viewpoint of theory, but also from that of the technological process.

Special attention was given to the chemical transformations of $SnCl_2 \cdot 2H_2O$ upon reaction with the various components making up the mixture being vulcanized. $SnCl_2 \cdot 2H_2O$ is one of the most effective known accelerators of butyl rubber vulcanization, but its practical application involves some difficulties connected with its insufficiently homogeneous distribution through the stock and with its adverse effect on the process equipment.

Christov et al. (1966) analyzed the changes in the Mössbauer spectra of $SnCl_2 \cdot 2H_2O$ and $SnCl_2$ used as accelerating additives in the course of butyl rubber vulcanization with alkylphenol–formaldehyde resins. It was found that with longer vulcanization times the $SnCl_2$ line lost some of its intensity, whereas the $SnO_2$ line of the spectrum increased. The authors succeeded in identifying the line of the intermediate product of this transformation, $Sn(OH)Cl$. It was shown that carbon black played a significant part in these transformations, particularly under conditions of heating. The stannic oxide formed in the vulcanizate was characterized by a much stronger temperature dependence of the Mössbauer effect than the ordinary polycrystalline $SnO_2$. The authors also found interesting correlations between the quantity of unreacted $SnCl_2$ and a number of physicomechanical parameters of the vulcanizates. Christov et al. (1971a) studied the role of the water of crystallization in the process of vulcanization. It was shown that, subjected to heating at $170^\circ C$, the mixtures of $SnCl_2 \cdot 2H_2O$ with carbon black, stearin, or machine oil showed in their Mössbauer spectra the singlet of anhydrous stannous chloride, and in the mixture with stearin, 60 min heating caused the initial doublet of $SnCl_2 \cdot 2H_2O$ to disappear altogether. On the basis of these data, an important conclusion of practical value was made as to the possibility of using anhydrous stannous chloride as an accelerating additive whose introduction into the compositions requires a fairly simple procedure. Subsequently, this assumption has been borne out.

Christov *et al.* (1970a) presented the results of Mössbauer experiments on the chemical transformations of $SnCl_2 \cdot 2H_2O$ mixed with bisphenolic resin (Vulcaresol 315E) and carbon black; these experiments were conducted with a view to testing the assumption that the reaction of these substances had a great effect on the butyl rubber vulcanization process. Figure 19A presents the Mössbauer spectra of the mixtures of $SnCl_2 \cdot 2H_2O$ with the resin and carbon black before and after being heated at 170°C for 1 hr; Fig. 19B presents the spectra of similar mixtures made with anhydrous $SnCl_2$. It can be seen in Fig. 19A that the prepared mixture composed of $SnCl_2 \cdot 2H_2O$ and the resin has a spectrum devoid of the doublet typical of this salt but shows two lines instead, one line close to the zero velocity region and the other at $2.0 \pm 0.2$ mm/sec. Upon heating of the mixture, the line of anhydrous

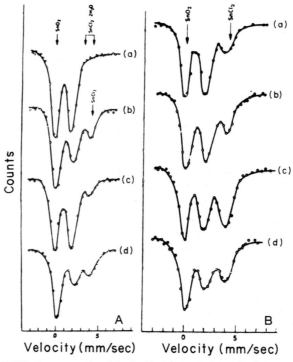

**Fig. 19.** (A) Mössbauer spectra of the $SnCl_2 \cdot 2H_2O$ mixtures with resin and carbon black; (a) $SnCl_2 \cdot 2H_2O$/resin prior to heating, (b) $SnCl_2 \cdot 2H_2O$/resin after heating, (c) $SnCl_2 \cdot 2H_2O$/resin/carbon black prior to heating, (d) $SnCl_2 \cdot 2H_2O$/resin/carbon black after heating. (B) Mössbauer spectra of similar compositions with $SnCl_2$. (Christov *et al.*, 1970a.)

stannous chloride appears [Fig. 19A(b)]. Addition of carbon black has a telling effect on the intensity of this line. It is observed in the Mössbauer spectrum even before the sample is heated [Fig. 19A(c)]. At the same time, carbon black was found to have practically no effect on the Mössbauer spectra of the heated compositions [Fig. 19A(b) and (c)]. In mixtures with anhydrous stannous chloride, some tin remains in the form of the initial compound giving an intense line in the Mössbauer spectrum [Fig. 19B(a)], while the rest of tin is transformed into the forms described above. In this case, too, carbon black additives cause a marked redistribution of the line intensities of the evolving tin compounds.

Analyzing these data, Christov et al. (1970a) came to the conclusion that both hydrated and anhydrous stannous chlorides formed a complex with the resin, with the isomer shift in the Mössbauer spectrum of 2.0 mm/sec. It was also found that dehydration of $SnCl_2 \cdot 2H_2O$ proceeded through a stage of formation of such a complex, and not by a purely thermal route. The authors made it clear that further studies were needed to elucidate the nature of the complex and the mechanism of formation of the compound having a zero isomer shift. They also called attention to the difference in the transformations of $SnCl_2 \cdot 2H_2O$ in the compositions studied (Christov et al., 1970a) and in its reactions with alkylphenol–formaldehyde resins (Christov et al., 1966).

With a view to systematically studying the processes of butyl rubber vulcanization in the presence of a variety of accelerators, Christov et al. (1970b) measured the Mössbauer spectra of a series of tin and iron salts and also carried out extensive Mössbauer studies into the chemical transformations of these salts in compositions with butyl rubber and various ingredients (Christov et al., 1970c).

The following compositions were studied: salt/butyl rubber, salt/phenol–formaldehyde resin, salt/stearin, salt/butyl rubber/resin, salt/carbon black, salt/butyl rubber/zinc oxide, and salt/butyl rubber/resin/zinc oxide, where the salt may be $SnF_2 \cdot 2HF$, $SnCl_2 \cdot 2H_2O$, $SnCl_2$, $SnCl_4 \cdot 5H_2O$, $SnBr_2$, $SnI_2$, $SnI_4$, $SnSO_4$, $Sn(PO_4)_2$, $Sn(CHO_2)_2$, $Sn(C_2H_3O_2)_2 \cdot 2H_2O$, Sn(II) propionate, Sn(II) monochloroacetate, Sn(II) trichloroacetate, $SnC_2O_4$, $FeF_2 \cdot 4H_2O$, $FeCl_2 \cdot 4H_2O$, $FeCl_3 \cdot 6H_2O$, $FeSO_4 \cdot 7H_2O$, $Fe_2(SO_4)_3 \cdot 9H_2O$, or $Fe_3(PO_4)_2 \cdot 8H_2O$. From the Mössbauer spectra, it was determined that all the tin salts under study, except Sn(II) propionate, did not react with butyl rubber on heating. The Mössbauer spectra of the heated compositions of tin halides with resin showed an intense line at zero velocity, which according to the authors, was associated with the transformation of the initial salts into finely divided stannic oxide. It was also found that tin sulfate, phosphate, and oxalate did not react with butyl rubber or with the ingredients either in the course of composition preparation or during the heat treatment stage.

Christov *et al.* (1970c) gave a detailed description of the changes in the Mössbauer spectra of the various compositions listed above, but we are not going to dwell on them except to point out that the transformations of tin and iron salts did not fall into overall patterns. Some compositions demonstrated a sizeable effect of carbon black on the Mössbauer spectra; this was attributed by the authors to the high concentrations and activity of carbon black.

Among the iron salt compositions, the most pronounced spectral changes were registered for $FeCl_3 \cdot 6H_2O$. It is a noteworthy fact that this compound is the most effective vulcanization accelerator of all the iron salts studied.

The data of Christov *et al.* (1966, 1970a, b, c, 1971a) constitute a big contribution toward understanding the processes behind butyl rubber vulcanization. Using the results of these works, the authors continued studies into similar systems, without recourse to Mössbauer spectroscopy (Christov *et al.*, 1971b, 1972, 1973).

## IX. Additional Remarks

This section is a brief summary of the works that have been published as only short reports as well as of the papers where Mössbauer spectroscopy has been only used as an auxiliary method.

Vértes and Nagy (1972) studied the reaction of polyvinyl alcohol, polyacrylamide, and polyvinylpyrrolidone with $FeCl_2$. Films were prepared from the aqueous solutions of the above polymers and $FeCl_2$; they were composed of 98% polymer, 1% water, and 1% $FeCl_2$. Analysis of the changes in the Mössbauer spectra against the parameters characteristic of pure $FeCl_2$ led the authors to conclude that iron formed coordination bonds with the polymers. It was further noted that in the composition with polyvinylpyrrolidone, about 40% of the iron was oxidized to a trivalent form, suggesting that this polymer was the strongest oxidation promoter of the compounds studied.

Blow (1969) noted that in a sample prepared from finely divided $PuFe_2$ pressed with polymethylmethacrylate powder, the iron reacted with the polymer, as evidenced by a change in the Mössbauer spectrum 5 months after the sample had been prepared. The author assumed that methylmethacrylate combined with divalent iron.

An attempt to study the regions of crystal ordering in cellulose using $^{57}Fe$ as a Mössbauer tracer was made by Tanakieva *et al.* (1968). Judging from the preliminary results of this work, Mössbauer spectroscopy may be successfully employed in research of this kind.

R. Larsson *et al.* (1972) discussed some catalytic processes on porous electrodes prepared from polyfluoroethylene resin mixed with activated charcoal and activated by means of iron polyphthalocyanine. Since the Mössbauer spectra of all the samples studied remained unchanged, the authors maintain that the iron atoms take no part in the electrode catalytic reactions, a valuable insight into processes of this kind.

The mechanism of adhesion of a number of polymers to shale rock surfaces was studied by Raff *et al.* (1973). Using Mössbauer spectroscopy, the authors showed that the iron contained in the shale formed no bonds with the polymers. The results obtained with the help of other methods used for investigating this problem go beyond the purposes of this review.


## X. Conclusion

The examples cited above suffice to convince anyone of the great advantages offered by Mössbauer spectroscopy as applied to a variety of problems related to the physics and chemistry of polymers. At this stage, one can only guess about the lines of future research and the amount of effort that will be spent on it. It should only be observed that the best results have invariably been achieved by those authors who compare their Mössbauer data with the results of other physicochemical methods used to study the same objects.

Undoubtedly, Mössbauer spectroscopy may prove of enormous value for solving the problems of stabilization of polymers, for understanding the thermooxidation processes, for studying the effects of ionizing radiations and light, and for studying the mechanism of catalytic polymerization with a view to developing modified polymers and improving their properties with the aid of additives. The number of problems that lend themselves to investigation by Mössbauer spectroscopy in this field is truly limitless. Very interesting research challenges are offered by metallopolymers containing colloidal particles of tin, iron, and other Mössbauer elements.

Apart from chemical studies, such problems as dynamics of molecular motions in polymers and polymerizing systems present great interest in view of the tasks mentioned in Section VII. It would be a real challenge to continue studying such objects and try to relate the Mössbauer parameters with the solid-phase polymerization data, preferably experimenting with differently oriented structures (unidimensional, two-dimensional, and three-dimensional). It is also desirable to conduct experiments to study the anisotropy of vibration of atoms or molecular groups in polymer chains and structures such as liquid crystals, experimenting with specimens both oriented in a natural way in the course of production and oriented through mechanical, electrical, or magnetic effects.

The list of possible applications of Mössbauer spectroscopy in the field of physics and chemistry of polymers can be easily extended on the understanding that as a rule, Mössbauer experiments with high-molecular compounds, although capable of yielding valuable information, are still primarily of an auxiliary nature. For this reason the way such experiments are conducted and their usefulness are determined by the effectiveness of their combination with other research techniques. No doubt, even in the near future, Mössbauer spectroscopy will find many new and successful applications for solving theoretical and practical problems in the field of physics and chemistry of high-molecular compounds.

## References

Aleksandrov, A. Yu. (1964). Candidate dissertation. Inst. of Petroleum Chem. Synthesis, Acad. Sci. USSR, Moscow.

Aleksandrov, A. Yu., Delyagin, N. N., Mitrofanov, K. P., Polak, L. S., and Shpinel, V. S. (1962). *Zh. Eksp. Teor. Fiz.* **43**, 2074.

Aleksandrov, A. Yu., Okhlobystin, O. Yu., Polak, L. S., and Shpinel', V. S. (1964a). *Dokl. Akad. Nauk SSSR* **157**, 934.

Aleksandrov, A. Yu., Berlyant, S. M., Karpov, V. L., Leshchenko, S. S., Okhlobystin, O. Yu., Finkel', E. E., and Shpinel, V. S. (1964b). *Vysokomol. Soedin.* **6**, 2105.

Aleksandrov, A. Yu., Berlyant, S. M., Karpov, V. L., Korytko, L. A. and Finkel', E. E. (1967a). *Vysokomol. Soedin.* **11A**. 2695.

Aleksandrov, A. Yu., Baldokhin, Yu. V., Braginskii, R. P., Gol'danskii, V. I., Korytko, L. A., Leshchenko, S. S., Finkel', E. E. (1968a). *Khim. Vys. Energ.* **2**, 331.

Aleksandrov, A. Yu. *et al.* (1968a). *Khim. Vys. Energ.* **2**, 331.

Aleksandrov, A. Yu., Gol'danskii, V. I., Korytko, L. A., Mal'tsev, V. A. and Plate, N. A. (1968b). *Vysokomol. Soedin.* **10b**, 209.

Aleksandrov, A. Yu., Gol'danskii, V. I., Zavarova, T. B., and Korytko, L. A. (1971a). *Vysokomol. Soedin.* **13B**, 784.

Aleksandrov, A. Yu., Gol'danskii, V. I., Zavarova, T. B., Korytko, L. A., and Plate, N. A. (1971b). *Vysokomol. Soedin.* **13B**, 76.

Aliyev, L. A., Vishnyakova, T. P., Paushkin, Ya. M., Sokolinskaya, T. Stukan, R. A., and Pendin, A. A. (1970a). *Izv. Akad. Nauk SSSR Ser. Khim.* 306.

Aliyev, L. A., Gol'danskii, V. I., Stukan, R. A., Paushkin, Ya. M., Vishnyakova, T. P., and Sokolinskaya, T. A. (1970b). *Dokl. Acad. Nauk SSSR* **194**, 843.

Belov, V. F., Vishnyakova, T. P., Gol'danskii, V. I., Makazov, E. F., Paushkin, Ya. M., Sokolinskaya, T. A., Stukan, R. P., Trukhtanov, V. A. (1964). *Dokl. Akad. Nauk SSSR* **159**, 831.

Berlyant, S. M., Karpov, V. L., and Finkel', E. E. (1968). *Vysokomol. Soedin.* **10A**, 1899.

Blow, S. (1969). *Phys. Lett.* **29A**, 676.

Bryukhanov, V. A., Goldanskii, V. I., Delyagin, N. N., Makarov, E. F., and Shpinel, V. S. (1962). *Zh. Eksp. Teor. Fiz.* **42**, 637.

Champeney, D. C., and Sedgwick, D. F. (1971). *J. Phys. C* **4**, 2220.

Champeney, D. C., and Sedgwick, D. F. (1972a). *J. Phys. C* **5**, 1903.

Champeney, D. C., and Sedgwick, D. F. (1972b). *Chem. Phys. Lett.* **15**, 377.

Champeney, D. C., and Woodhams, F. W. D. (1968). *J. Phys. B* **1**, 620.

Charlesby, A. (1960). "Atomic Radiations and Polymers." Pergamon, Oxford.

Christov, D., and Bontschev, Cv., (1973). *Kaut. Gummi Kunsts.* **26**, 199.
Christov, D., Bontschev, Cv., Skortsev, B., Dimov, D., and Ormandjiev, S. (1966). *Kaut. Gummi Kunsts.* **19**, 418.
Christov, D., Bontschev, Cv., and Dimov, D. (1970a). *Kaut. Gummi Kunsts.* **23**, 15.
Christov, C., Bontschev, Cv., Manuschev, B., and Ormandjiev, S. (1970b). *Kaut. Gummi Kunsts.* **23**, 207.
Christov, D., Bontschev, Cv., Manuschev, B., and Tschakarov, Chr. (1970c). *Kaut. Gummi Kunsts.* **23**, 556.
Christov, D., Bonchev, Zw., Manouchev, B., Dimov, D. and Nenon, N. (1971a). *Proc. Conf. Appl. Mössbauer Effect, Tihany, 1969* (I. Dézsi, ed.), pp. 795–797. Akademial Keado, Budapest.
Christov, D., Bontschev, Cv., and Dimov, D. (1971b). *Kaut. Gummi Kunsts.* **24**, 647.
Christov, D., Bontschev, Cv., Dimov, D., and Nenon, N. (1972). *Kaut. Gumi Kunsts.* **25**, 403.
Dézsi, I., Keszthelyi, L., Molnár, B., and Pócs, L. (1965). *Phys. Lett.* **18**, 28.
Finkel', E. E., Leshchenko, S. S., Braginskii, R. P. (1968). "Radiatsionnaya khimiya i kabel'naya tekhnika" ("Radiation Chemistry and Cable Engineering"). Atomizdat Publ., Moscow.
Gashnikova, N. P. et al. (1963). *Tr. NII Kabel'noi promyshlennosti (Proc. Sci.-Res. Inst. Cable Ind.)* issue 8, p. 143. Gosenergoizdat, Moscow.
George, M. H., and Hayes, G. F. (1973). *J. Polym. Lett. Ed.* **11**, 471.
Gladkova, G. I., Yegorova, Z. S., Karpov, B. L., Leshchenko, S. S., Mitrofanova, L. V., Slovokhotova, N. A., Finkel', E. E., Chernetsov, S. M. (1965). *Plastmassy*, No. 9, 8.
Gol'danskii, V. I. (1963). "Effekt Mossbauera i Yego primeneniye v khimii ("The Mössbauer Effect and Its Application in Chemistry"). Acad. Sci. Press, Moscow.
Gol'danskii, V. I. (1971). *Vysokomol. Soedin.* **13**, 311.
Gol'danskii, V. I., and Herber, R. H. (eds.) (1968). "Chemical Applications of Mossbauer Spectroscopy." Academic Press, New York.
Gol'danskii, V. I., Gorodinskii, G. M., Karyagir, S. V., Korytko, L. A., Krizhanskii, L. M., Makarov, E. F., Suzdalev, I. P., Khrapov, V. V., (1962). *Dokl. Akad. Nauk SSSR* **147**, 127.
Gol'danskii, V. I., Makarov, E. F., Stukan, R. A., Trukhtanov, V. A., and Khrapov, V. V. (1963). *Dokl. Acad. Nauk SSSR* **151**, 357.
Gol'danskii, V. I., Rochev, V. Ya., and Khrapov, V. V. (1964). *Dokl. Acad. Nauk SSSR* **154**, 709
Gol'danskii, V. I., Plate, N. A., Purinson, Yu. A., and Khrapov, V. V. (1968b). *Vysokomol. Soedin.* **11B**, 498.
Gordon, G. Ya. (1963). "Stabilizatsiya sinteticheskikh polimerov ("Stabilization of Synthetic Polymers"). Goskhimizdat, Moscow.
Greenwood, N. N., and Gibb, T. C. (1971). "Mössbauer Spectroscopy." Chapman and Hall, London.
Gusakovskaya, I. G., and Gol'danskii, V. I. (1967). *Vysokomol. Soedin.* **9B**, 390.
Gusakovskaya, I. G., and Gol'danskii, V. I. (1968). *Khim. Vys. Energ.* **2**, 46.
Gusakovskaya, I. G., and Larkina, T. I. (1973). *Fiz. Tverd. Tela* **15**, 1329.
Gusakovskaya, I. G., Larkina, T. I., Trukhtanov, V. A., Shcherbinin, Yu. S., and Gol'danskii, V. I. (1972a). *Vysokomol. Soedin.* **14A**, 1390.
Gusakövskaya, I. G., Larkina, T. I., and Gol'danskii, V. I. (1972b). *Fiz. Tverd. Tela* **14**, 2631.
Harrison, P. G., and Zuckerman, J. J. (1969). *J. Amer. Chem. Soc.* **91**, 6885.
Herber, R. H., and Chandra, S. (1970). *J. Chem. Phys.* **52**, 6045.
Herber, R. H. Chandra, S., and Hazony, Y. (1970). *J. Chem. Phys.* **53**, 3330.
Holzinger, H. W., and Tiller, M. J. (1972). *Plaste Kaut.* **19**, 656.
Holzinger, H. W., Meyer, K., and Tiller, H. J. (1973). *Z. Chem.* **13**, 32.
Ingham, G., Rosenberg, S., and Gillman, H. (1960). *Chem. Rev.* **60**, 459.

Kabanov, V. A., Papisov, I. M., Gvosdetskii, A. N., and Kargin, V. A. (1965). *Vysokomol. Soedin.* **7**, 1787.

Karakozova, E. I., Ratner, P. M., Paushkin, Ya. M., Stukan, R. A., Karmileva, L. V., Vishnyakova, T. P., Enikolopyan, N. S. (1972). *Dokl. Akad. Nauk SSSR* **205**, 97.

Karyagin, S. V. (1963). *Dokl. Acad. Nauk SSSR* **148**, 1102.

Kargin, V. A. (ed.) (1973). "Radiatsionnaya khimiya polimerov" ("Radiation Chemistry of Polymers"). Atomizdat Publ., Moscow.

Kargin, V. A., and Kabanov, V. A. (1961a). *J. Polym. Sci.* **52**, 71.

Kargin, V. A., Kabanov, V. A., and Zubov, V. P. (1959). *Vysokomol. Soedin.* **1**, 265.

Kargin, V. A., Kabanov, V. A., and Papisov, I. M. (1961b). *Vysokomol. Soedin.* **3**, 426.

Kargin, V. A., Kabanov, V. A., Papisov, I. M., and Zubov, V. P. (1961c). *Dokl. Acad. Nauk SSSR* **141**, 389.

Kargin, V. A., Kabanov, V. A., and Papisov, I. M. (1964). *J. Polym. Sci.* **C4**, 767.

Kessler, G. E., Rochev, V. Ya., Stukan, R. A., and Romanov, L. M. (1973). *Vysokomol. Soedin.* **15B**, 159.

Khrapov, V. V., Rochev, V. Ya., Aztemova, Yu. V., Viznik, A. P., Zemlyanskii, N. N., Gol'danskii, V. I., Rogovin, Z. A. (1970). *Vysokomol. Soedin.* **12B**, 145.

Korytko, L. A. (1972). Candidate dissertation. Inst. Chem. Phys. Acad. Sci. USSR, Moscow.

Kritskaya, D. A., Ponomarev, A. M., and Tal'roze, V. A. (1968). *Khim. Vys. Energ.* **2**, 61.

Larsson, R. Mrha, J., and Blomqvist, J. (1972). *Acta Chem. Scand.* **26**, 3386.

Markova, Ye, I., Paushkin, Ya. M., Polak, L. S., and Selezneva, E. I. (1969). *Zh. Fiz. Khim.* **43**, 2408.

Matalygina, Zh. I., and Moshkovskii, Yu. Sh. (1972). *Zh. Fiz. Khim.* **46**, 2474.

Minsker, K. S., and Fedoseeva, G. T. (1972). "Destruktsiya i stabilizatsiya polivinilkhlorida" ("PVC Destruction and Stabilization"). Khimiya Publ., Moscow.

Minsker, K. S., Zavarova, T. B., Purinson, Yu. A., Platé, N. A., Fedoseeva, G. T., and Kargin, V. A. (1968). *Vysokomol. Soedin.* **10A**, 1336.

Minsker, K. S., Zavarova, T. B., Fedoseeva, G. T., and Krats, E. O. (1971). *Vysokomol. Soedin.* **13A**, 2265.

Morozova, I. S., Tarasova, G. M., Ivanov, V. V., Bryukhova, Ye. V., and Enikolopyan, N. S. (1971). *Dokl. Acad. Nauk SSSR* **199**, 654.

Morris, D. R., and Rockett, B. W. (1972). *J. Organomet. Chem.* **35**, 179.

Neiman, N. B. (ed.) (1964). "Stareniye i Stabilizatsiya polimerov" ("Ageing and Stabilization of Polymers"). Nauka, Moscow.

Papisov, I. M., Kabanov, V. A., and Kargin, V. A. (1965). *Vysokomol. Soedin.* **7**, 1779.

Paushkin, Ya. M., Burova, L. M., Voronina, M. A., Vishnyakova, T. P., Sokolinskaya, T. A., and Aliyev, L. A. (1969). *Dokl. Acad. Nauk SSSR* **186**, 108.

Platé, N. A., and Mal'tsev, V. V. (1970). *Vysokomol. Soedin.* **12A**, 1533.

Platé, N. A., Mal'tsev, V. V., Davydova, S. L., and Kargin, V. A. (1966). *Vysokomol. Soedin.* **8**, 1890.

Platé, N. A., Zavarova, T. B., Mal'tsev, V. V., Minsker, K. S., Fefoseeva, G. T., and Kargin, V. A. (1968). *Vysokomol. Soedin.* **11A**, 803.

Purinson, Yu. A., Plate, I. A., Davydova, S. L., Nurkeeva, Z. S., and Kargin, V. A. (1968). *Vysokomol. Soedin.* **10B**, 257.

Raff, R. A., Haygen, J. A., and Mostafa, M. F. (1973). *J. Appl. Polym. Sci.* **17**, 1315.

Reich, S., and Michaeli, I. (1972). *J. Chem. Phys.* **56**, 2350.

Sanaya, I. F., Rochev, V. Ya., Kedrina, N. F., Stukan, R. A., Gol'danskii, V. I., and Yenikolopyan, N. S. (1971). *Dokl. Acad. Nauk SSSR* **197**, 869.

Seishi, Y., and Mamoru, O. (1972). *Chem. Lett.* No. 10, 843.

Seishi, Y., and Mamoru, O. (1974). *Chem. Lett.* No. 3, 277.

Stöckler, H. A., and Sano, H. (1967). *Phys. Lett.* **25A**, 500.

Stöckler, H. A., and Sano, H. (1968). *Phys. Rev.* **165**, 406.
Stöckler, H. A., and Sano, H. (1969). *J. Chem. Phys.* **50**, 381.
Tanakieva, M., Quiles, J. P., Chene, M., Christov, T., Chevalier, R., and Belakhovsky, M. (1968). *C. R. Acad. Sci. Paris* **C267**, 1013.
Vértes, A., and Nagy, M. (1972). *Radiochem. Radioanal. Lett.* **9**, 221.
Vértes, A., Csakvasi, B., Gomory, P., and Komor, M. (1972). *Radiochem. Radioanal. Lett.* **9**, 303.
Von Joachim Voigt (1966). "Die Stabilisierung der Kunststoffe gegen licht undwarme." Springer-Verlag, Berlin and New York.
Voronina, M. A., Vishnyakova, T. P., Paushkin, Ya. M., Bozhilova, M. A., and Aliyev, L. A. (1969). *Vysokomol. Soedin.* **11B**, 862.
Zuckerman, J. J. (1970). *Advan. Organomet. Chem.* **9**, 22.

# Index

# RETURN TO: CHEMISTRY LIBRARY
### 100 Hildebrand Hall • 510-642-3753

| LOAN PERIOD | 1 | 2 *1 Month* | 3 |
|---|---|---|---|
| 4 | | 5 | 6 |

## ALL BOOKS MAY BE RECALLED AFTER 7 DAYS.

Renewals may be requested by phone or, using GLADIS, type inv followed by your patron ID number.

## DUE AS STAMPED BELOW.

| | | |
|---|---|---|
| ~~FEB 24~~ | | |
| ~~MAY 02~~ | | |
| | | |
| | | |
| | | |
| | | |
| | | |
| | | |
| | | |

FORM NO. DD 10
3M 7-08

UNIVERSITY OF CALIFORNIA, BERKELEY
Berkeley, California 94720–6000